CLIMATE CHANGE: AN INTEGRATED PERSPECTIVE

ADVANCES IN GLOBAL CHANGE RESEARCH

VOLUME 1

CLIMATE CHANGE: AN INTEGRATED PERSPECTIVE

Edited by

Pim Martens
International Centre for Integrative Studies (ICIS),
Maastricht University,
Maastricht, The Netherlands

and

Jan Rotmans
International Centre for Integrative Studies (ICIS),
Maastricht University,
Maastricht, The Netherlands

Co-editors:

Darco Jansen
Koos Vrieze

KLUWER ACADEMIC PUBLISHERS
DORDRECHT / BOSTON / LONDON

A C.I.P. Catalogue record for this book is available from the Library of Congress.

ISBN 978-90-481-5331-2 ISBN 978-0-306-47982-3 (eBook)
DOI 10.1007/978-0-306-47982-3

Published by Kluwer Academic Publishers,
P.O. Box 17, 3300 AA Dordrecht, The Netherlands.

Sold and distributed in North, Central and South America
by Kluwer Academic Publishers,
101 Philip Drive, Norwell, MA 02061, U.S.A.

In all other countries, sold and distributed
by Kluwer Academic Publishers,
P.O. Box 322, 3300 AH Dordrecht, The Netherlands.

Printed on acid-free paper

CONTENTS

Chapter 10: Epilogue: scientific advice in the world of power politics 357

(S. Boehmer-Christiansen)

CONTRIBUTORS

M. van Asselt
International Centre for Integrative Studies (ICIS)
Maastricht University
Maastricht, The Netherlands

S. Boehmer-Christiansen
Department of Geography, Faculty of Science and the Environment,
University of Hull,
Hull, UK

K. Chatterjee
Development Alternatives
Global Environment Systems Branch
New Delhi, India

M. den Elzen
Dutch National Institute of Health and the Environment
Bilthoven, The Netherlands

J. Gupta
Institute for Environmental Studies
Free University
Amsterdam, The Netherlands

D. Jansen
Faculty of Natural Sciences
Department of Environmental Science and Engeneering
Open University
Heerlen, The Netherlands

J. Kinter
Center for Ocean-Land-Atmosphere Studies
Calverton, USA

P. Martens
International Centre for Integrative Studies (ICIS)
Maastricht University
Maastricht, The Netherlands

M. Parry
Jackson Environment Institute
School of Environmental Sciences
University of East Anglia
Norwich, UK

J. Rotmans
International Centre for
Integrative Studies (ICIS)
Maastricht University
Maastricht, The Netherlands

E. Schneider
Center for Ocean-Land-
Atmosphere Studies
Calverton, USA

J. Shukla
Center for Ocean-Land-
Atmosphere Studies
Calverton, USA

D. Straus
Center for Ocean-Land-
Atmosphere Studies
Calverton, USA

K. Vrieze
Department of Mathematics
Maastricht University
Maastricht, The Netherlands

PREFACE

Several years ago the Open University in Heerlen and Maastricht University decided to launch a course on '*Climate and the Environment*', with a diverse team of authors. Both natural and social scientists, from several regions of the world, contributed to this book. Initially, the book was intended as a textbook within this course for students of Environmental Sciences programmes at the Open University and Maastricht University. As the book developed it became clear that it would be an excellent source to anyone professionally engaged in the wide area of the enhance greenhouse effect.

This notion and new developments at the Open University at the time when the first draft was finished, caused a change of plan: the book should not only be written at the student-level, but should also reach the diverse group of policy-makers and scientists. Also the title of the book changed into '*Climate Change: An Integrative Perspective*'. It then took another few years before we could complete the manuscript as it lies in front of you.

This book aims to give you, the reader, a clear understanding of the nature of global climate change. It also makes clear what is known about the problem, and what is unknown or uncertain. It furthermore advocates the need for an integrative perspective to analyse and understand the complexity of the climate phenomenon. We are convinced that an integrated perspective can provide a useful guide to the problem of global climate change, and complement detailed analyses that cover only some parts of this complex problem.

Given the long history, many people contributed directly or indirectly to this book. First of all, we would like to thank the authors, for continuously updating their chapters as the book developed, and for their patience all throughout. Furthermore, we would like to thank all colleagues at the Open University and at the International Centre for Integrative Studies (ICIS) at Maastricht University for their help in finalising this book. Special thanks to Debby Jochems and Janneke Hogervorst for their efforts in proof-reading the final version of this book.

Pim Martens & Jan Rotmans

Maastricht, June 1999

Chapter 1

CLIMATE CHANGE: AN INTEGRATED PERSPECTIVE

P. Martens and J. Rotmans

1.1 Introduction

The consequences of rapid and substantial human-induced global climate change on life on Earth could be far-reaching. The impact on society of stringent emission control programs could be enormous, and the efficiency of such action may be highly debatable. The climate change issue's characteristic of prompt costs and delayed benefits has resulted in early policy research being focused on analysis of the cost-effectiveness of various greenhouse gas abatement strategies. These studies do not help decision-makers to identify climate change policy objectives, they only address the costs of meeting various abatement targets and the efficacy of different strategies. Consequently, scientific climate research has focused on explorations of the Earth's environment assuming that the atmospheric concentration of greenhouse gases continues to increase. Little effort has been expended on the exploration of the interactions among the various elements of the climate problem, on a systematic evaluation of climate stabilisation benefits or on the costs of adapting to a changed climate.

Because there is an immediate need for policy decisions on how to prevent or adapt to climate change, and how to allocate scarce funds for climate research, we need to move beyond isolated studies of the various parts of the climate problem. Many historical cases of isolated environmental problems, e.g. the decline of the Aral sea, teach us again that the interactions among the various social-economic, social-cultural, physical and political processes at play in the anthropogenic enhancement of the greenhouse effect can only be underestimated. Analysis frameworks are needed that incorporate our knowledge about precursors to, processes of, and consequences of climate change. These frameworks also need to represent the reliability with which the various pieces of the climate puzzle are understood and should be able to propagate uncertainties through the analysis and reflect them in the conclusions.

In this book we provide an integrative perspective by presenting an Integrated Assessment framework of climate change (Integrated Assessment is an iterative, continuing process, where integrated insights from the scientific community are communicated to the decision-making community, and experiences and learning effects from decision-makers form one input for scientific assessment – see for more details Chapter 7). This enables us to structure present scientific knowledge of global climate change. Furthermore, Integrated Assessment encompasses a variety of methods and tools which provides the structure for the interlinkages among the various chapters of the book. However, the analytical side of Integrated Assessment is elucidated rather than the process side. This means that the reader has to make its own integrated assessment based on the various pieces of the climate change puzzle we offer the reader.

1.2 Climate change in perspective

Global climate change is closely related to the natural resilience and buffer capacity of the biosphere in relation to anthropogenic disturbances. On a global scale, these disturbances can be represented by a set of interrelated cause-effect chains, here chosen as starting-points. The inextricably interconnected cause-effect chains form an organised whole, the properties of which are more than just the sum of their constituent parts. The cause-effect chains may be aggregated in terms of systems and subsystems, and one way of doing this is by distinguishing between pressure subsystems, environmental subsystems, impact subsystems and response subsystems, as denoted in Figure 1.1:

- Pressure system: social and economic forces underlying the pressure on the biosphere;
- State system: physical, chemical and biological changes in the state of the biosphere;
- Impact system: social, economic and ecological impacts resulting from human activities;
- Response system: human interventions as response to ecological and societal impacts.

Figure 1.1 represents the framework as a simple pressure-state-impact-response diagram that represents many interactions between the human system and the climate system. Figure 1.1 shows that changes unleashed in one subsystem result in a cascade of events that may eventually find their way back to the starting point. More specifically, while demographic and consumption pressures lead to changes in land use and energy consumption, such activities lead to environmental and economic impacts that may play significant roles in the future evolution of demographic and consumption patterns.

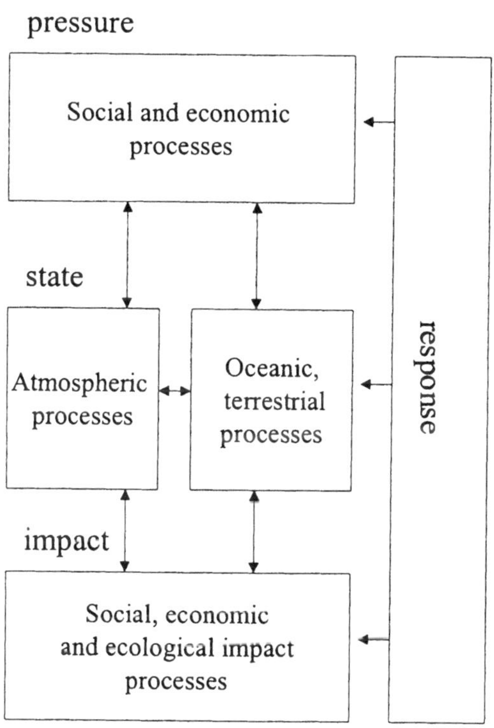

Figure 1.1 A simple pressure-state-impact-response diagram for the various elements of the climate change issue

Two dimensions of integration can be usefully distinguished. First, the vertical integration captures as much as possible of the cause-effect relationships (causal chain) of a phenomenon. Vertical integration closes the pressure-state-impact-response loop, linking a pressure to a state, a state to an impact, an impact to a response, and a response to a pressure. Second, horizontal integration addresses the cross linkages and interactions among various pressures, among various states, and among various impacts. Total integration means that various pressures are linked to various states, to various impacts and to various responses. In practice, many hybrid forms of integration exist, where vertical and horizontal integration routes are mixed.

Whereas the arguments for integrated assessment are intellectually compelling, current understanding of the climate problem by the natural and social sciences is highly incomplete. The problem of having not enough knowledge about both the human and natural system is further complicated by a lack of knowledge about the effects and feedbacks within and between the systems. In the realm of the natural system the dynamics of the climate system is far from well understood because of a dearth of knowledge about the internal dynamics of this system. The nascent nature of climate science is typified by a continuing stream of surprise findings and disappointments in solving what were once thought to be tractable problems. For example:

- The problem of balancing the global carbon cycle, made more difficult with recent findings that the role of nitrogen fertilisation as carbon sink is lesser than thought;
- Chlorofluorocarbons, once thought to be the most potent greenhouse gases, are now believed to have a negligible net warming effect;
- Fuel and biomass burning as well as biogenic sources lead to emission of greenhouse gases and aerosols. The former leads to long-wave radiation being trapped in the atmosphere, the latter leads to reflection of short-wave radiation. Estimating the magnitude of this warming and cooling effect continues to be a challenge.

Knowledge of key dynamics of social systems is also limited. In addition, understanding social dynamics suffers from a dearth of basic data, in particular outside industrialised countries. A few examples of the limitations of our knowledge of social dynamics are:

- What brings about the demographic and epidemiological transition?
- What drives environmentally related human behaviour?
- What are the roots and dynamics of technological innovation?
- How are preferences formed and do they evolve through time?

- What drives processes such as urbanisation and migration?

As indicated above, exploring global climate change and its consequences for human society is beset with many uncertainties, some of which are structural and irreducible by nature. This form of fundamental uncertainty in the climate problem provides room for different perspectives. A perspective is here defined as a consistent and coherent description of how the world functions and how decision-makers should act upon it. Therefore, perspectives can be characterised by two dimensions: (i) worldview, i.e. coherent conception of how the world functions and (ii) management style, i.e. signifying policy preferences and strategies. An ultimate challenge is then to cast the problem of global climate change from a selective number of representative perspectives involved in the climate debate.

1.3 This book

The integrated systems approach concentrates on the interactions and feedback mechanisms between the various subsystems of the cause-effect chains of global climate change rather than focusing on each subsystem in isolation. Therefore, in this book a multi-disciplinary systems approach, based on the integration of knowledge gleaned from a variety of scientific disciplines, will be used as a guiding principle.

The framework we present in this book is a simplified conceptual model of the extremely complex problem of global climate change. Nevertheless, we think that even a simplified but integrated model can provide a useful guide to global climate change, and complement detailed analyses that cover only some parts of the phenomenon of global climate change. In general, this book aims to give the reader a clear understanding of an integrated perspective of global climate change, which acts as a guideline for this book. The underlying objective is to acquire basic knowledge of what is known about the complex climate system and its interrelations with societal dynamics, but also what the basic uncertainties are in current knowledge.

In **Chapter 1: Climate change: an integrated perspective** we explain the basic rationale behind this book, and describe in general terms the coherence among the various chapters.

The main issue of **Chapter 2: The climate system** is to gain insight in these different interactions of the complex system called climate. Because the climate system is comprehensive, it is not possible to give a sharp definition. In reality the climate system contains all processes that are related to the mean weather. The circulation in the atmosphere and ocean plays an essential role. Also changes in terrestrial system, as far as they alter the composition of the atmosphere and ocean, are important factors. To better understand the Earth's climate, climate models are constructed by expressing the physical laws, which govern climate mathematically, solving the resulting equations, and comparing the solutions with nature. Given the complexity of the climate, the mathematical models can only be solved under simplifying assumptions, which are a priori decisions about which physical processes are important.

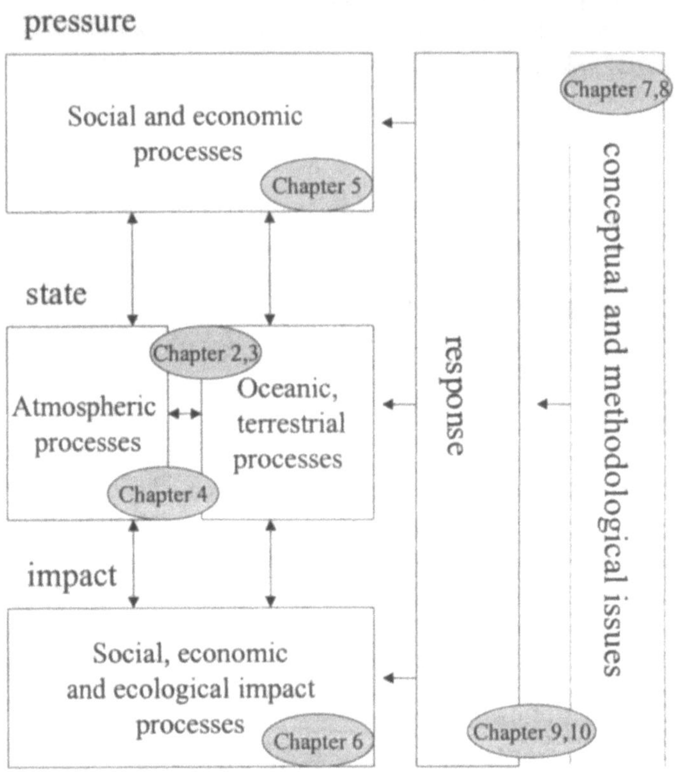

Figure 1.2 The pressure-state-impact-response diagram as starting point for the structure of this book

In **Chapter 3: Modelling of the climate system** current climate models are described that are used to project how the Earth's climate will respond to changes in external conditions, as well as their advantages and limitations. Although climate models have improved considerably over the past decades, they still have many deficiencies. Current climate models do not have an adequate treatment of solar cycles and glacial cycles. They also do not take into account several geophysical and biogeochemical feedbacks. Therefore, there is a need to improve our knowledge of these feedback processes, so that models can be developed that combine the fast components of climate, i.e. atmosphere, land and upper oceans, with the slow ones, such as the biogeochemical cycles.

In order to demonstrate how global climate change could be reconsidered in terms of human disturbances of the major global element cycles - the global cycles of carbon, nitrogen, phosphorus and sulphur - **Chapter 4: Global biogeochemical cycles** discusses the current and future state of the global element cycles. It is shown how the major global element cycles determine life conditions on Earth, and to what extent these global cycles are in unbalance. The range of anthropogenic perturbations responsible for this unbalance is discussed, varying from consumer behaviour to transport and construction. It is demonstrated how the global budget method can be used to determine the disturbance in the global element cycles, and what that means in terms of global climate change. In particular the interrelations between these global cycles are treated, by discussing common driving forces as well as common impacts.

Chapter 5: Causes of greenhouse gas emissions assesses the human activities resulting in anthropogenic emissions of greenhouse gases. Main developments of human-induced greenhouse gas emissions in both developed and developing countries are sketched. The two types of human activities that have lead to significant increases in greenhouse gas emissions are fossil fuel burning due to generation of electricity, cooking, heating, industrial activities, transportation, etc., and land use changes, such as deforestation and agricultural activities. Generation of electricity by using fossil fuels and transportation are among the key sectors, showing a considerable growth of emissions over the past decades. Both population growth and the level of energy and materials consumption in general are responsible for the rapid, unprecedented increase in greenhouse gas emissions.

Climate change is associated with a multitude of effects: climate change will shift the composition and geographic distribution of many ecosystems, with likely reductions in biological diversity. Climate change will lead to an intensification of the global hydrological cycle and may have impacts on regional water resources. Additionally, climate change and the resulting sea-level rise can have a number of negative effects on energy, industry, and transportation infrastructure, agricultural yields, human settlements, human health and tourism. Although there is a wide range of systems and sectors, upon which changes of climate may have an effect (e.g. from mountain glaciers to manufacturing), the approaches applied in climate impact assessment are broadly similar. The purpose of **Chapter 6: Impacts of climate change** is to give an overview of the current state of knowledge of potential impacts of climate change and to outline the array of methods available for assessing these impacts.

The complexity of the global climate change issue demands an integrated approach to ensure that key interactions, feedbacks and effects are not inadvertently omitted from the analysis. The various pieces of the complex puzzle can no longer be examined in isolation. **Chapter 7: Integrated Assessment modelling** describes how Integrated Assessment models of climate change try to capture the whole cause-effect chain of global climate change. IA models aim to fit the pieces of the climate puzzle together, thereby indicating priorities for climate policy at different levels. Integrated Assessment is considered as a cyclic process, in which models are used as a tool to gather insights. The IA-cycle encompasses the following steps: problem definition, conceptual model design, model development, model experiments, uncertainty and sensitivity analysis, the assessment itself, and the resulting policy recommendations. Three critical methodological issues are identified in IA modelling: the aggregation issue, treatment of uncertainty, and representation of qualitative knowledge. Suggestions are made to overcome the communication barrier between the rather technocratic world of IA modellers and the decision-making arena. Further, research suggestions are done to improve the overall quality of current IA models, and challenging research topics that could be addressed by IA models are given. Finally, new and promising research paradigms are discussed that could be used by the next generation of IA models.

Because of the uncertainties in the various parts of the cause-effect chains of global climate change, there are many valid perspectives according to which one may describe and evaluate phenomena of global climate change. Choosing a particular social-cultural perspective automatically includes a specific way of describing global climate change. This implies that each scientific description of global climate change is value-based. Therefore, various perspectives will be elaborated in this book, as described in **Chapter 8: Perspectives and the subjective dimension in modelling**. This book intends to incorporate this challenge by concentrating on different sources and types of uncertainty, and how these uncertainties could be coloured according to the biases and preferences of various perspectives. This Chapter aims to give a better understanding of the major role of uncertainty in climate change modelling, and to demonstrate why the subjective dimension should not be neglected in this modelling field. Furthermore, it aims to offer the reader but one way to deal with subjectivity in a systematic manner. After having studied the Chapter, the reader should understand how such a multiple-perspective approach could be applied in other modelling efforts. The uncertainty that surrounds climate change legitimates different interpretations of the problem, resulting in different kinds of solutions. This is among the book's essential ingredients.

Evaluation of the implications of various climate strategies considering future risks, and the development of preventive policies that reduce risks to tolerable levels needs a comprehensive view of past and current global decision initiatives and of the available tools for decision-making (**Chapter 9: Global decision making: climate change politics**). The future is inherently uncertain and thus unpredictable. Nevertheless, people in general, and decision-makers in particular, are interested in exploring future developments in order to make plans. One of the roles of science has been, and will continue to be, to assist decision-makers by sketching images of the future of the planet and of humankind. This Chapter examines the science policy relationship with respect to the climate change problem. It first looks at the role of scientific uncertainty and controversy on climate change. Then different theories on how science influences policy-making are explained, as well as an integrated model on the relationship between science and policy-making. Then this integrated science-policy model is applied to the formation and development of the climate change regime.

The book concludes with a reflective Chapter, which addresses fundamental issues with regard to the role of scientific knowledge in the policy-making arena, thereby focusing on the climate change policy process. **Chapter 10: Epilogue: scientific advice in the world of power politics** discusses the function of scientific advice in public policy (and hence law making) since the mid-1980s, concluding the book with a reflective chapter which addresses fundamental issues with regard to the role of scientific knowledge in the policy-making arena, thereby focusing on the climate change policy process.

This book is meant to be used by a diverse group of people. In general, we hope this book is a useful source to anyone professionally engaged in the wide area of global climate change. More specifically we aim to reach:

Policy-makers and policy analysts: this book intends to give a clear understanding of the uncertainties in our current knowledge of the climate system, and of the causes and impacts of future changes in the climate system as a result of human activities. In particular we hope to contribute to the continuous dialogue between scientists and policy-makers.

Natural and social scientists: this book intends to give a good understanding of both the physical and biogeochemical aspects of global climate change and the social-economic, social-cultural and institutional backgrounds. Mostly, global climate change is approached from either of the two scientific worlds, so this is one of the few attempts to bridge the two arenas.

Educationalists in developing countries: this book may help in the advancement of capabilities in developing countries to keep up with, to judge and to apply climate information, scientific results and their social-economic consequences.

Students: We think this book is highly interesting for students in the field of environmental science, but also for students with a multidisciplinary focus or background. It is particularly useful to them to learn more about integrated assessment as a promising research field to tackle complex societal problems.

In general, this book challenges the reader to study the complexity of the climate problem from different perspectives. The complexity of global climate change can not be straightjacketed into a single-disciplinary harness. The essentials of global climate change can not at all be captured by a simple cost-benefit analysis, just as much as reducing global climate change to a simple emissions reduction game, as viewed in the Kyoto-process, does no right to the complexity of the policy dimension of it.

A much more sensible way of coping with climate change is the integrative approach. Climate change as such will most likely have no disastrous effects on society, because our rapidly changing society is able to adjust to changes in climate. However, the integrative approach shows that global climate change needs to be considered as the spider in the web, as triggering factor for a whole range of other problems that are highly related to it: land use changes, water supply and demand, food supply, energy supply, human health, air pollution, stratospheric ozone depletion, etc.

Such an integrative approach, taking account of all known but also uncertain knowledge pieces, does not produce definitive, straight and popular answers. However, what it does provide is useful insights, based upon which comprehensive and effective climate strategies can be put into action.

Chapter 2

THE CLIMATE SYSTEM

D. Jansen

2.1 Introduction

Climate is a comprehensive system and therefore it is not possible to give a sharp definition. In its simplest way climate is related to the average weather experience of everyday. Meteorologists describe climate as the average weather over a period of 30 years. As such climate is related to mean values of surface temperature, rainfall, wind speed, and so on. However, it is not only the average weather condition that determines climate but also the number of extreme events that some regions meet. Especially temperature and precipitation and their seasonal variations influence for example the vegetation pattern. Therefore, the geographical distribution of climate is strongly related to the distribution of the major types of natural vegetation.

In reality the climate system contains all processes that are related to the mean weather. The circulation in the atmosphere and ocean plays an essential role. Also changes in the vegetation and in the geophysical and biogeochemical cycles (Chapter 4), as far as they alter the composition of the atmosphere and ocean, are important factors. On longer time scales the position of the continents influences the local as well as the global mean climate.

From a physical point of view climate is a giant thermodynamic machine with many interactions but eventually only driven by a single force: nuclear fusion in the Sun. The main issue of Chapter 2 is to gain insight in these different interactions of the complex system called climate.

2.2 Radiation budget

The interplay between the Earth and space essentially has to do with three physical quantities: energy, mass and linear momentum (the amount of movement). The radiation energy from the Sun is the main force of climate but it also drives many biological, geophysical and biogeochemical processes that may ultimately influence climate itself. The sun radiates all kind of radiation but for the climate system only electromagnetic radiation is of direct importance. The solar constant *S* is defined as the amount of radiation energy per unit time per unit area (i.e., energy flux) intercepted by Earth at the average distance from the Sun. The value of *S* is estimated at 1368 Wm^{-2} and is sometimes referred to as the solar irradiance. The spherical Earth intercepts an amount of radiation energy proportional to its sectional plane (πr^2 where *r* is the radius). This energy has to be divided over the whole surface area of the Earth ($= 4\pi r^2$). So per area unit of a spherical Earth: $(S/4) = 342 Wm^{-2}$. However, not all radiation energy from the Sun is absorbed. Seen from space, the Earth 'shines with light' because it reflects back some of the incoming solar radiation. The planetary albedo α defines the ratio of energy flux, which is reflected or scattered to that which is absorbed. The global mean value of the α as measured by satellites is 0.3. So the Earth absorbs an amount of $(1-\alpha)\, S/4 = 239$ Wm^{-2} of the incoming solar radiation

Not only the Sun but all objects with non-zero temperature (in degrees Kelvin) emit electromagnetic radiation. The rate at which an object radiates energy in all directions depends on absolute temperature *T*. A blackbody radiator is an idealised physical model that describes the radiation characteristics of an object. A blackbody radiates energy at the maximum rate at each temperature and wavelength. According to the Stefan-Boltzmann law, the energy flux at which an object radiates energy away from itself is proportional to T^4, the proportionality factor σ is called the Stefan-Boltzmann constant ($5.67*10^{-8}$ $Wm^{-2}K^{-4}$).

If an object is in radiation balance, the rate at which radiation energy is absorbed must balance the rate at which it is radiated. If the two rates fail to match, the temperature of an object will change until the balance is restored. Hence, the radiation balance for the Earth is given by $(1-\alpha)\ S/4 = \sigma T^4$. Insertion of appropriate values yields to a temperature of 255K or −18°C. This temperature is often referred to as the earth's radiation temperature. This is much lower than the observed mean surface temperature of 15°C. In fact a temperature of −18°C is only seen in the atmosphere at an altitude of 5km. Apparently, a large part of the radiation that leaves the climate system is emitted by the atmosphere and not by the surface itself. To put it differently, most radiation emitted by the Earth's surface is absorbed by the atmosphere, which in return radiates energy in all directions. Hence, the surface receives an additional source of energy from the atmosphere (the so-called downward backradiation) and therefore heats up: this is referred to as the (natural) greenhouse effect.

2.2.1 The greenhouse effect

Figure 2.1 gives an illustration of the greenhouse effect. Figure 2.1a states the radiation balance without an atmosphere; the surface temperature is only determined by the amount of solar radiation absorbed ($(1-\alpha) S/4 = 239 Wm^{-2}$). The temperature of –18°C calculated from the above radiation balance is called the effective radiating temperature. This temperature differs from the global mean surface temperature because the Earth does have an atmosphere.

The atmosphere resembles a greenhouse only because of the analogy with radiation trapping by the panes of glass. Most radiation of the Sun is not absorbed by the atmosphere, and hence the surface is heated. The surface also emits radiation but, in contrast, this radiation is for a large part absorbed by the atmosphere. Like all objects, the atmosphere (essentially the molecules and atoms in the atmosphere) emits electromagnetic radiation. The downward backradiation is an additional source of energy for the surface. In Figure 2.1b such a hypothetical atmosphere is included. The atmosphere is seen as just one layer that only absorbs all radiation emitted by the surface. The top of the atmosphere, the atmosphere itself as well as the surface must be in radiation balance. The surface absorbs as much as $478 Wm^{-2}$ corresponding to a surface temperature of 30°C. Hence, such a hypothetical atmosphere heats the surface by an additional 48°C.

In reality the atmosphere consists of many layers which all have their own characteristics, i.e., the amount of radiation of the Sun, of the surface and of other atmosphere layers that is absorbed. These characteristics strongly depend on another important aspect of radiation, namely its wavelength.

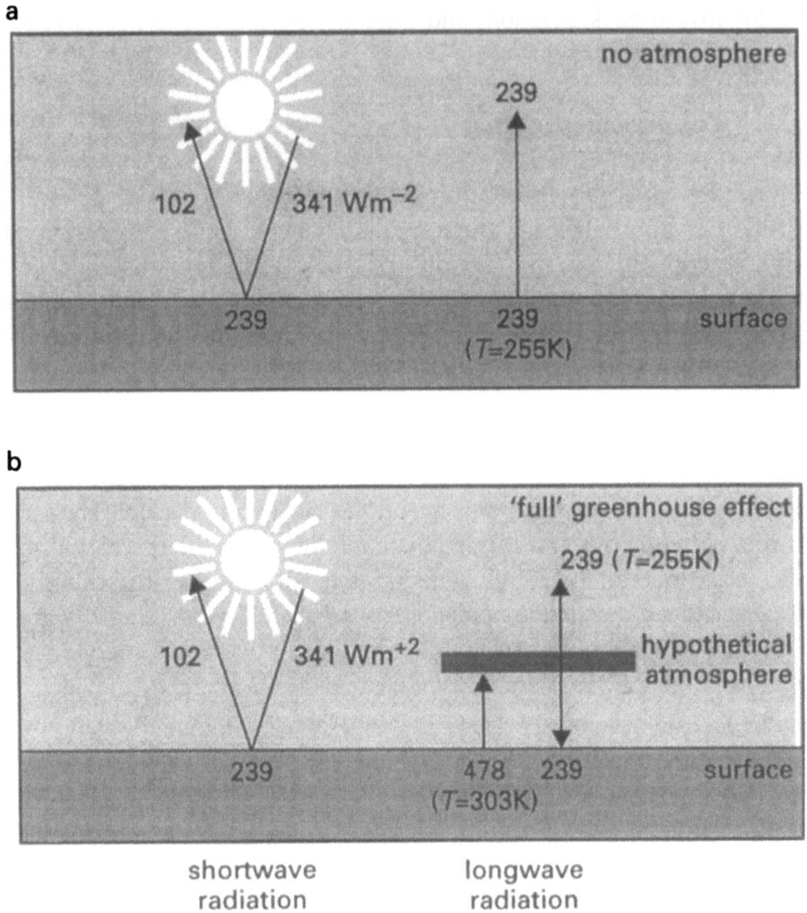

Figure 2.1 Schematic representation of the global annual averaged radiation balance without (2.1a) and with an atmosphere (2.1b). The hypothetical atmosphere in Figure 2.1b is represented by just one layer that is opaque to long-wave radiation and transparent to short-wave radiation.

2.2.2 Greenhouse gases

The radiation temperature not only determines the rate of radiation energy flow from an object but also the distribution of wavelengths ($=\lambda$) emitted. The part of the electromagnetic spectrum relevant for climate can be divided in ultraviolet (λ<400nm), visible (400nm<λ<700nm) and infrared (λ>700nm) radiation. A basic relationship exists between the wavelength of the radiation and the energy E of photons[1]: $E = hc/\lambda$ where h is Planck's constant and c is the speed of light. Thus, photons with short wavelengths have higher energies than photons with longer wavelengths. The energy of ultraviolet radiation is high enough to break down many molecules in the atmosphere (e.g., ozone). By contrast, infrared radiation is only capable to bring an electron in a molecule or atom in a higher energy state. As the molecule loses energy again, that energy may be emitted as radiation too.

Planck's law describes that for an object with a certain temperature, some radiation is emitted at all wavelengths. However, the quantity emitted falls off steeply both below and above the peak emission wavelength λ_{peak}. The product of λ_{peak} with the radiating temperature is, according to Wien's displacement law a constant value of $2.898*10^{-3}$ mK. The theoretical emission spectrum of the Sun and Earth is shown in Figure 2.2a and 2.2b, respectively. The Sun, with an effective radiation temperature of about 5900K, has a maximum in its radiated energy in the visible band (at about 491nm) but there are also contributions in the ultraviolet band and at longer wavelengths (often called the near infrared). The emission spectrum of the Earth (with T=255K) lies entirely at longer infrared wavelengths. Notice that the emission spectrum of the Sun barely overlaps that of the Earth. For that reason one speaks generally of incoming short-wave and outgoing long-wave radiation.

As stated before the atmosphere absorbs both short-wave and long-wave radiation. The observed spectrum from the Sun at Earth's surface (Figure 2.2a) and from the Earth at the top of the atmosphere (Figure 2.2b) clearly differ from that emitted by the Sun and Earth, respectively. The absorption spectrum in Figure 2.2c summarises these differences. Apart from the 'wing' at the ultraviolet side of the solar spectrum, the atmosphere is almost transparent to the incoming short-wave radiation. On the other hand the atmosphere strongly absorbs the outgoing long-wave radiation. Only in the 'atmospheric window' (i.e., between about 8 and 12mm) the absorption of the atmosphere is relatively weak.

Part (d) of Figure 2.2 explains the individual contribution of different atmospheric gases to the absorption spectrum (part (c)). Oxygen (O_2) and ozone (O_3) are the dominant absorbers in the ultraviolet region. Absorption by oxygen occurs at wavelengths lower than 240nm while ozone mainly absorbs solar radiation with wavelengths between 200 and 320nm. As such life on Earth is protected by this high energetic Sun radiation.

[1] One could easily read photon as well as particle. Light behaves either as a wave or as a particle, i.e., light has a dualistic character.

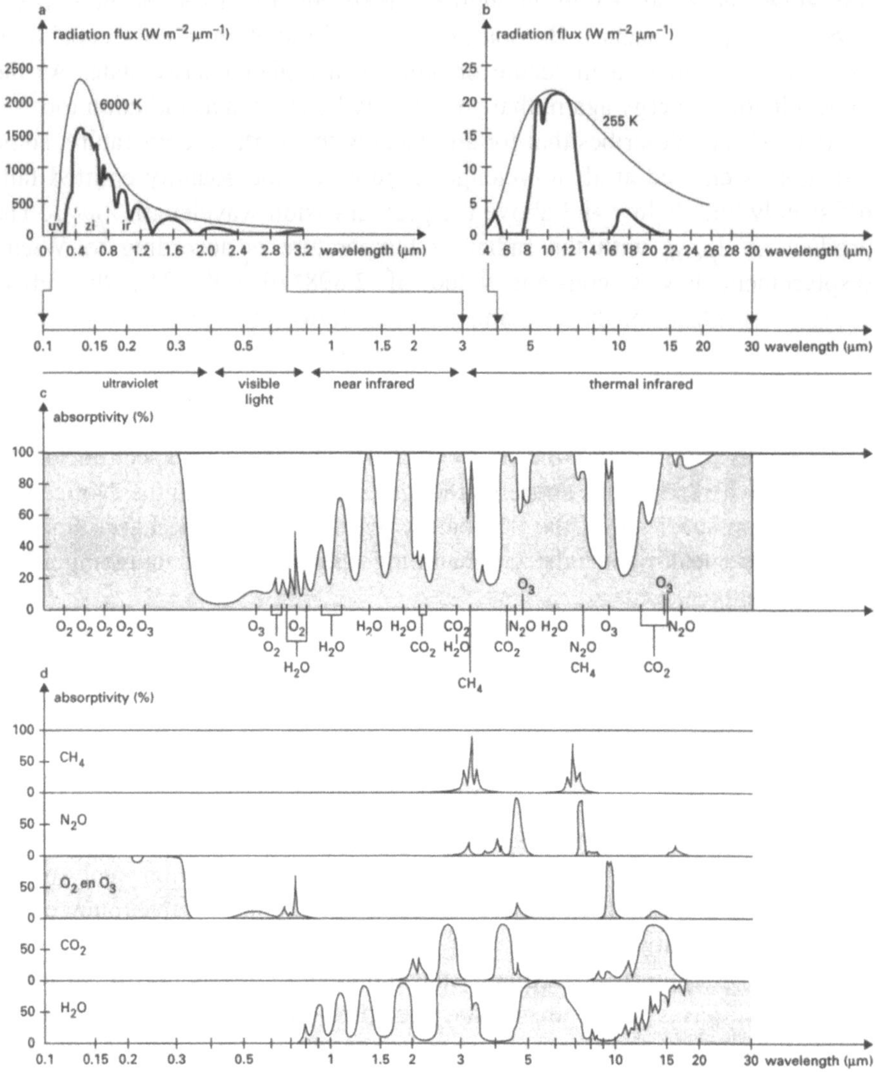

Figure 2.2 (a) The blackbody emission spectrum of the Sun with an effective radiating temperature of 5900K (black curve) and the spectrum as observed at the Earth's surface (thin line). (b) The blackbody emission spectrum of the Earth with an effective radiating temperature of 255K (black curve) and the spectrum as observed in outer space by satellites (thin line). (c) The overall absorption spectrum for the entire vertical extent of the atmosphere. (d) The absorption spectrum of the major greenhouse gases as occurring in their natural concentrations.

Long-wave radiation is mainly absorbed by water vapour (H_2O), carbon dioxide (CO_2), ozone, methane (CH_4) and nitrous oxide (N_2O). Infrared radiation emitted by the Earth's surface is absorbed by these gases and partially re-emitted back to the planetary surface. So, for that reason, atmospheric constituents that absorb long-wave radiation, and hence contribute to the greenhouse effect, are known as greenhouse gases. Surprisingly these greenhouse gases are only minor constituents of the atmosphere (Table 2.1).

Table 2.1 Average concentration of different trace gases in the atmosphere. Nitrogen gas (78%) and oxygen gas (21%) are the major constituents leaving about 1% to all trace gases. Remember that not all trace gases are greenhouse gases. Ppmv is parts per million by volume, ppbv is parts per billion per volume, pptv is parts per trillion by volume (relative measures of concentration).

Trace gas	**Chemical formula**	**Atmospheric lifetime (years)**	**Pre-industrial concentration**	**Concentration in 1994**
Water vapour	H_2O	Week	Varies strongly	Varies strongly
Carbon dioxide	CO_2	50-200	280 ppmv	358 ppmv
Methane	CH_4	12	0.7 ppmv	1.72 ppmv
Nitrous oxide	N_2O	120	275 ppbv	312 ppbv
Perfluorocarbon	CF_4	50.000	0	72 pptv
CFC-11	CCl_3F	50	0	268 pptv
CFC-12	CCl_2F_2	102	0	503 pptv
CFC-13	$Cl_2FC\text{-}CClF_2$	85	0	82 pptv
HCFC-22	$CHClF_2$	12	0	110 pptv

These trace constituents or 'trace gases' are sometimes highly variable in their concentration in the atmosphere. For example, the total amount of water vapour in the tropics is much higher than in subtropical desert areas. The contribution that each of the greenhouse gases make to the total greenhouse effect varies, depending on two main factors: the wavelength it actually absorbs (Figure 2.2d) and its atmospheric concentration (Table 2.1). The combined influence of these two factors and the resulting radiation trapping ensure that H_2O and CO_2 (Figure 2.2d) are the dominant contributions to the natural greenhouse effect.

One way to estimate the intensity of the natural greenhouse effect is to make use of the simple radiation balance of Section 2.1.1: $(1-\alpha)\ S/4 = \sigma T^4$. For this purpose the outgoing long-wave radiation is related to the planetary surface temperature T_S. Hence, the global radiation balance can be written as

$$(1-\alpha)\ S/4 = \sigma T^4 = \tau_e \sigma T_S^{\ 4} \qquad (2.1)$$

where τ_e is the effective transmittance of the atmosphere. Hence, τ_e is determined by the ratio of the long-wave radiation flux emitted to outer space to the emitted radiation flux by the planetary surface. For an Earth with no greenhouse effect (Figure 2.1a) and with a 'full' greenhouse effect (Figure 2.1b) the effective transmittance of the atmosphere is 1 and 0.5, respectively. This is consistent with our intuition: the more long-wave radiation of the surface is transmitted through the atmosphere, the higher the value of τ_e and the less intense is the greenhouse effect. The present value of τ_e for Earth's atmosphere is 0.61. Mankind may affect climate through one of the three climate parameters in (2.1). The on-going increase in the atmospheric burden of greenhouse gases is central to the anthropogenic or enhanced greenhouse effect (often falsely referred to as the greenhouse effect itself). The amount of infrared radiation that a specified increase in a greenhouse gas causes to be retained in the earth system is called radiative forcing.

2.2.3 The enhanced greenhouse effect

If the atmosphere behaves as in Figure 2.1b, than an anthropogenic input of greenhouse gases will hardly have any effect on the global mean surface temperature. The above statement clarifies the importance of the absorption spectrum of different greenhouse gases (Figure 2.2d) in relation to the overall absorption spectrum of the atmosphere (Figure 2.2c). For example, anthropogenic input of a hypothetical greenhouse gas that only absorbs radiation with wavelength larger than 30mm does not affect the overall intensity of the greenhouse effect. The atmosphere already absorbs all outgoing long-wave radiation at those wavelengths (Figure 2.2c). Hence, the global-mean surface temperature is hardly[2] modified by anthropogenic input of such a hypothetical gas.

Natural quantities of carbon dioxide are so large that the atmosphere is very opaque near its 15mm band. The anthropogenic input of carbon dioxide is expected to increase the percentage of atmospheric absorption at the edges of the 15mm band, and in particular around 13.7 and 16mm. Although especially H_2O and CO_2 are responsible for the natural greenhouse effect, also other gases in Figure 2.2d can make a significant contribution to the enhanced greenhouse effect. They strongly absorb at wavelengths within the atmospheric window (ozone) or close to it (nitrous oxide, methane and CFCs: chlorofluorocarbons).

[2] At first, one would expect that, in this hypothetical case, the surface temperature is not modified. However, concentration variations in the vertical of such a greenhouse gas may adjust the radiation budget in the vertical (see also Section 2.3).

Estimation of the change of radiative forcing out of changes in the concentration in greenhouse gases requires detailed radiative transfer models. These models simulate, among others, the complex variations of absorption and emission with wavelength, the overlap between absorption bands and the effects of clouds on the transfer of radiation. To make it even more complex, changes in radiative forcing may affect atmospheric chemical processes. These interactions also change the radiation balance and as such are called indirect radiative effects.

By now we only have looked at climate with a very simple, zero dimensional model (i.e., equation 2.1). With this model, climate is only characterised by a single parameter, the global mean surface temperature. However, to understand climate in its full complexity, we must add its temporal variability both horizontally and vertically.

2.3 Circulation of energy

In Section 2.2 we assumed that the Earth is in radiative balance or radiative equilibrium, i.e., the short-wave radiation flux absorbed equals the long-wave radiation flux emitted. However, observations show that this assumption does not hold. But before the radiation budget in the vertical can be estimated, the variation of temperature and the variation of the concentration of greenhouse gases have to be discussed.

Due to gravity, atmospheric pressure and density increase very rapidly with decreasing altitude, i.e., near the surface the total number of molecules in a volume is higher, and so is the pressure (see Figure 2.3). In practice 90% of the total atmospheric mass is found within some 16km of the surface, and 99% lies below 30km. Atmospheric chemistry may cause departure from this general picture (e.g., ozone). However, most greenhouse gascs are more abundant close to the ground level, and increasingly scarce at higher altitudes. This would tend to produce the highest temperatures near the surface, close to where the molecules that absorb long-wave radiation are more abundant. This picture is confirmed in Figure 2.3, however, the temperature variation with height is more complex.

Inflection points on the temperature profile in Figure 2.3 are used to distinguish different regions. Beginning at the Earth's surface, these regions are called the troposphere, the stratosphere, the mesosphere, and the thermosphere. Their boundaries are referred to as the tropopause, the stratopause and mesopause, respectively. The observed temperature profile is so complex because it is determined by the interplay of absorption by the atmosphere of both short-wave and long-wave radiation. The temperature increases from the tropopause to the stratopause mainly because of increasing absorption of shorter wavelength, i.e., higher energy, ultraviolet radiation. This energy is sufficient to break chemical bounds. This process, called photolysis, is central to the ozone budget in the stratosphere. As such, some 90% of ozone is found in the stratosphere.

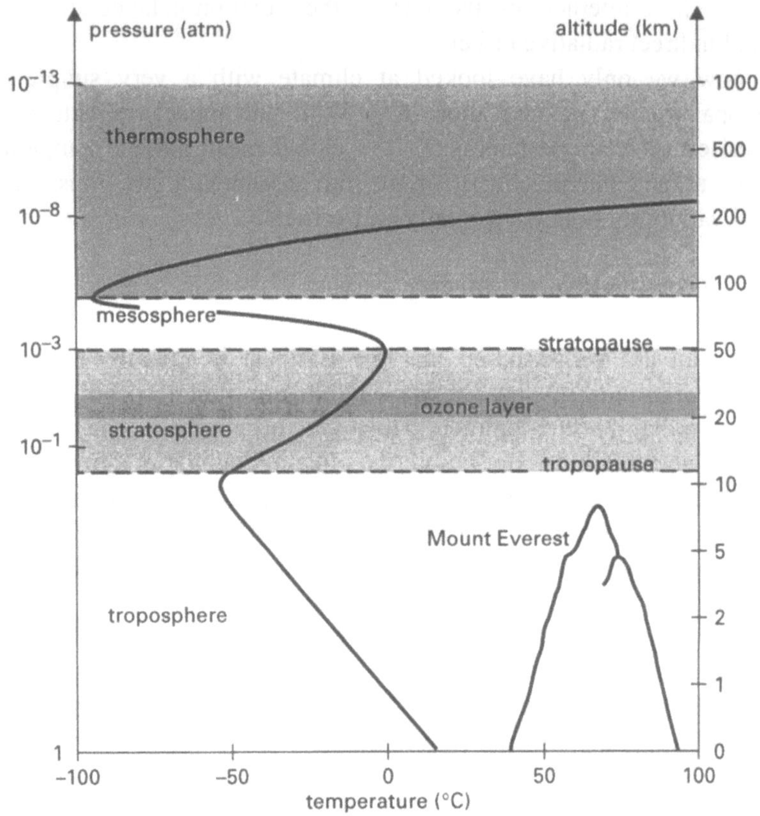

Figure 2. 3. Average vertical temperature variation with height and pressure.

The most important regions for weather and climate are the troposphere and the stratosphere. Based on the atmospheric temperature and greenhouse gases variations with height, and the different emission, reflection and absorption characteristics, a radiation budget can be determined. For simplicity only the global yearly mean budget of the surface and the mean troposphere are considered (Table 2.2).

Table 2.2 Yearly mean radiation budget averaged over the globe for the mean troposphere and the surface (negative values correspond to lost energy of the 'object').

	Short-wave	Long-wave	Net budget
Troposphere	+65	-176	-111
Surface	+174	-63	+111

Both the surface and mean troposphere are not in radiation balance. However, every object strives for balance. If both the troposphere and the surface strive for radiation balance, temperature must decrease and increase, respectively (the emittance of long-wave radiation must balance the absorption of radiation). However, this situation is not stable and only through transport of energy can balance be achieved.

To get hold of the term stability in the vertical direction of the atmosphere, the following reflection is useful. Consider a parcel of air in rest, near the surface, that is heated. The parcel expands, the density and pressure decrease and it starts to rise. As the parcel of air moves up adiabatically, i.e., without exchanging heat with the air outside, its temperature decreases. The rate at which the temperature falls with height, due to expansion, is called the dry adiabatic lapse rate and has a value of about $10 Kkm^{-1}$. If the temperature of the surroundings fell off less with height, a rising parcel would find itself colder than its surroundings, and therefore further rising is stopped. In other words, the atmosphere would be stable and the parcel of air returns to a height where the surroundings have the same density. If conversely the temperature of the surroundings fell off more quickly with height, the temperature and hence the density differences would increase and the parcel will undergo further acceleration. In that case the atmosphere is not stable.

Therefore, an atmosphere in radiative balance in the vertical induces a lapse rate that is unstable. Hence, heat is transferred from the surface upward into the atmosphere by convection (i.e., vertical air motion). Today's weather as we experience it, is restricted to the troposphere; the stratosphere, with an increasing temperature with height, is very stable. However, the radiation properties of molecules in the stratosphere are important for forcing the climate system. Moreover, it is the change in radiative flux at the tropopause that expresses the radiative forcing of the troposphere.

Water vapour, produced by evaporation at the surface, introduces another complexity. Air at a given temperature and pressure can only hold a certain amount of water vapour. The amount of water vapour relative to this saturation value is called relative humidity. When the relative humidity reaches 100%, water droplets condense out of the air, thereby forming clouds. The condensed water ultimately returns to the Earth's surface as precipitation. Thus, the amount of water vapour a parcel of air rising adiabatically can hold decreases with height. When a rising parcel cools down, the air saturates with water vapour, whereupon condensation takes place. If the parcel becomes saturated with water vapour, latent heat will be released as the parcel rises, so that the rate of decrease of temperature with height will be less than for dry air. This rate of decrease with height is called the moist adiabatic lapse rate and has a value that depends on the temperature and pressure. Near the surface, the value is about 5Kkm^{-1} at 10°C. Hence, whether an air parcel is unstable relative to its surrounding, depends on the lapse rate of the surrounding and the temperature and the amount of water vapour in the parcel itself.

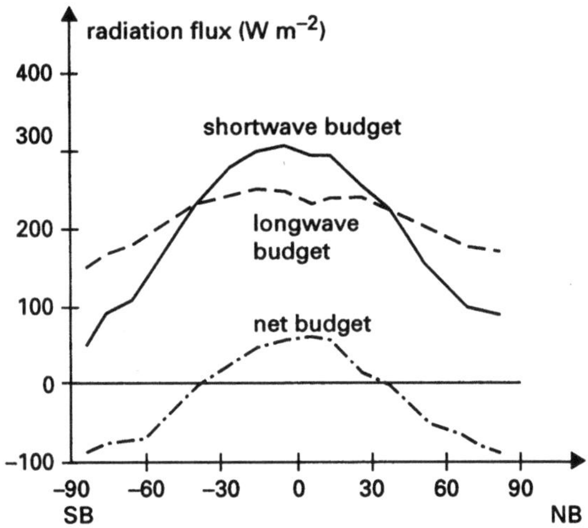

Figure 2.4 Variation with latitude of the net solar radiation absorbed (black line), the net long-wave radiation emitted (stippled line), and the total radiation budget of the Earth-atmosphere system (net budget). All Figures are averaged over height, over longitude and over year.

In summary, the large vertical temperature gradients that would be produced by radiation acting in insolation, result in convection that tends to reduce these gradients. Heat is transferred into the atmosphere by dry convection (sensible heat) and by latent heat released when condensation takes place in clouds (moist convection). The latter accounts on average for about 80% of the convective transport. Radiative-convective models try to represent the effects of convection without bothering about horizontal variations (see Section 3.2.2). Climate, however, is also associated with large horizontal variations.

Also in the horizontal direction climate is not in radiative balance. Figure 2.4 shows that the tropics (poles) absorb more (less) of the incoming short-wave radiation than it emits as long-wave radiation.

The variation with latitude of the absorbed radiative flux would lead to large horizontal temperature gradients if radiation acts in isolation. Again as in the vertical, such gradients are so large that the dynamics of the atmosphere and ocean induce motions that tend to reduce these gradients. Now heat must be transferred from the tropics to the poles. A variety of dynamical-circulation-processes both in atmosphere and in ocean contributes to this redistribution of energy.

2.3.1 Atmospheric circulation

One might expect the excess heating at the tropics (Figure 2.4) to cause rising motion; the heated air rises until it reaches to tropopause where its density becomes the same as the surrounding air, whereupon it moves polewards. In principle then, a single cell would be set up, which is known as Hadley cell. Due to the rotation of the Earth and to the large temperature differences with such a single cell, the Hadley cell is confined to low latitudes, less than about 30° (Figure 2.5).

As the equatorial air rises, air from higher latitudes moves towards the equator to take its place (trade winds). However, trade winds do not blow directly to the equator. The Earth's rotation induces a fictitious force, the Coriolis force, that results in the deflection of all moving objects not at the equator to the right of the direction of motion in the northern atmosphere and to the left in the southern Hemisphere. Hence, the trade winds on the northern and Southern Hemisphere blow from the Northeast and Southeast, respectively (Figure 2.5). They are not coming together (converging) exactly at the equator but at a region called the Inter Tropical Convergence Zone (ITCZ, see Figure 2.5). Moist air from the trade wind zone, where evaporation exceeds precipitation, is drawn into the areas of rising motion. The strong upward motion of air in the ITCZ is characterised by heavy precipitation in convective thunderstorms. On the other hand, the amount of water vapour a parcel of air can hold increases as the parcel descends, so the relative humidity in a descending parcel decreases. Hence, the regions with subsiding air (and as such with diverging surface winds) are dry, and include in particular the desert regions between latitudes 20° and 30° (Figure 2.5).

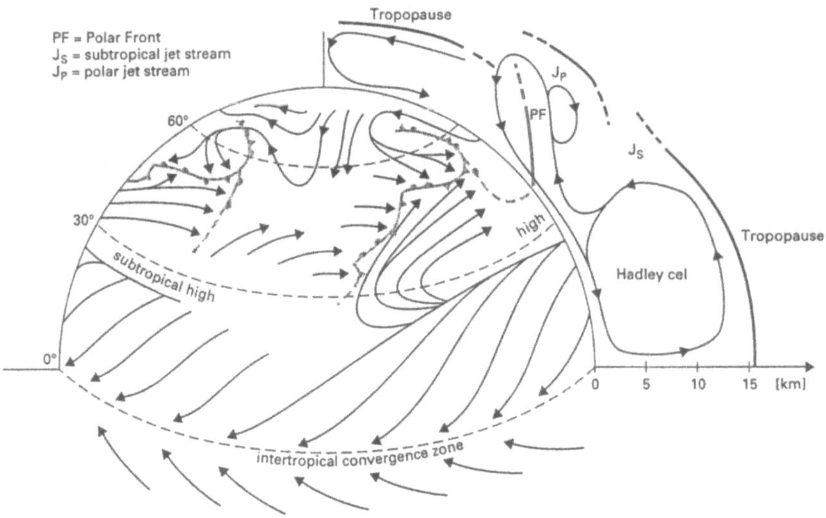

Figure 2.5 Idealised atmospheric circulation on the Northern Hemisphere. Three basic features can be distinguished: the (sub) tropical Hadley cell, a zone with relative strong westerlies at mid-latitudes and the polar anticyclone. The mid-latitude westerlies are most strongly developed at higher altitude, often referred to as 'jet stream'. The polar jet stream (Jp) is highly variable but is generally stronger than the subtropical jet stream (Js). Note that the tropopause extends to higher altitudes in the tropics. Near the polar front (PF) the tropopause makes break whereupon air exchange between troposphere and stratosphere takes place.

Also both poles are referred to as (polar) desert areas because air descends, and moves towards lower latitudes at the surface. The Coriolis force deflects these surface winds westwards, and these are therefore called polar easterlies.

In mid-latitudes, the picture is quite different. Because of the rotation of the earth, the motion produced by the horizontal density gradients is mainly West-East, and there is relatively little meridional circulation. This relative strong horizontal temperature gradient at mid-latitudes is referred to as polar front and separates relative cold polar air from warm subtropical air. However, the situation is not a stable one, and large transient disturbances (which appear as anticyclones and cyclones or depressions on the weather map) develop. Figure 2.6 shows how a small instability in the Polar front (left side), gradually grows to a fully developed cyclone.

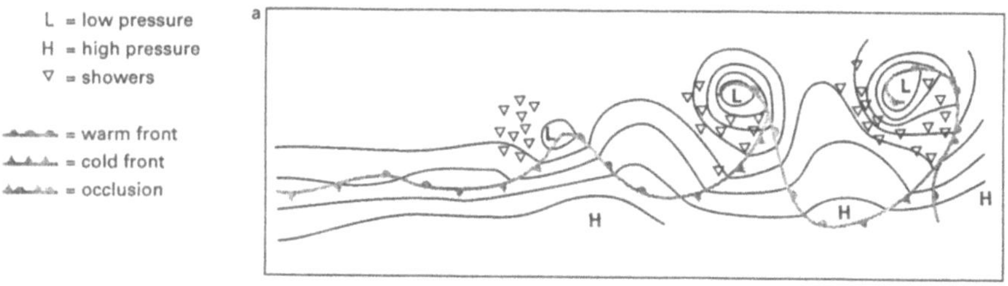

Figure 2.6. From the left different phases of development of a mid-latitude cyclone near the surface are shown. Black lines mark isobars (lines connecting points with same pressure value) and thin lines are fronts. An undulation in the polar front (entirely left) intensifies when it moves to the east with the westerly flow. In general the cold front moves faster than the warm front.

How efficient the atmosphere is in redistributing energy, strongly depends on the various mechanisms of atmospheric circulation. Cyclones and anticyclones at mid-latitudes are most effective at transporting heat polewards. Hence, the transport of heat in the atmosphere varies with latitude with a maximum at 50° (Figure 2.7).

2.3.2 Oceanic circulation

The ocean is as important as the atmosphere in transporting heat. Also the effectiveness of doing so strongly depends on the specific mechanisms. There is a large difference between atmosphere and ocean itself. Ocean currents that carry warm water from the tropics towards higher latitudes are very efficient in transporting heat at latitudes of about 20° (Figure 2.7).

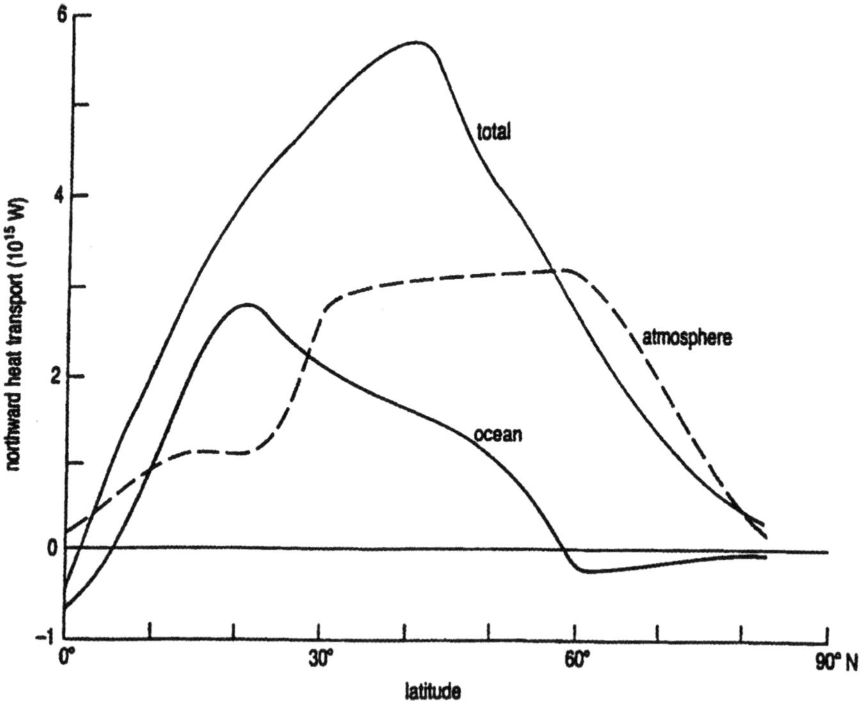

Figure 2.7. The northward transport of energy by ocean and atmosphere as a function of latitude (averaged at each latitude around the globe over a year). Negative values denote southward transport.

Displaced water warms the air and, indirectly, the land over which the air blows. As such land areas near oceans have maritime climates, which are less extreme than continental ones, with smaller day-night and winter-summer differences. The surface currents in the ocean are forced by atmospheric surface winds. Although there is a large similarity in patterns, the connection is not as obvious as it might first appear. One difference is due to the shapes of the ocean basins. Hence, the tendency for circular (gyral) motion is even more noticeable in the oceans, e.g., the North Pacific gyre, which includes the Kuroshio, the North Pacific current, the California current and the North Equatorial current (Figure 2.8).

Another outstanding difference is the complex current system near the equator. The equatorial counter-currents, i.e., currents against the prevailing wind direction, are small and sometimes not very good developed. Note that Figure 2.8 represents average conditions and that differences between ocean currents and wind fields are most evident during different seasons. The oceanic counterparts of transient weather systems (e.g., cyclones) are called mesoscale eddies. They last much longer than atmospheric (anti)cyclones and are also smaller.

So far, we have looked at the surface of the ocean. However, vertical circulation cells, like the Hadley cell in the atmosphere, are also possible in the ocean. Vertical motion in the atmosphere is related to density differences. However, these differences are only effective in producing motion because of gravity. In reality differences in buoyancy, i.e., the weight per unit volume, are important. A particle is said to be more buoyant when it has less weight. The ocean also moves because of buoyancy contrasts, but these differences are due to salinity (i.e., concentration of dissolved salt) as well as to temperature. The ocean circulation that is driven by buoyancy differences is called the thermohaline circulation.

Salinity is expressed in ‰ or concentration in parts per thousand by weight. The major dissolved constituents are chloride, sodium, sulphate and magnesium ions, respectively. Salinity in surface waters of the open ocean ranges from 33 to 37. The way salinity varies throughout the oceans depends almost entirely on the balance between evaporation and precipitation (Figure 2.9). The salinity of the surface waters is at a maximum in latitudes of about 20°, where evaporation largely exceeds precipitation. Remember that these regions are close to the descending part of the Hadley cell. Salinities decrease both towards higher latitudes and towards the equator.

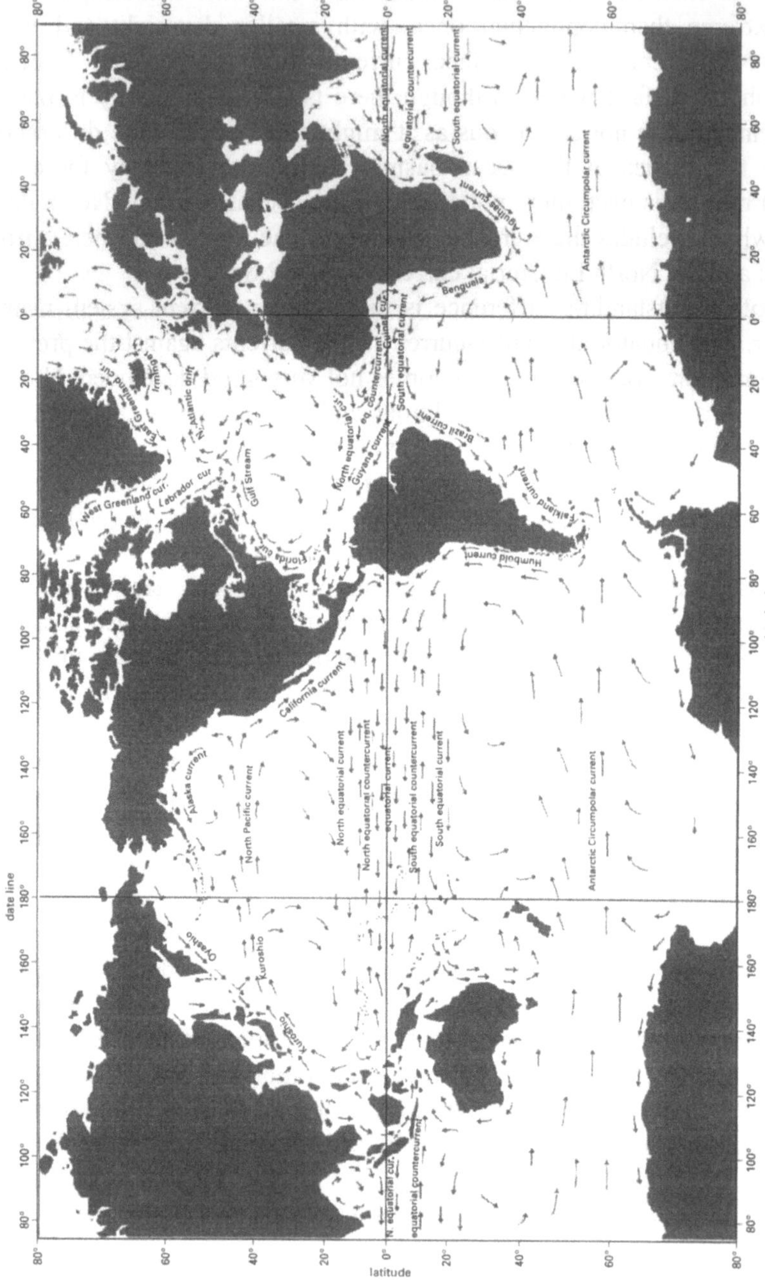

Figure 2.8 The annual mean surface currents in the oceans. There are seasonal differences in most areas, particularly in areas that experience strong seasonal variations in wind directions. Currents on the Western side of the ocean are 50-100km wide, the Eastern currents are even wider but also slower.

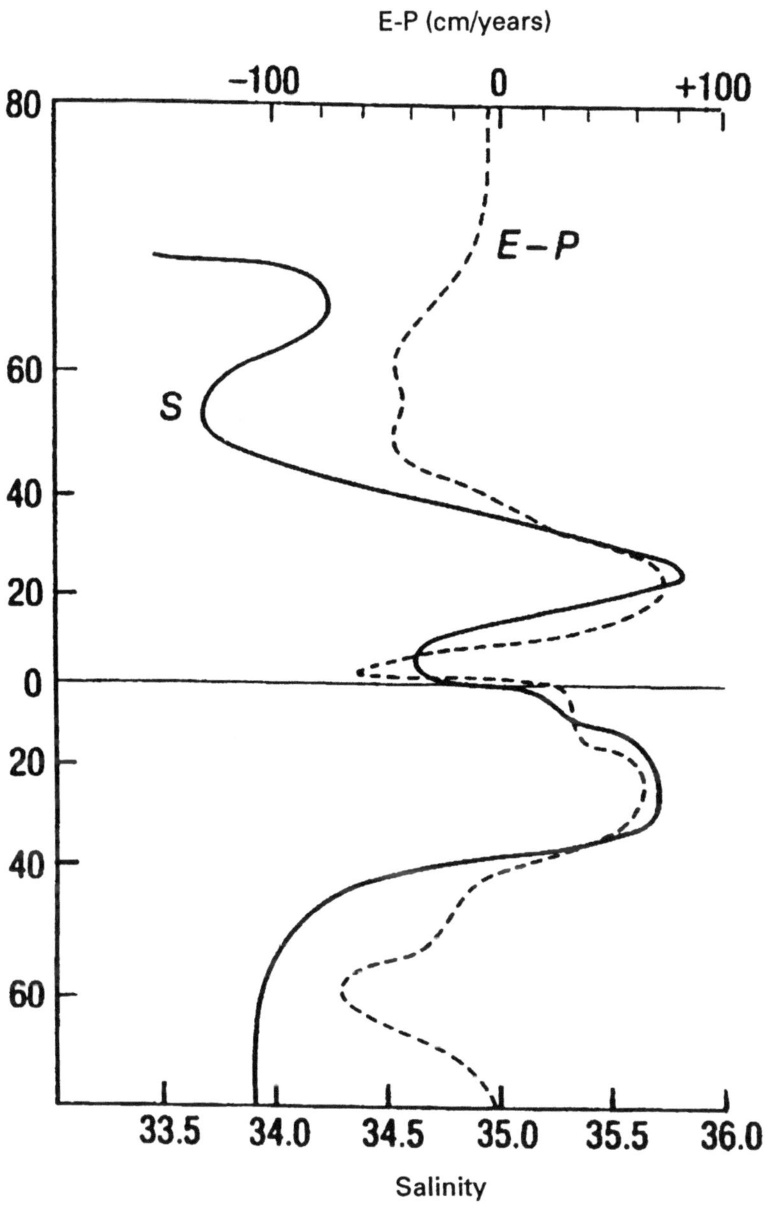

Figure 2.9 Average values of salinity and the differences between annual mean evaporation and precipitation (E-P) as function of latitude.

The way both temperature and salinity influence buoyancy is very important to the thermohaline circulation. Figure 2.10 shows that the freezing point and the temperature of maximum density are the same when salinity of water reaches about 25. Hence, the density of seawater increases with increasing salinity and falling temperature right down to the freezing point. Evaporation at the ocean surface, for example, decreases buoyancy by cooling and by increasing salinity. Precipitation, meltwater and rivers increase buoyancy locally by decreasing salinity.

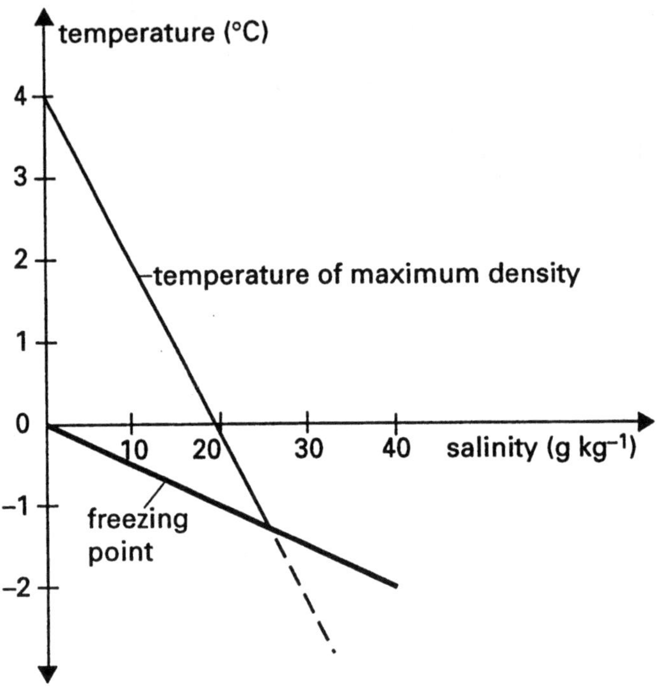

Figure 2.10 Temperature of freezing point and maximum density of water as functions of salinity.

Also the formation of sea-ice influences buoyancy locally. When sea-ice is formed, almost all the salt is left in the remaining seawater, which thus becomes more saline. This cold seawater with higher salinity is very dense, and sinks to greater depths. In this way some of the densest water is formed near Antarctica. This Antarctic Bottom Water flows along the bottom of the ocean around Antarctica and out into the Atlantic, Pacific and Indian Oceans (Figure 2.11). Near the North Pole, North Atlantic Deep Water (NADW) is formed mainly between Greenland and Norway, and flows south along the bottom of the Atlantic Ocean. NADW flows over Antarctic Bottom Water because it is less dense (Figure 2.11).

Measurements of the characteristics of water masses reveal a gigantic ocean 'conveyor belt' of deep water driven by the dense water sinking in the

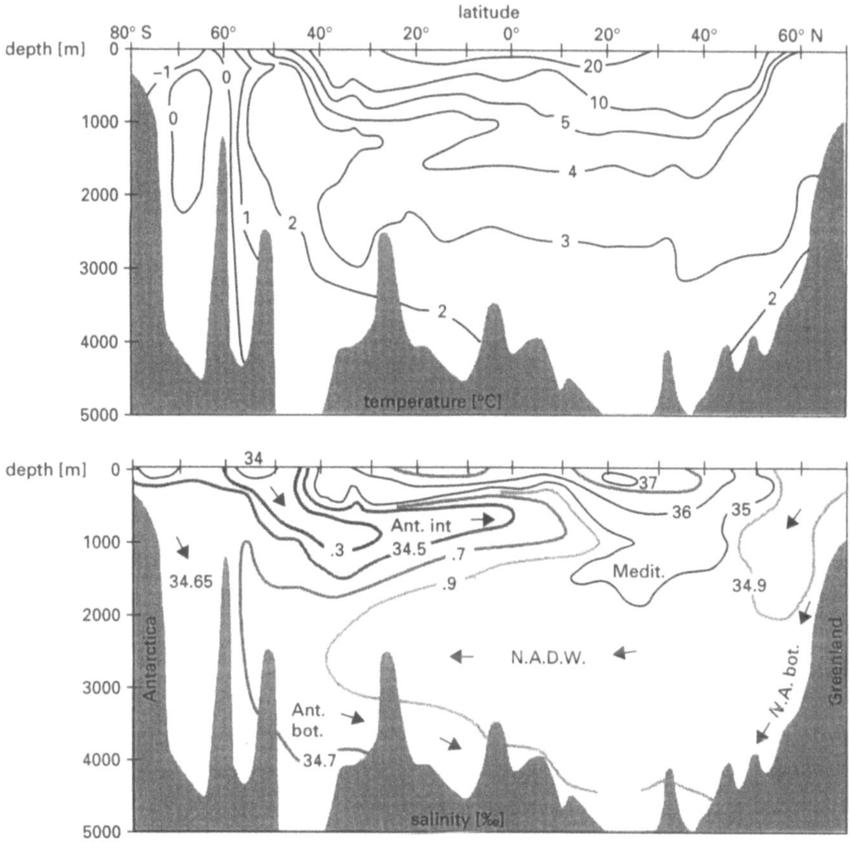

Figure 2.11 Meridional cross-section of the Atlantic Ocean, showing temperature- and salinity distribution. Bodies of water are identifiable by, among others, temperature and salinity for the very reason that mixing in the deep ocean occurs very slowly. Water masses that form in semi-enclosed seas provide particularly clear examples of bodies of water with recognisable temperature and salinity characteristics. Mediterranean Water (Medit.) is distinguished from other water masses by its relative high temperature and high salinity, and therefore can be recognised throughout much of the Atlantic Ocean at a depth of about 1000m, where it is neutrally buoyant. Intermediate water (e.g., Antarctic Intermediate Water; Ant. int.) is formed in subpolar regions where precipitation exceeds evaporation, and its salinity is therefore low. NADW and Ant. bot. refer to North Atlantic Deep Water and Antarctic Bottom Water, respectively.

North Atlantic region (Figure 2.12). On timescales of about a millennium this conveyor includes the transport of deep water from the North and South Atlantic into the Indian and Pacific Oceans, and the return flow of relative warm water near the surface (Figure 2.12). This part of the thermohaline circulation transfers heat from the North Pacific and Indian Ocean into the North Atlantic. As such it has large influences on local climate over much of the Northern Hemisphere.

Note that the thermohaline and ocean surface circulation are interconnected. When the NADW spreads to the bottom of the sea, and begins to flow towards the equator, this must necessarily intensify the surface current in the opposite direction. Hence, NADW warms Northwest Europe in two ways, first by transporting 'cold' to the deeper ocean, second by intensifying warm water transport of the North Atlantic drift.

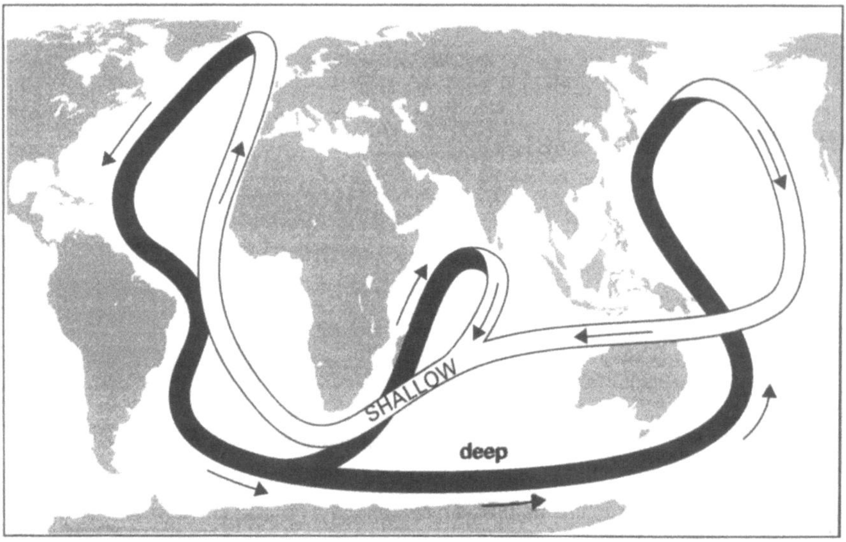

Figure 2.12 The long term thermohaline circulation or 'conveyor belt'. NADW transports cold water towards the North Pacific and Indian Oceans. The narrow Agulhas Current near South Africa (Figure 2.9) is critical for the return flow of heat through surface currents from the Indian Oceans to the Atlantic Ocean. The heat and fresh water fluxes between these two oceans take place largely by Agulhas rings that pinch-off from the Agulhas Current and penetrate the Atlantic. This transport of heat associated with this conveyor belt fluctuates substantially over time-scales ranging from years to millennia.

2.4 Changing climate

Equation (2.1) is a too simple formula but it is very illustrative to indicate how climate may change. Variation of the effective transmittance of the atmosphere (τ_e), albedo (α) and the solar constant (S) are the basic parameters that change climate.

2.4.1 Solar radiation and Milankovich

The total amount of radiation incident on the earth in 1 year, i.e., the solar "constant" (S) is variable. The most logical factor is the variation of the intensity of the Sun. The radiation output from the Sun does vary on a wide range of time-scales, from days to millions of years. Beyond the very slow evolution of the Sun, i.e., as a hypothesis of stellar evolutionary theory, there is direct observational evidence for shorter-term variations in solar irradiance. Scientists have long tried to link sunspots to climatic changes. Sunspots are huge magnetic storms that show up as cooler (dark) regions on the Sun's surface. They occur in cycles, with their number and size reaching a maximum approximately every 11 years (Schwabe cycle). However, they are thought to have relatively little effect on Earth's climate. First, these variations are very small: less than 0.1%, and second, they are also too short-term to influence the more slowly responding parts of the climate system like ice-sheets, glaciers, ocean, etc. (see also Section 2.5.1).

The length of the Schwabe cycle (defined through the interval between successive sunspot maxima) varies between 8 and 12 years over a period of about 80 years (Gleissberg cycle). Statistical analysis shows a good match between the average surface temperature and the length of the Schwabe cycle. Lower-than-normal surface temperatures tend to occur in years when the sunspot cycle is longest, and visa versa. This close correlation could account for the average surface temperature changes from 1940 back to the 16th century, and could partly explain the slightly cooling phase between 1940-1970. The period known as the Little Ice Age corresponds to a minimum level of sunspot activity (the Maunder Minimum, 1645-1715), the estimated change in solar irradiance is a 70-year-long reduction of about 0.14%. Studies with climate models (Chapter 3) suggest that such a drop would neither be large enough, nor long enough to explain the observed cooling during the Little Ice Age. However, these climate models do not include the observed match between surface temperature and the length of the Schwabe cycle. But as long as the mechanism behind this correlation is not well understood, it cannot be incorporated in these climate models. Moreover, close correlation without a realistic mechanism does not prove

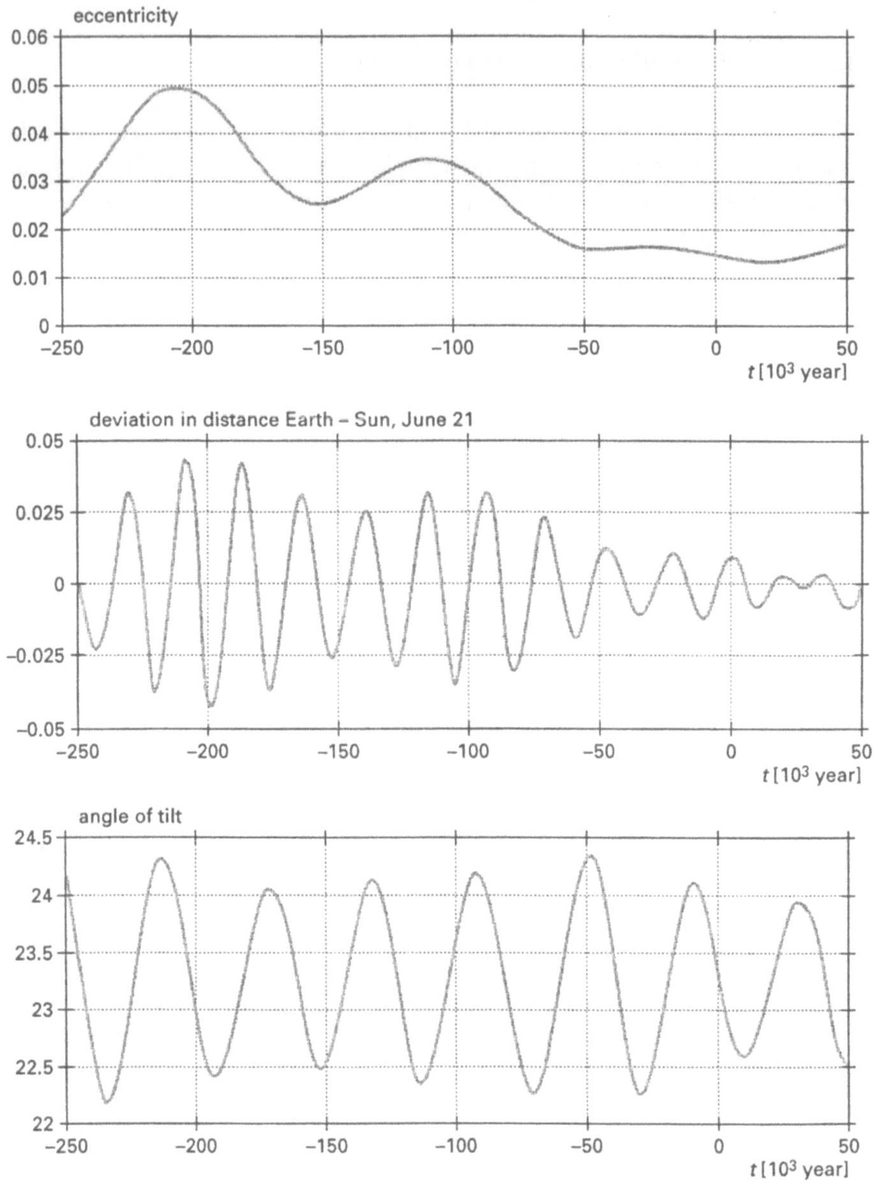

Figure 2.13 Long-term variations of the three Milankovich orbital parameters: eccentricity (a), precession (effectively the Earth-Sun distance on June 21st (b)) and tilt (c) of the Earth.

anything. Recent studies about a possible mechanism focus on a connection between the global cloud cover and cosmic rays.

Another reason why S varies is because of changes in the average distance Earth-Sun. The eccentricity expresses to what extent the Earth's orbit around the Sun differs from a circle. The orbit is somewhat elliptical, and a higher eccentricity corresponds to a more elliptical orbit. The eccentricity varies regularly with periods of about 100.000 and 400.000 years (Figure 2.13).

The maximum change in S associated with variation in eccentricity is about 0.1%. However, the eccentricity has a stronger influence on the seasonal variation of short-wave radiation. When eccentricity is relatively high, the Earth receives more radiation on days when it is closer to the Sun (see Figure 2.14). This intensifies the seasons in one Hemisphere but moderates them in the other.

The eccentricity is just one of three orbital parameters that influence seasonal and latitudinal changes in short-wave radiation reaching the Earth. The angle of tilt of the Earth's axis of rotation (Figure 2.14) varies between 22° and 24.5° with a periodicity of about 40 000 years (Figure 2.13). The greater the tilt, the more extreme the seasons in each Hemisphere become. For example, the duration of the winter-darkness near a pole is determined by the tilt only.

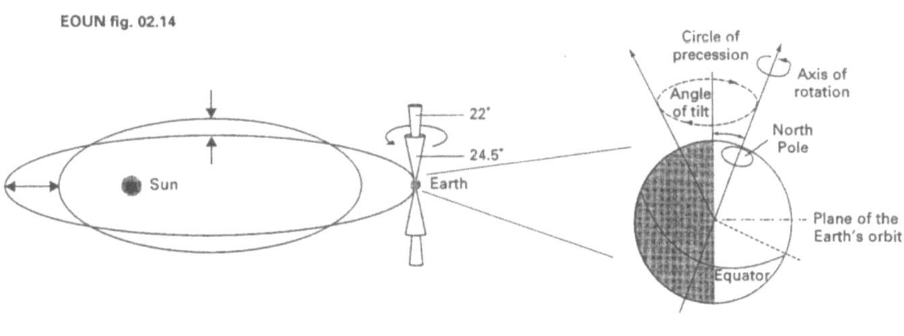

Figure 2.14 Variation of the orbital parameters as seen from space. The ellipse of the Earth's orbit is somewhat exaggerated.

Precession is the change of direction in space of the Earth's axis of rotation. The axis changes direction with a periodicity of about 22 000 years (Figure 2.13). There is some correspondence between a top; the axis of rotation swings according to an imaginary cone perpendicular to the plane of the Earth's orbit (Figure 2.14). The precession affects the seasonality in each Hemisphere because it determines when the distance between the Earth and Sun is at a minimum or maximum. At present, the distance is a minimum in the Southern Hemisphere winter, so the Southern Hemisphere has on the whole slightly warmer summers and colder winters. The distance Earth-Sun in June 21st, as shown in Figure 2.13, is controlled by both precession and eccentricity. Active calculation of the solar radiation on an arbitrary place on Earth and time of year is not so easy. In practice the tilt of the Earth's axis of rotation becomes more important polewards.

All the above mentioned components of the orbit vary because of the gravitational attraction between the Earth and the other planets. The orbital changes are known as Milankovich orbital changes. He did put forward the theory that the periodic changes of climate between glacial and interglacial are related to the orbital changes of the Earth. They are only important for climate changes on very long timescales. Hence, when modelling the enhanced greenhouse effect, the solar constant can probably be handled as a true constant.

2.4.2 Albedo and albedo-temperature feedback

The value of reflectivity strongly varies with the object and depends on wavelength. The reflectance averaged out over all wavelengths is called the albedo of an object. The planetary albedo refers to the time-averaged value of the Earth. Measurements with satellites show large latitudinal variation of both the planetary albedo and the albedo under clear skies (Figure 2.15).The minimum albedo is close to the value without clouds, ice and snow. On land, the value is usually about 0.15, with higher values in desert (up to 0.3) and in icy regions, reaching about 0.6 in parts of Antarctica (Figure 2.15). Comparison of the albedo under clear skies with the planetary albedo shows the effect of clouds. For instance, most of the ocean within 40° has a surface albedo of about 0.12, but the average albedo is close to 0.3. Also the solar elevation is important for both the albedo of clouds and of oceans. The albedo of liquid water strongly increases to about 0.7 when the Sun is close to the horizon. For example, around sunset the disc of the Sun is clearly visible in water. The highest albedo of 0.95 is measured above virgin snow, but the albedo of snow decreases to about 0.6 when it gets contaminated.

It is clear from these figures that the factors that influence albedo are very important in determining the energy balance of the Earth. In fact, as temperature changes, the planetary albedo changes with it. These associate changes even amplify the initial temperature change, i.e., feedback process. The climate system can respond to imposed changes by either reinforcing the initial change (a positive feedback) or by dampening their magnitude (negative feedback). A relative strong positive feedback associated with the albedo of snow and ice is known as the albedo-temperature[3] feedback.

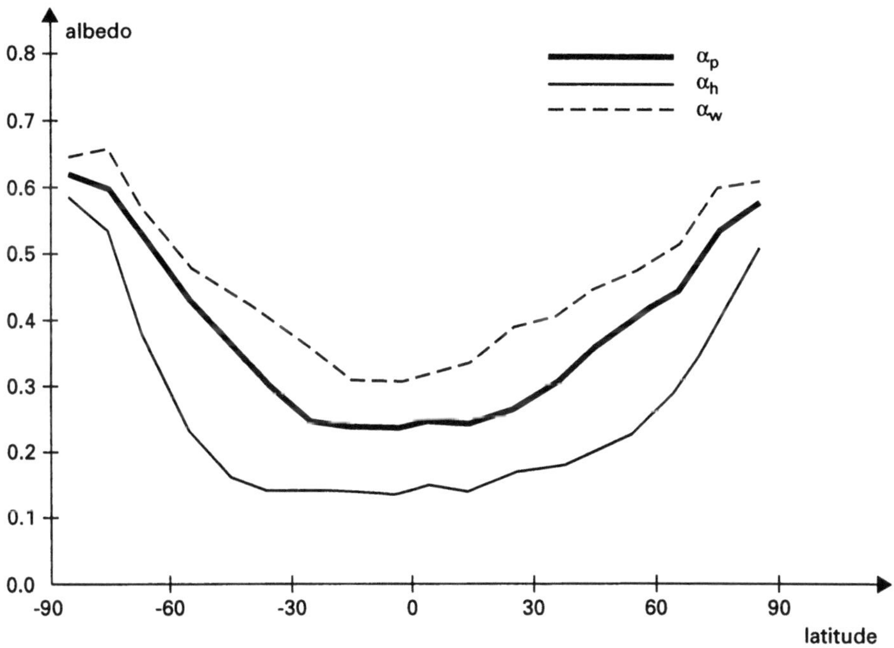

Figure 2.15 The latitudinal variation of the planetary albedo (α_p), the albedo under clear skies (α_c) and the deduced albedo for clouds (α_w). Values are averaged over height, over longitude and over year.

[3] Different expressions exist in literature concerning this feedback process: ice-albedo feedback and snow-climate feedback..

When global-mean surface temperature decreases, e.g., due to a lower value of the solar constant, the average snow- and ice cover increases, reflecting more of the solar radiation and reducing the absorption of solar radiation and, hence, decreasing the effective radiating temperature. This creates a positive feedback whenever the lower radiating temperature also implies a lower surface temperature[4]. The strength of this albedo-temperature feedback also depends on the average temperature, e.g., a temperature decrease in a completely ice-covered planet is not enhanced by the albedo-temperature feedback. At present, a rise in temperature near the equator will not modify the snow- and ice-cover. As a consequence, the strength of the albedo-temperature feedback varies both seasonal and geographical.

2.4.3 Greenhouse gases and the water vapour-temperature feedback

Variation of the effective transmittance of the atmosphere (τ_e) is related to the changing concentration of greenhouse gases in the atmosphere. The concentration of these gases, however, is not constant to a changing temperature. Hence, they may be involved in different feedback processes. A clear example is the positive feedback that involves water vapour. As temperature rises the evaporation of liquid water increases and the amount of water vapour air can hold, before it saturates, increases. Because water vapour is a strong greenhouse gas, an increase in its atmospheric concentration will amplify the initial temperature rise. This positive feedback called water vapour-temperature feedback works equally well whatever the direction of temperature change is. In the above reflection it is assumed that the relative humidity hardly changes with temperature. This assumption strongly influences the strength of the water vapour-temperature feedback. However, concerning the strength of the water vapour-temperature feedback, there is a remarkable agreement between large climate models and with observation data. It is important to notice that the enhanced greenhouse effect is not a positive feedback by itself, but its effect can be amplified or dampened by different feedbacks.

2.4.4 Crucial role of aerosols and clouds

Clouds not only contribute to the greenhouse warming; they are also highly reflective, thus contributing to the planetary albedo. The planetary albedo and the albedo under clear skies are measured by satellites. Together with the estimated cloud cover, the albedo of clouds can be determined

[4] This conclusion is not directly evident because properties of the atmosphere in the vertical may change.

indirectly (Figure 2.15). The albedo of clouds strongly varies with latitude because the albedo of liquid water strongly increases with a decreasing solar elevation. This effect is related to the cloud optical depth, i.e. the path of a sunray in a cloud. In this respect also the amount of liquid water and the amount of ice-crystals in a cloud are important. Ice-crystals are formed at very low temperature and, hence, depend on the altitude of the cloud. Also the downward backradiation varies with the cloud-level because long-wave radiation emitted by clouds depends on temperature. The net effect under annual mean conditions is that the cooling through reflection dominates the greenhouse contribution. Hence, on net, climate is cooler with clouds than under clear-sky conditions. More important is the role of clouds when temperature changes. In other words, can clouds reduce the global warming (i.e., produce a negative feedback) due to the enhanced greenhouse effect? Feedback mechanisms related to clouds are extremely complex. Predictions of the enhanced greenhouse effect with large atmosphere models differ mostly because the strength of the cloud feedback[5] and even its sign is highly uncertain. Processes that determine the strength of the cloud feedback involve cloud cover, cloud-level, cloud droplet size distribution and the interplay between temperature and the presence of ice crystals or liquid droplets. In this respect also the amount of condensation nuclei is important (see hereafter).

Looking from the top of a mountain into a valley, one will notice a hazy atmosphere that reflects some sunlight. The effect is caused by the scattering of sunlight on solid and liquid particles in the atmosphere. All these floating particles are called aerosols. The total effect of aerosols on the planetary albedo is estimated at 0.05. Like clouds, aerosols have a dualistic character in determining the radiation budget. The outgoing long-wave radiation is hampered by aerosols, such that the effective transmittance of the atmosphere (τ_e) decreases. The net effect of aerosols on climate is very difficult to estimate because of the general lack of observational data, together with the extraordinary diversity of tropospheric aerosols (in size, optical properties, atmospheric lifetimes and chemical composition). However, stratospheric aerosols are an exception. For these particles it is estimated that changes in the short-wave radiation budget dominate those in the long-wave radiation budget. Hence, aerosols in the stratosphere tend to cool Earth's climate.

Again aerosol concentrations in the lower troposphere as well as in the stratosphere are not constant. Concentration of aerosols fluctuates strongly by infrequent occurrence of dust storms, volcano eruptions, biomass combustion and anthropogenic input in industrial, urban and agricultural

[5] This feedback includes, among others, cloud cover-albedo feedback and cloud height-temperature feedback.

areas. The latter is the main cause of the haze that reduces visibility over industrialised areas. Once particles are levitated into moving air currents, they can be transported over long distances. Convection may also transport aerosols higher into the troposphere. However, the relative high concentration of aerosols in the stratosphere is not directly related to the above sources because the air exchange between troposphere and stratosphere is very slow. The perturbations of stratospheric aerosol concentration are strongly related to high energetic eruptions of volcanoes. There are two types of particles volcanoes inject into the stratosphere. Volcanic ash is typical silicate dust created from the fragmentation of solid rock. Absorption of solar radiation by volcanic dust prevents the short-wave radiation from reaching the Earth's surface. More important, however, is the back scattering on sulphate aerosols that increases the planetary albedo. Most of the volcanic ash falls out of stratosphere within a few months, but sulphate aerosols may remain there for several years. On average, major volcanic eruptions are thought to reduce the global mean temperature by 0.5°C for up to two years (e.g., the eruption of Mt. Pinatubo in June 1991).

Stratospheric sulphate aerosols are produced out of carbonyl sulphide (COS) and out of one of the volcanic gases released during an eruption: sulphur dioxide (SO_2). The responsible process, called gas-to-particle-conversion, is another important source of very small particles in the troposphere. These very small particles occur by direct nucleation from the gas phase but are mostly created by accretion onto pre-existing particles. Sulphur-containing gases are major participants in gas-to-particle-conversion. Sulphuric acid is formed out of reactive gases, like dimethylsulphide (CH_3SCH_3), hydrogen sulphide (H_2S), and sulphur dioxide. Sulphuric acid quickly condenses out onto cloud droplets and aerosol surfaces. Under natural conditions the main source is dimethylsulphide emitted into the atmosphere as a biological waste product from the ocean surface. However, today, the anthropogenic source of sulphur dioxide induces a severe increase of sulphate aerosols in the troposphere. Coal and oil both contain significant quantities of sulphur. Hence, anthropogenic input of sulphur dioxide is related to that of carbon dioxide. Therefore, this anthropogenic climate effect can be estimated reasonable well. Recently, calculations with climate models suggest that this increase of sulphate aerosols in the troposphere may temporarily and partly offset the enhanced greenhouse effect.

This direct cooling effect of tropospheric sulphate aerosols may be enhanced by an indirect effect. Most aerosols can act as so-called cloud condensation nuclei (CCN), i.e., by condensation of water vapour on aerosols cloud droplets are formed. More aerosols can cause an increase of the number of small droplets whenever the amount of cloud water stays the same. Smaller droplets increase the reflectivity of clouds. Hence, the cloud albedo over continents, especially over more polluted areas, is somewhat higher than over oceans. Because of the lack of observational data and due to the complexity of processes involved, the possible influence of anthropogenic SO_2-emission on cloud albedo and subsequent on climate is difficult to quantify. Moreover, the direct and indirect cooling effect by tropospheric sulphate aerosols is not a solution to offset the enhanced greenhouse effect for two reasons. First, the anthropogenic SO_2-emission is the main source of acid rain. Second, the residence time of tropospheric sulphate aerosols is very short (a few weeks at most) compared to that of most greenhouse gases (decades to centuries). Moreover, radiative forcing by tropospheric sulphate aerosols is highly regional, while the atmospheric CO_2-concentration increases globally.

2.5 Changing climate interacting with the different spheres

Climate changes by variation of climate parameters like albedo, the effective transmittance of the atmosphere and the solar constant. These climate parameters determine the radiative forcing of the climate system. But even when the magnitude of a changing radiative forcing is known, it is still very difficult to quantify such changes into predictions of future climate both global and regional.

First of all general circulation in the atmosphere and ocean provides the transfer of heat necessary to balance the system. As such, they are interconnected with each other. The ocean circulation is driven by the atmosphere through surface winds and the fresh water flux (i.e., mainly evaporation – precipitation). The oceans store and redistribute heat, before releasing it to the atmosphere, mainly by evaporation. Latent heat, released when water vapour condenses, forces the atmosphere, and, hence, the surface winds and the amount of precipitation. Moreover, the atmosphere and ocean strongly differ in their effectiveness of transporting heat polewards (see Figure 2.7). The prediction of especially local climate requires detailed description of both systems and how they influence one another (i.e., coupled atmosphere-ocean models, see Section 3.2 and further).

Second, the climate system can respond to the imposed changes by different feedback processes that influence the radiative forcing itself. Processes within the climate system both influence and are influenced by a multitude of interactions between the relevant processes. Hence, without knowledge of how the whole climate system responds, predictions of the enhanced greenhouse effect will be limited to quantitative statements or, at the most, to give the magnitude of the warming within a range (e.g., 1.5°C-4.5°C).

In the previous Section we have looked at the most evident feedback processes. In general they are referred to as geophysical feedbacks. To reveal other relevant feedback processes, it is essential to look at the whole climate system. Climate is generally regarded to be influenced by the operations of five regimes. The influence of the atmosphere and ocean on climate are already extensively discussed in the previous Sections. The cryosphere is that part of the Earth that is covered with ice and snow. The biosphere contains all forms of life on Earth and the geosphere is the soil bearing part of Earth's surface.

2.5.1 Cryosphere

The cryosphere is found near the poles and on top of high mountains on all continents. It influences the radiative forcing of climate mainly because of the large difference in albedo with ice- and snow-free regions. Ice sheets and mountain glaciers contain about 80% of the fresh water on the globe. They therefore also influence both the sea level and the thermohaline circulation.

Sea-ice prevents atmospheric surface winds to force ocean currents. Also the exchange of heat- (e.g., latent heat) and fresh-water between the ocean and the atmosphere is influenced. Hence, through sea-ice, the cryosphere tends to decouple ocean and atmosphere. During summer months, the ice-free ocean at high latitudes absorbs more solar radiation, primarily due to the less reflective surface. This heat is again released to the atmosphere during autumn and winter, which delays the onset of sea-ice formation. In a warmer climate more heat is stored in the ocean, which delays the sea-ice formation even further. In return, the ice-free ocean can then continue to release heat over a longer period. Over continental regions, the reduced extent and earlier melting of high albedo snow cover similarly acts to amplify the surface warming. However, the land surface losses heat much quicker than the ocean. Hence, sea-ice effects climate on timescales of seasons, while snow cover over land effects the radiative budget on timescales of days. Sea-ice also influences the buoyancy of surface waters by salt extrusion during the freezing period and as such, it influences the thermohaline circulation.

Information about past climates is inferred from a variety of biological, chemical and geological indicators. Examples are fossil pollen grains, the shells of marine micro-organisms, sedimentary records, landforms associated with glaciation, and so on. They reveal that Earth's climate experienced different periods of glaciation. Figure 2.16a gives an estimate of global-mean temperature variation derived from proxy data over the past million

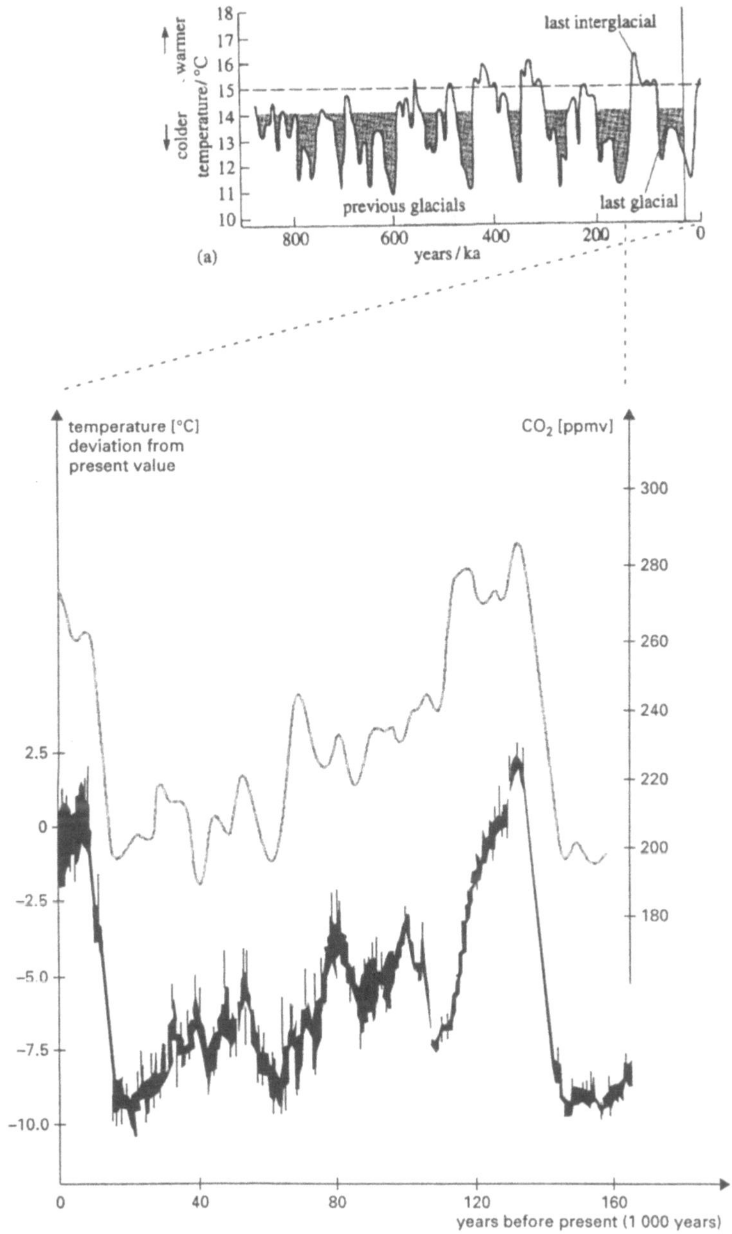

Figure 2.16 Global-mean surface temperature variations over the past million years (a) and during the last glacial-interglacial period (lower half of b). Part (a) expresses the mean value derived from proxy date (e.g., ice-cores and deep-ocean sediment) while part (b) is from the Vostock ice-core only. The latter also reflects the carbon dioxide concentration during the last glacial-interglacial period (upper half of b).

years. The glacial-interglacial cycles manifest themselves on a global scale: data from ice cores in Greenland and Antarctica and sedimentary records over different deep oceans are revealing the same pattern.

The Milankovich orbital changes are thought to at least initiate these periodic changes of climate between glacial and interglacial. However, the formation and subsequent removal of substantial ice cover is not that simple. Figure 2.16a shows an asymmetric periodicity of the glaciation cycle. The formation of an ice sheet occurs relatively slowly because the growth is limited to the precipitation of snow. On Antarctica the average precipitation conditions yield to growth of 0.15m ice each year. However, by melting and evaporation most of it disappears again. The average thickness of the present ice-sheets is about 2000m. So, large ice-sheets are only formed after at least 20.000 years. Due to gravity, ice formed in areas, where accumulation of ice dominates the melting (accumulation area), will flow to regions where melting dominates or calving in occurs (ablation area). The melting of ice-sheets may happen very fast as air-temperatures may rise far above zero. As such, an ice-sheet may collapse within some millennia. Therefore, summer temperatures are far more important than winter temperatures to determine whether an ice-sheet will develop or collapse. The summer radiation on higher latitudes is largely determined by the tilt of the Earth's axis of rotation and therefore experiences a periodicity of at most 40.000 years (see Section 2.4.1). Glacial periods, however, typically seem to occur every 100.000 years (Figure 2.16a). Hence, the Milankovich hypothesis is not the whole picture of glacial cycles. In practice, it is the interplay of different complex (feedback) processes that makes the cryosphere react in a non-linear way (i.e., the response characteristics are different from the forcing pattern).

Even during a glacial or interglacial period climate is not constant. Relative rapid climate shifts are observed in glacial periods as well as during the transition between a glacial and interglacial. One example of the latter is the period called the Younger Dyras, a cold spell-period of about 1.000 years, some 11.000 or 12.000 years ago, with rapid onset (700 years) and an even more rapid termination (in some places only 50 years). The cooling appears to have been the strongest in the North Atlantic region but it also shows up in palaeoclimatic records from New Zealand and Antarctica. This and other relative rapid climate shifts are thought to be connected to the North Atlantic's thermohaline circulation. Massive iceberg discharge into the North Atlantic Ocean reduced the density of the North Atlantic surface water and, hence, the production of NADW. Evidence comes from changes in North-Atlantic surface temperature that closely mirror high frequency temperature changes recorded in Greenland ice cores over the past 90.000 years. There is even a close correspondence to deep-water temperature and as such, to the history of NADW production. Hence, the climate influence of

NADW variability is widespread. NADW production decreased during glacial times and also declined during the Younger Dyras cool period at the end of the last glaciation.

Figure 2.16b shows that temperature variations largely coincide with variations of carbon dioxide concentration during the last glacial cycle. Carbon dioxide concentrations in the past can be measured from the air in ice pores. However, different pores remain interconnected for 100 to 1.000 years after snow deposition. Together with uncertainties in the measurement of temperature and carbon dioxide concentration, the time differences between the two records must be greater than about 3.500 years, before lags can be determined. Hence, this dataset cannot reveal whether carbon dioxide concentration changed before temperature did or whether changes in carbon dioxide were caused by temperature changes. However, there are some indications that carbon dioxide concentration lags behind the temperature change during the transition from the interglacial to the last glacial (i.e., during a cooling trend).

Changes in solar radiation induced by orbital changes are not large enough in themselves to cause the size of temperatures changes deduced from the proxy data. To trigger the onset or the termination of glacial periods, this direct effect must be amplified by positive feedbacks within the Earth's climate system. Studies with climate models indicate that changes in the strength of the greenhouse effect (as a result of changes in atmospheric CO_2) are at least as important as the geophysical feedback processes to amplify the initial change and can almost explain the derived temperature change. Why CO_2 concentration changed during an interglacial-glacial transition is still an open question. However, there is some evidence that feedback processes within the oceanic part of the biosphere are at play.

2.5.2 Biosphere and biogeochemical feedbacks

The biosphere is defined as the spherical shell encompassing all forms of life on Earth. The biosphere essentially interacts with climate by the albedo and by the concentration of greenhouse gases in the atmosphere. The surface albedo differs for different types of land cover. The albedo for snow decreases in a forest to about 0.35. A forest will temper the cooling effect associated with high reflective snow. Desert areas, on the other hand, have a higher albedo than other land types. Hence, the biosphere may act as a thermostat, such that Earth's climate is kept into bounds to support the maintenance of life. This concept is central in the Gaia hypothesis (see Chapter 8.3) developed by James Lovelock, who demonstrated this with the well-known Daisyworld concept. The cooling effect of sulphate aerosols as discussed in Section 2.4.4 is central to another possible example.

Dimethylsulphide produced by planktonic algae in seawater oxidates to sulphate particles that reflect sunlight both directly and indirectly (CCN). According to this proposed negative feedback, a warmer climate will lead to enhanced oceanic productivity and to larger populations of dimethylsulphide-emitting plankton, thus to higher concentrations of sulphate aerosols and, hence, to dampening of the initial warming.

The biosphere fulfils an important role in the strength of the greenhouse effect because, e.g., plants use the two major greenhouse gases (CO_2 and H_2O) to produce simple organic compounds. The energy for this process, called photosynthesis, comes from sunlight and makes plants autotrophs ('selffeeders'). A proportion of the CO_2 that plants take up is returned to the air through their own respiration, i.e., the reverse of photosynthesis. The fixed carbon not released back into the atmosphere is called the net primary production (NPP), i.e. carbon taken up by photosynthesis minus carbon lost during plant respiration.

Photosynthesis in the oceans is limited to the surface, i.e., the photic zone, because water strongly absorbs sunlight. The photic zone rarely extends more than 200m. Microscopic phytoplankton drives the biological cycle in the photic zone and deeper in the oceans. They take their CO_2 from that dissolved in the seawater around them. The phytoplankton are grazed by zooplankton that package most of their fast products into faecal pellets, which in return are consumed and decomposed by other organisms. A small fraction (about 10%) of the carbon fixed by photosynthesis escapes from the photic zone by gravity, downward mixing and/or downwelling associated with the formation of deep water. Hence, these sinking particles represent a net and direct transfer of carbon from the surface into the deep ocean: the *biological pump*. In an unpertubated world this downward transfer of carbon is balanced by a net upward flux (e.g., the upward transport in the Indian and Pacific Oceans as part of the large 'conveyor belt', see Section 2.3.2.). This pumping around of carbon is a part of the carbon cycle discussed in Chapter 4.

Careful analyses and dating of marine sediments have shown that the biological pump was stronger during the last glacial period. Thus, more fixed carbon settled out of surface waters, i.e., carbon lost from the atmosphere went into the deep ocean. The very mechanism behind this enhanced marine productivity is still unclear. Changes in atmosphere and, thus, oceanic surface currents or even in the thermohaline circulation could themselves have initiated these changes. However, the interplay of physical, chemical and biological processes is very complex and makes it difficult to reveal the sequence of processes. More observational data will perhaps clarify this subject.

Coupling between climate and the terrestrial biosphere is controlled by the extent of plant growth that depends on, among other factors, temperature, precipitation and carbon dioxide concentration. Under *ideal* conditions, an increased CO_2 level will enhance net photosynthesis and therefore stimulates plant growth, the so-called CO_2-fertilisation effect. Plant growth will remove CO_2 from the atmosphere, thus potentially dampening its atmospheric rise. Thus, this fertilisation feedback may hamper the anthropogenic input of CO_2. However, the strength of this feedback strongly depends on the type of plants. Many tropical crops, like maize and sugar, are less stimulated by raised CO_2 levels. The runner bean and the beach pine, for example, even reduce photosynthesis in response to elevated CO_2. On the other hand, higher surface temperatures may influence both the NPP and the respiration rate. A global temperature increase enhances soil and litter respiration rates, thereby causing release of additional CO_2 to the atmosphere and, thus, stronger greenhouse forcing. This respiration feedback amplifies the anthropogenic influence on climate. The NPP-feedback reflects a net increase of the NPP as temperature increases and, as such, is a negative feedback. The sign of the total temperature effect again strongly depends on the plant-species involved and the average temperature itself.

Under natural conditions, growth is limited by the availability of nutrients, light, water and temperature (limiting factors). In general only one of these factors limits photosynthesis. In the photic zone it is often nitrate and on land it is often phosphorus. On land the rate of photosynthesis also strongly depends on the availability of water. Diffusion of gases through leaves typically depends on pores (stomata) on the lower leaf surface (the rest of the leaves acts as a barrier because of a waxy layer called cuticle). To ensure uptake of CO_2, these stomata are open during daylight. At the same time, however, this causes severe loss of water by evaporation, especially in bright, warm and windy conditions. Water lost by this process of transpiration has to be replaced from the soil through roots. Plants may suffer from water stress if the water content of the soil is reduced and therefore they have to (partially) close their stomata. Hence, the opening and closing of the stomata reflects the continuing conflict of maximising photosynthesis and avoiding an excessive loss of water. The partial closure of stomata is also triggered when the CO_2 level increases. There is even some evidence from palaeoclimatic research that the number of stomata decreased when CO_2 levels increased. The advantage is that precious water is conserved without compromising photosynthesis. Thus, when CO_2 concentration is increasing, plants tend to use water more efficiently.

In summary, living organisms participate in climate regulation in an active and responsive way. There exists a tightly coupled system, which include the biota, the atmosphere, the ocean and the geosphere. Within such a system, the growth of organisms changes environmental conditions and environmental change feeds back on growth. The overall response of the biosphere in a climate with an enhanced greenhouse effect is subject to many factors. Different plant species already react differently to the enriched CO_2 levels. In a (possible) future environment that is warmer and richer in CO_2, the soil moisture content will be lower. This means that we have to look at the combined effect of all these changes. How a plant will react to all these factors is still uncertain, but the overall response of all species in an ecosystem is subject to even larger uncertainties (see also Chapter 6.3.2 and 6.3.4).

2.5.3 Geosphere

The geosphere is defined as the solid portion of the Earth's surface. Its influence on climate on relatively short timescales (days up to months) is known as land processes. Groundwater processes determine the interaction with, mainly, the biosphere. The amount of fresh water available for the terrestrial biosphere depends on the storing capacity, the evaporation rate and the amount transported as run-off to other locations. These factors are influenced by soil type and the terrestrial biosphere itself (e.g. vegetation). Run-off by rivers also may influence the buoyancy of the surface water in the oceans. Interaction with the atmosphere depends on exchanges of aerosols (see Section 2.4.4), gases (e.g., as a medium to vegetation) and water (latent heat).

On a much longer time scale (up to millions of years) geological processes are important. The continental structure changes because of the tectonic motions (± 5cm per year). Continents are more reflective than oceans, and snow only accumulates to large ice-sheets over land (compare, for example, the present situation: there is only sea-ice at the North Pole and a large-ice-sheet over Antarctica). Hence, cooler periods primarily were met when continents were located at higher latitudes. The location of the continents also strongly influences local climate because the continents act as boundaries to ocean currents and, hence, determine the effectiveness of the polewards transport of heat. When different tectonic plates are moving together, features like folded mountain ranges (± 1mm per year) may occur (mainly if the plates have the same densities). These mountains influence atmospheric circulation (e.g., forcing winds to move polewards or increasing precipitation by forcing air to rise) and, hence, local climate. Volcanoes mainly occur where plates with different densities collide (i.e., subduction zones). Because volcanic eruptions are essentially random and their effects are short-lived, volcanic aerosols are not expected to significantly influence long-term changes in climate. Only if eruptions are large enough, volcanoes may influence climate for several years.

2.6 Discussion

In the above Sections we restricted the analysis of a changing climate to the interaction with the different spheres involved. Central to these analyses stood the changing concentration of greenhouse gases in the atmosphere and the changing albedo by cloud cover, aerosols and snow cover. The dominant processes described, however, only involve five chemical compounds: carbon (greenhouse gases: CO_2, CH_4, CFC), nitrogen (greenhouse gas, aerosols, limiting nutrient for the photic zone), phosphorus (limiting nutrient for terrestrial biosphere), sulphur (aerosols) and water (greenhouse gas, different phases and albedo, water stress). Another method to understand changing climate is to quantify the flows into, the transformations within, and the removal from different reservoirs (e.g., spheres). This concept is central to the construction of a budget. The radiative budget has some similarities with this method. The connection of all reservoirs that conserve and transport a specific element is called a cycle. Hence, by definition, a cycle is closed, i.e., no material is lost.

Predictions of future climate strongly depend on our knowledge of the budget and cycles involved (see Chapter 4). At present, uncertainties are so large that specific assumptions have to be made. One assumption is that we know some factors that determine the radiative forcing of climate, e.g., the varying concentration of greenhouse gases with time. Another assumption for example is that the influence of solar activity over the next decades is not a dominant factor in climate dynamics (Section 2.4.1). Climate models (Chapter 3) may tell us then how climate may look in such a hypothetical world. However, as climate changes, all reservoirs and the flows between them will also change. Different cycles influence and are influenced by climate and continually interact with the different reservoirs. Moreover, also the social-economical aspects interact with a changing climate. Modelling these interdisciplinary processes is subject to integrated assessment models, as will be described in Chapter 7. However, because of limited computer capacity, our limited knowledge of most subsystems (e.g., climate) and lack of consistent and long-term datasets, these models are also subject to many assumptions.

References

European Environmental Science: Towards Sustainability (N.09.1.1.1). Course Open University of the Netherlands.

Environmental Policy in an International Context (N.22.2.1.2). Course Open University of the Netherlands.

Graedel, T.E. and Crutzen, P.J. (1993). Atmospheric Change: An Earth System Perspective. Freeman and Company, New York.

Intergovernmental Panel on Climate Change (IPCC) (1990). Climate Change. The IPCC Scientific Assessment. Editors: J.T. Houghton, G.J. Jenkins, J.J. Ephraums, Cambridge Univ. Press.

Intergovernmental Panel on Climate Change (IPCC) (1994), Climate Change 1994: radiative forcing of climate change and an evaluation of the IPCC 1992 emission scenarios, Cambridge University Press, Cambridge, U.K.

Intergovernmental Panel on Climate Change (IPCC) (1995), The Science of Climate Change, Cambridge University Press, Cambridge, U.K.

Peixoto, P.J. and A.H. Oort. Physics of Climate, American Institute of Physics, NY, 1992.

Chapter 3

MODELLING OF THE CLIMATE SYSTEM

J. Shukla, J.L. Kinter,
E.K. Schneider and D.M. Straus

3.1 Introduction

To better understand the earth's climate, climate models are constructed by expressing the physical laws, which govern climate mathematically, solving the resulting equations, and comparing the solutions with nature. Given the complexity of the climate, the mathematical model can only be solved under simplifying assumptions, which are a priori decisions about which physical processes are important. The objective is to obtain a mathematical model, which both reproduces the observed climate and can be used to project how the earth's climate will respond to changes in external conditions.

There are several factors, which must be taken into account. The earth and its enveloping atmosphere have a spherical geometry, the atmosphere and oceans are gravitationally attracted to the centre of the earth, the earth rotates on its axis once per day and revolves about the sun once per year, and the composition of the earth's atmosphere includes several radiatively active gases, which absorb and emit energy. All of these factors introduce effects on the climate, which may vary with longitude, latitude, altitude, and time of the day or season of the year. Additionally, some of these factors may feedback on other processes (see Chapter 2 and 4), making the climate system non-linear in the sense that feedbacks among diverse physical processes make it difficult to predict the collective response to the processes from their individual influences.

3.2 Simple climate modelling

For simplicity, existing climate models can be subdivided into two main categories: (1) General Circulation Models (GCMs), which incorporate three dimensional dynamics and all other processes (such as radiative transfer, sea-ice processes, etc.) as explicitly as possible, and (2) simple models in which a high degree of parameterisation of processes is used. Both types exist side by side and have been improved during the past several years. Both types of models have been able to benefit from each other: GCM-results provide insight into climate change processes, which allow useful parameterisations to be made for the simple models; simple models allow quick insight into large-scale processes on long time scales since they are computationally fast in comparison with GCMs. Moreover, due to computational limits, only simple climate models can currently be used to study interaction with processes on long time scales (e.g., the slow adjustment of the biosphere, and the glacial cycles).

There are several choices of simplifying assumptions, which may be applied. For example, one can integrate the mathematical equations either horizontally or vertically or both in order to simplify the system. In the simplest possible climate model, a single number is obtained to describe the entire climate system. In the two sections that follow, two simple climate modelling schemes; energy balance and radiative-convective balance, are described.

3.2.1 Energy balance climate models

In the case of an energy balance climate model, the fundamental laws, which are invoked, are conservation of total energy and total mass. No appreciable mass is assumed to escape from the top of the earth's atmosphere, and the earth and its atmosphere are assumed to be in thermal equilibrium with the space environment. These are robust assumptions, which can be validated by observations. It is also possible to assume that the energy flux is in equilibrium at the earth's surface, although this is not strictly true since there may be considerable heat storage in the ocean on millennial and shorter time scales. Such models have been used to determine the sensitivity of the earth's climate to variations in the solar radiation at the top of the atmosphere. Analytic solutions to the energy balance equations have been obtained in some classes of models.

The simplest possible model, a zero dimensional model in which the global average, time average fluxes at the top of the atmosphere are in balance, may be solved for the equilibrium temperature (Chapter 2). By assuming that the energy flux from the sun is a constant, and that the earth conforms to the Stefan-Boltzman "black body" law for radiative emission, the energy balance may be written as:

$$\frac{S_0}{4}(1-\alpha)=\sigma T^4 \qquad (3.1)$$

where S_0 is the energy from the sun (solar constant), α is the planetary albedo which is the ratio of energy flux which is scattered to that which is absorbed, σ is the Stefan-Boltzman constant, and T is the effective temperature of the earth-atmosphere system. The factor of 4 on the left hand side represents the ratio of the surface area of the spherical earth (emitting surface) to the surface area of the circular disk of solar radiation intercepted by the earth (absorbing surface). Given a measurement for the solar constant (1,372 Wm^{-2}), the model may be solved for the effective temperature at the top of the atmosphere up to the parameter α. Measurements from space indicate that the earth's radiant temperature is 255 K and the albedo is 0.3 (Chapter 2).

This simple model can be used as a means to test the sensitivity of the earth's climate to changes in either the solar energy flux reaching the top of the atmosphere or the planetary albedo, which is a function of the cloud cover and the snow and ice cover at the surface. For example, a one percent change in the solar energy reaching the top of the atmosphere results in a 0.65 K change in the earth's effective temperature. In order to establish a quantitative relationship between the radiative energy flux at the top of the atmosphere and the climate near the surface, it is necessary to take into account the effects of the atmosphere, particularly its vertical structure, and the effects of surface conditions, particularly feedbacks associated with snow, ice and clouds (chapter 2).

Figure 3.1 shows the earth's radiation energy balance with the incoming solar energy flux normalised to 100 units (100 units = 343 W m^{-2} = 1,372/4 W m^{-2}). As may be seen in the figure, the solar energy is scattered to space by clouds or by the surface (28%), absorbed by the atmosphere (25%) or by the earth's surface (47%). In order to preserve the thermal equilibrium, the energy absorbed at the surface must be transported to the atmosphere, where it can be re-emitted to space. This is accomplished by surface radiative emission and sensible and latent heat transfers. The surface emits 391 W m^{-2}, primarily in the infrared portion of the electromagnetic spectrum, to the atmosphere, which absorbs 374 W m^{-2} and allows 17 W m^{-2} to pass into space. The atmosphere, in turn, emits 229 W m^{-2} to space and 329 W m^{-2} back to the earth's surface. This downward emission by clouds and radiatively active atmospheric gases is termed the "greenhouse effect" by analogy to a greenhouse, which glass walls permit solar radiation to pass through, but inhibit the transmission of infrared radiation from inside.

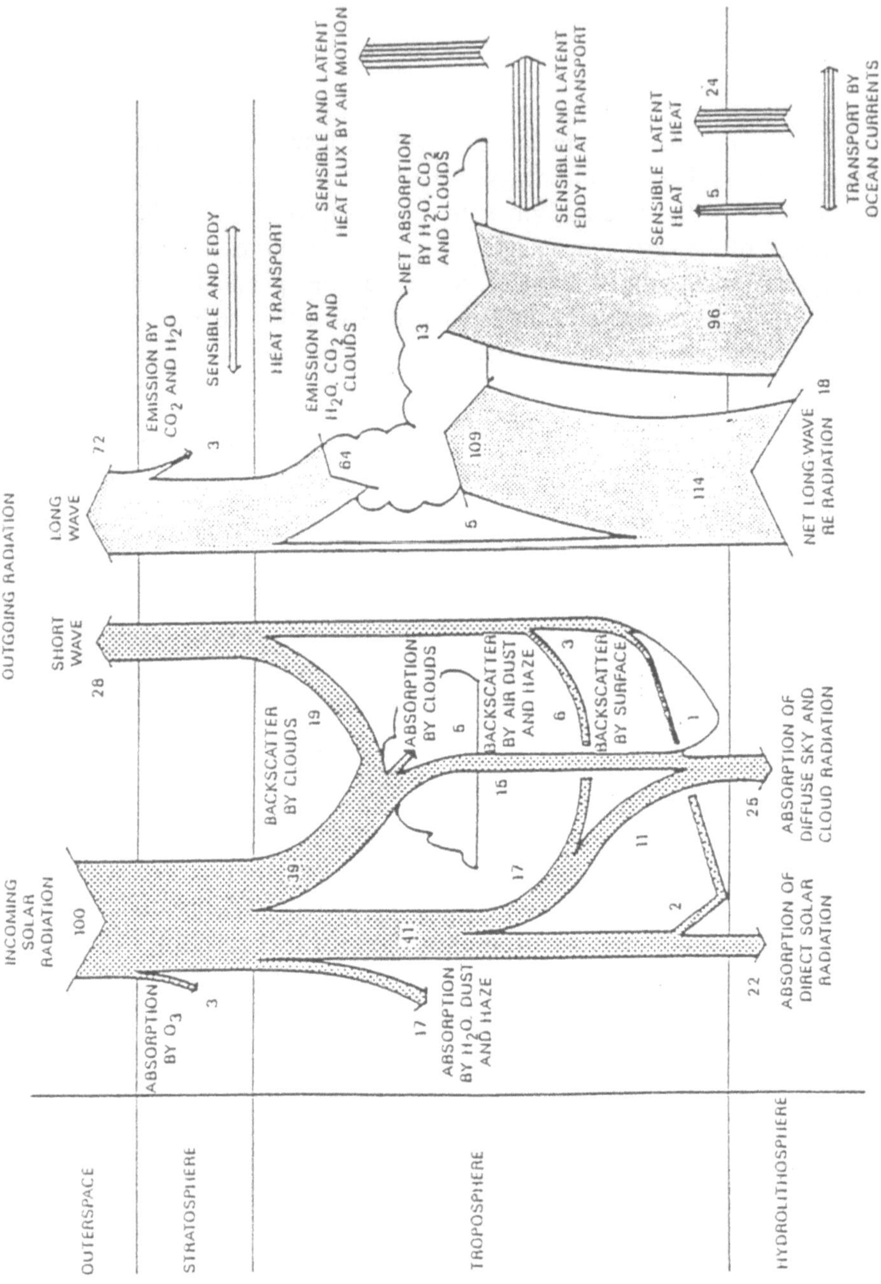

Figure 3.1 Schematic diagram of earth radiation budget components. Incoming solar energy normalised to 100 units. Adapted from "Understanding Climate Change", U.S. National Academy of Sciences, Washington, D.C., p. 14, 1975.

Thus, the surface energy balance is strongly influenced by the composition of the atmosphere, the amount of cloudiness, and the transport of water vapour (latent heat).

The effects of surface conditions can also profoundly influence the surface energy balance, primarily by the variations of the snow and ice cover at the surface and their feedback on the climate. Since snow and ice are bright, they contribute to the planetary albedo by scattering solar radiation back to space before it is absorbed. If snow or ice cover were to increase for some reason, then the scattering of solar radiation would increase, the planetary albedo would increase and it may be seen that the effective temperature of the earth would decrease. If that lower temperature at the top of the atmosphere were related to a similarly reduced surface temperature, then there would be a resultant increase in snow and ice, creating a positive feedback with the albedo effect.

3.2.2 Radiative-convective models

The second simplest climate model is one in which the effect of the vertical structure of the atmosphere is considered. Since the atmosphere is a fluid, the physical mechanism, which is absent in energy balance climate models, but present in a model with vertical structure, is the vertical motion of the air. The relevant forces in such a motion are the gravitational attraction of the atmosphere toward the centre of the earth and convection.

As was shown in the previous section, the atmosphere absorbs 86 Wm^{-2} of the solar energy and 374 Wm^{-2} of the terrestrial energy it receives, and it emits 229 Wm^{-2} to space and 329 Wm^{-2} back to the surface of the earth. The latter is referred to as the "greenhouse effect" and is primarily due to water vapour and clouds, with smaller contributions by other radiatively active gases, such as carbon dioxide, ozone and methane. The atmosphere is a net exporter of radiant energy at a rate of 98 Wm^{-2}. Therefore, there is a radiative cooling of the atmosphere with a corresponding radiative heating of the earth's surface.

When a fluid is heated from below and cooled internally, the result is convection (Chapter 2). Convection is the destabilisation of fluid stratification by heating and the resultant overturning circulation of the fluid to restore stable stratification. The overturning of the fluid may take place by large-scale circulation or by small-scale turbulent transfers of heat and water vapour. Given the radiative heating of the atmosphere from below by emission from the earth's surface and the radiative cooling of the atmosphere by emission to space and back to the earth, the earth's atmosphere is prone to convective overturning. The temperature of the atmosphere tends to have its maximum near the earth's surface and to decrease with altitude. The declining temperature of the atmosphere with height above the surface is called the lapse rate (Chapter 2). It is possible to determine from the lapse rate whether the atmosphere is stably, neutrally, or unstably stratified. It is also possible to construct a mathematical climate model on the basis of balancing the two atmospheric processes of radiative cooling and convection.

By assumption, the convective overturning of the atmosphere is assumed to be efficient, so that the equilibrium state of the atmosphere is a neutrally stable lapse rate. Convection dominates the lower portion of the atmosphere, called the troposphere, and radiation dominates the balance in the upper portion of the atmosphere, called the stratosphere. A radiative-convective climate model, then, is one in which a radiative balance is assumed in the stratosphere, a convectively neutral lapse rate is assumed in the troposphere and the surface temperature may then be determined. The radiative equilibrium may be quite complicated, due to the diversity of absorbing and emitting radiative gases.

The most important advantage that radiative-convective models have over energy balance climate models is that they can be used to quantify the cloud albedo feedback mechanism under various assumptions about cloud formation. The climate sensitivity to variations in cloudiness may then be examined critically using such models.

3.3 General Circulation Models (GCMs)

3.3.1 Introduction

Climate models may be organised into a hierarchy, based on the complexity of the models, which also bears upon the simplifying assumptions, which must be made. The simplest model is the zero dimensional energy balance model described in Chapter 2 and Section 3.2.1. Next in the hierarchy are one and two dimensional energy balance models (Section 3.2.1) in which the atmosphere is treated as a single layer, and the one and two dimensional radiative-convective models (Section 3.2.2) in which deviations from the global or zonal area mean are neglected, but vertical structure within the atmosphere is considered. At the top of the hierarchy are three dimensional general circulation models (GCM). A GCM is a model in which all horizontal and vertical motions on scales larger than a chosen "resolved" scale are included (see Section 3.3.2). Motions, which take place on scales smaller than the resolved scale, are represented parametrically in terms of the large-scale climate variables. Parametric representation (or parameterisation) involves devising a set of mathematical rules, which relate phenomena occurring on unresolved scales to the large-scale variables that are computed directly. In general, such parameterisations are based on a combination of empirical (i.e., drawn from observations) and theoretical studies. Also included are the effects of radiative heating and cooling, convective overturning (both in the resolved large scales and in the unresolved or parameterised scales), thermodynamic conversions of water vapour to liquid and back, and surface effects associated with surface ice, snow, vegetation, and soil.

General circulation models are used in place of energy balance models or radiative-convective models, when the horizontal and vertical structures or transient nature of the atmosphere are important considerations. Energy balance models can yield valuable insights into climate sensitivity and different feedback processes (Chapter 2) can be investigated very easily. However, the effects of clouds, aerosols, vertical heat transport, meridional heat transport and momentum transports can not be modelled adequately using energy balance models.

The starting point for a GCM is the set of governing laws. The laws of conservation of energy and mass are postulated, as is Newton's law (changes in momentum are related to the sum of external forces acting on a body), which applies with the slightly more restrictive assumption that all motions are hydrostatic (defined below). Newton's law for fluids is expressed mathematically in what are called the Navier-Stokes equations. With the hydrostatic assumption, changes in density are related to changes in pressure, and the downward gravitational force is balanced by the upward pressure gradient force, regardless of the motion of the fluid. The hydrostatic approximation was developed to filter sound waves, which have no importance on climate time and space scales. Mathematical equations may be written, which describe the conservation of atmospheric mass (also called the continuity equation), the conservation of energy (expressed by the first law of thermodynamics), and the changes in momentum, due to external forces, which include gravity, the pressure gradient force caused by differences in pressure from place to place, and the Coriolis force (Section 2.3.1). This set of equations, called the primitive equations of motion, is a set of non-linear, partial differential equations that have been known for centuries.

The spatial and temporal derivatives in the resulting equations, which are continuous in nature, are then approximated by discrete forms, which are suitable for a numerical treatment. The discrete equations are algebraic and may be solved by computer to determine the three dimensional distribution of temperature and winds. While various discrete forms of the primitive equations have been known for some time, only since the 1960s have computational resources become available to make their solution feasible. In addition, since the temporal dimension is also treated discretely and numerically, it is possible to solve the equations for their time dependent part, so that such processes as the annual cycle associated with the revolution of the earth about the sun, the inter-annual variation of climate, and the slow response of the climate to changes in external forcing, such as the Milankovitch orbital changes (Chapter 2) or the composition of the atmosphere, may be examined. The techniques of discretisation and numerical solution were originally developed for the problem of weather prediction. The first such successful application was attempted in the 1950s with a one layer atmospheric model.

3.3.2 Basic characteristics

Space and time are represented as continuous in the Navier-Stokes and the primitive equations. In order to allow solutions to be computed, space and time in the model world are each represented by discrete sets of points. The distance between these points defines the *resolution* of the model; high resolution represents the fields in finer detail, while low resolution can capture only the largest scale spatial or temporal structures. Those structures that can be seen at the given resolution are the *resolved scales*, while those structures that are too small to be seen are the *unresolved scales*.

The stable stratification in the atmosphere and oceans allows one to consider each as series of fluid layers, among which there is very little interaction. The representation of vertical derivatives selected, depends upon the problem being considered, but is typically effected by means of finite differences between layers or levels which are pre-selected. The choice of a co-ordinate to represent the vertical structure of the atmosphere or ocean can be complicated, because of the substantial irregularity of the earth's surface and ocean bathymetry. The hydrostatic approximation suggests that the most natural atmospheric vertical co-ordinate would be pressure, but the very steep topography at many places on the earth's surface make this a poor choice, since co-ordinate surfaces of some constant pressure are pierced by mountains. A more successful choice for the vertical co-ordinate is the s co-ordinate, which is pressure normalised by its value at the earth's surface. Ocean models make use of either distance from the sea surface (Z co-ordinate) or density (isopycnal or S co-ordinate) as the vertical co-ordinate.

The horizontal discretisation may be effected in a number of ways. The simplest formulation is an application of finite difference approximations to the continuous derivatives in both the longitudinal and latitudinal directions. Finite element approximations, which are useful in the vicinity of irregular boundaries, have been applied in limited domain models, as well as in sub-domains of global GCMs in order to more accurately simulate the atmospheric flow over and around mountains. Another class of discretisation techniques is the set of spectral methods in which the basic variables (temperature, moisture, wind speed, etc.) are expressed as series expansions in ortho-normal basis functions.

The temporal discretisation is typically effected by means of finite differences, but a complication arises because the atmosphere and oceans, being fluids, are capable of supporting waves. The complication is due to the fact that the speed, at which the phase of atmospheric and oceanic waves propagates, must be accurately resolved. This means that the time resolution (time step) and spatial resolutions are related.

Typically, the time step is chosen to be as large as possible, without causing the computational solution to become unstable, due to inaccuracies in representing the propagation speed of waves. A finer spatial resolution requires a proportionally smaller time step. For example, doubling the spatial resolution reduces the time step by a factor of two, so that doubled resolution in each direction increases the number of computations by a factor of 16 and the storage requirement by a factor of eight. An order of magnitude increase in computational resources supports only a modest increase in model resolution.

Once the continuous differential equations are transformed to a discrete set of algebraic equations, they may be solved computationally if boundary conditions and initial conditions are specified. Boundary conditions establish the values of model variables at the edges of the model domain. For example, since fluid cannot flow through a solid wall, the velocity component, normal to the earth's surface or the coasts and bottom of the oceans, is specified to be zero. Given the values of the variables at a specific moment in time and all points in the model domain, the values of the variables can be advanced one step. This process, known as time marching, is repeated time step by time step until the desired length of climate simulation is obtained. To start time marching, a set of initial conditions must be specified.

The specification of boundary conditions at the ocean-atmosphere interface is of particular importance. The appropriate air-sea boundary conditions to be specified for the oceanic GCM, as part of a coupled climate model, are the wind stress in the zonal and meridional directions, the net heat flux and the net fresh water flux. The ocean circulation is in large measure a response to these fluxes. Additionally, the flow and properties of water entering the ocean from land areas (i.e. river flow) need to be specified, as well as the heat flow across the solid boundaries (i.e. geothermal heating), is usually taken to be zero.

The result of the computational solution of the discrete primitive equations is a simulation of the three dimensional structure of the earth's atmosphere or oceans. Such a simulation should be capable of reproducing the observed characteristics of climate, including the global mean vertical structure of temperature and humidity, the zonal mean structure of the pressure and wind fields with subtropical jets near the top of the tropopause, and the longitudinally varying distribution of pressure, temperature and humidity, which is observed. In addition to the mean fields, the simulation should realistically represent the temporal variability of the main features, such as the degree to which the Polar front meanders, the seasonal shifts of major pressure belts, such as the subtropical highs, the annual cycle of tropical features, such as the Inter-tropical Convergence Zone and the Asian monsoon, and the progression of weather systems (Chapter 2). A reasonable ocean simulation reproduces the observed mean distributions of temperature and salinity, as well as the major currents. Important inter-annual variations such as El Niño (Sections 3.3.5 and 3.3.7) and the decadal oscillation of the North Atlantic should also be simulated.

3.3.3 Climate sensitivity

General circulation models are extremely useful tools for studying climate sensitivity. The procedure to study climate sensitivity is quite straightforward. First, the climate model is integrated to simulate the current climate. This simulation is referred to as the *control* run. Once a satisfactory simulation of the current climate has been obtained, an input parameter to the model is changed for the desired sensitivity experiment and the model is integrated again. This integration is referred to as the *experiment* run. The difference between the two model simulations (*experiment* minus *control*) is referred to as the model response (or sensitivity) to the particular parameter that was changed.

During the past 20 years, about 30 climate modelling groups in the world have conducted hundreds of climate sensitivity experiments using atmospheric GCMs. These numerical experiments are carried out to test a certain hypothesis about climate sensitivity, something that can not be done by analysing the past data alone. Although a detailed description of these experiments is beyond the scope of this chapter, they can be classified into the following broad categories:

a) Sensitivity to boundary conditions at the earth's surface:
A control integration is made with one set of values of sea surface temperature (SST), soil moisture, vegetation, albedo, snow cover, sea ice, height of mountains, etc. The integration is then repeated by changing one or more of the boundary conditions and the difference between the two simulations is interpreted as the effect of the change in that particular

boundary condition. The similarity between the observed and model simulated anomaly patterns validates not only the hypothesis but also the model.

Sensitivity experiments in a manner described above have been carried out by removing all the mountains (mountain - no mountain experiment); removing all the land masses (aqua-planet experiment); changing the configuration of the land masses (paleo-climatic experiments); replacing the forests by grass (deforestation experiment); expanding the deserts (desertification experiment); changing the extent and depth of snow and sea ice (snow, sea ice experiment) and changing the wetness of the ground (soil moisture experiment).

b) Sensitivity to changes in the chemical composition of the atmosphere: Of particular interest are the experiments to study the sensitivity of the earth's climate to changes in the concentration of greenhouse gases (see Sections 3.3.8 and 3.3.9). Sensitivity experiments have also been carried out to study the impact of a large number of nuclear explosions (nuclear winter experiment) and effects of volcanic eruptions.

c) Sensitivity to changes in physical parameterisations and numerical techniques:
The examples include sensitivity to parameterisations of convection, liquid water and ice crystals, cloudiness, radiation formulation, boundary layer schemes, surface roughness, land-surface processes, vertical mixing of momentum, heat, salt and water in the oceans and atmosphere, and numerical formulations for solving the mathematical equations that describe the climate model.

One of the major limitations of climate sensitivity studies is that the model simulated control climate, in many instances, has large errors, compared to the observed current climate, and therefore, the simulated anomalies (*experiment - control*) might be erroneous. It is generally assumed that deficiencies of the model and the control climate cancel in the subtraction of the *experiment* and *control*. However, this may not always be the case. Generally, it is up to the researcher to decide if a particular model, being used for a particular climate sensitivity study, is good enough to test the particular hypothesis.

3.3.4 Atmospheric modelling

In addition to the issues raised in Section 3.3.2, regarding transforming the continuous primitive equations to a discrete set of algebraic equations, there are several other problems, which must be addressed in modelling the

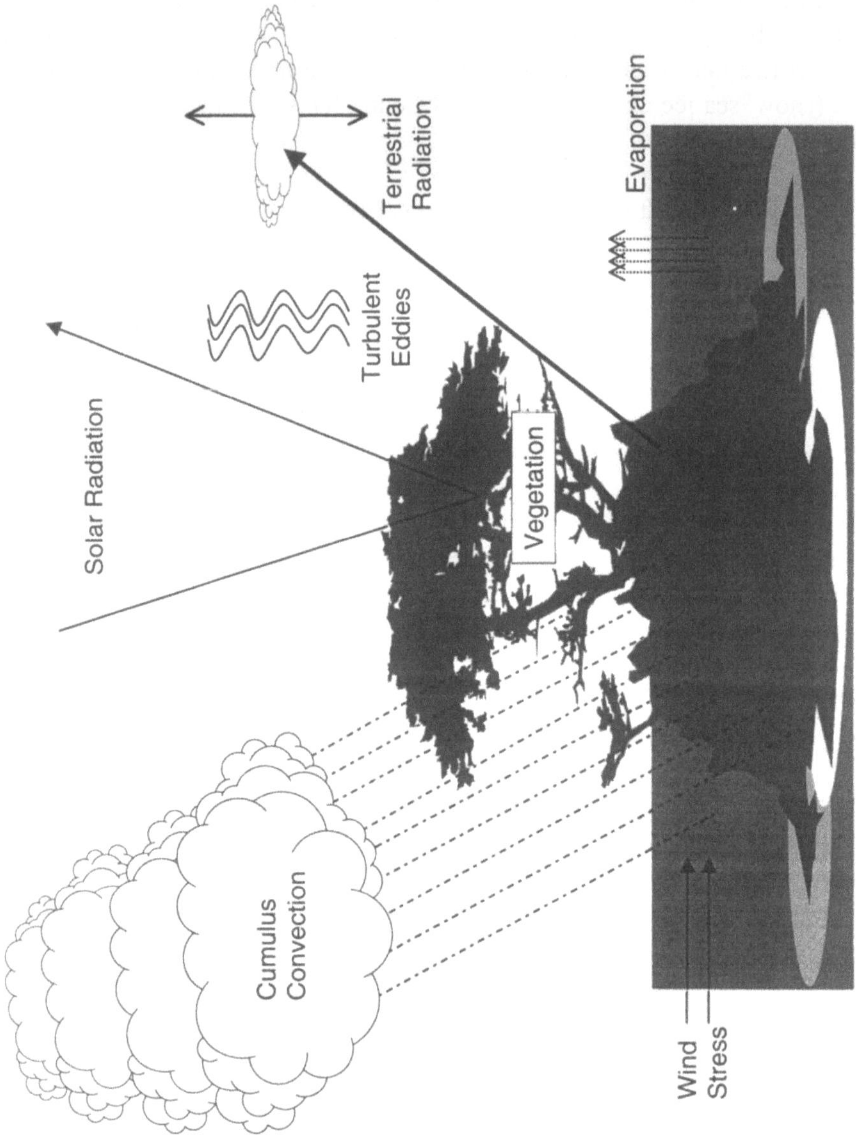

Figure 3.2 Schematic diagram of sub-grid scale processes, which must be parameterised in General Circulation Models

earth's atmosphere. These form a class of problems of parametrically representing the physical processes that act on scales smaller than the resolved scale. Such representations are referred to as sub-grid scale parameterisations. A schematic showing the various processes is given in Figure 3.2. In most cases, these phenomena involve small-scale structures or processes, which collective effect on the large-scale variables may not be simply related to those variables. For example, transports of heat or momentum are accomplished by turbulent wave motions, also known as eddies, at very different rates, depending on whether the large-scale flow is stably or unstably stratified. As a result, some assumptions, referred to as closure assumptions, must be made to establish the relationship between the sub-grid scale processes and the large-scale variables. The relationships so constructed may then be validated using observational data and incorporated into atmospheric GCMs (AGCMs). The most important sub-grid scale parameterisations - radiation, convection and the planetary boundary layer - are described below.

Radiation

In order to represent the energy balance at the earth's surface and to accurately simulate the heating by solar energy and the cooling by infrared emission to space, it is necessary to parameterise the absorption and emission of radiation in the atmosphere. The wide difference in temperature between the photosphere of the sun and the surface of the earth means that the radiation from the two bodies, if assumed to conform to the black body law, is in completely distinct frequency bands of the electromagnetic spectrum. The atmosphere is nearly transparent at the ultraviolet and visible frequencies, at which solar energy is emitted, and nearly opaque to the infrared radiation band in which the earth's surface emits. The typical AGCM radiative transfer model embodies three characteristics of the atmosphere: (1) the atmosphere is nearly transparent to solar radiation, (2) the atmosphere is nearly opaque to terrestrial radiation, and (3) the radiative flux may be characterised by a two stream approximation, in which all radiation propagates in either an upward or downward direction. The degree to which a column of air is cloudy, considerably complicates this model of radiative transfer, and, therefore, a GCM must include some model of sub-grid scale cloud formation in order to accurately take account of the effects of clouds. Cloud models based on relative humidity, upward air velocity, and other large-scale variables have been introduced into existing GCMs. More explicit cloud formation models, in which cloud condensation nuclei and droplet aggregation are included to simulate cloud liquid water content, have been formulated and are being incorporated into climate models at the present time.

Planetary Boundary Layer

Transfers of heat, moisture and momentum within the atmosphere, and exchanges of these quantities between the atmosphere and the underlying surface (land, ocean or ice), can be carried out by two types of motion. The "resolved" scales of motion are those, which are explicitly treated by the atmospheric model, while the "sub-grid" scale motions are those, which have a spatial scale too small to be resolved, but because of their importance must be "parameterised" in terms of the variables describing the resolved flow. In the context of a GCM the sub-grid scale motions are considered to be turbulence and are especially important in the layer just adjacent to the earth's surface, termed the "planetary boundary layer" (PBL). In this layer, in which friction plays a major role in the balance of forces, vertical turbulent transfers are highly significant.

There are currently two predominant classes of parameterisations of the PBL turbulent mixing. "Turbulence Closure" modelling is a general mathematical approach, in which the system of equations, describing the very small scale flow, is formulated in a non-convergent series of statistical moments. The system may then be closed by a set of consistent approximations. This approach is actually applied throughout the model atmosphere and allows the GCM to internally generate its own PBL, with a depth that varies in response to both the forcing from the surface and the influence of the resolved flow. In contrast, "Mixed Layer Theory" treats most of the PBL as a single layer. The basic idea here is that since heat and moisture are so well mixed in this layer, the transition between the top of the PBL and the atmosphere above is quite discrete, often accompanied by a discontinuity in temperature and moisture. The profiles of heat, moisture and momentum within the PBL, their fluxes into the free atmosphere and even the depth of the PBL itself can be approximated as functions of the resolved variables.

Convection

One of the very significant sources of heating in the interior of the atmosphere is the convection that takes place in many areas of the tropics. This "cumulus" convection is quite extensive in the vertical domain, often reaching to the top of the troposphere, and yet it occurs on horizontal scales of only 1 to 10 kilometres, far too small to be resolved (explicitly treated) by the atmospheric GCM (The word "cumulus" refers to clouds with a clumped structure). The transfer of heat, moisture and momentum by turbulence within the cumulus clouds is a very complex problem, analogous to the problem of parameterising the PBL discussed in the above section.

The approaches to this problem vary widely. The "Moist Convective Adjustment" scheme eliminates the gravitational instability of moist air by suitably adjusting the temperature and moisture in the vertical, whenever the instability occurs. The "CISK" class of parameterisations uses the horizontal convergence of moisture within the lowest layers of the atmosphere to determine when convection will occur. A parameterisation for the partition of sensible vs. latent heating (that is, heating, which increases the temperature vs. heating, which evaporates liquid water) is generally incorporated into this type of scheme, as is a treatment of the heat exchange between clouds and the environment. Yet a third class of parameterisation, the "Cloud-Buoyancy" or "Mass-Flux" schemes, attempt to explicitly model an ensemble of cumulus clouds. The basic kinematics of the clouds, including the "entrainment" (bringing in) of air near the cloud base, and the "detrainment" (letting air out) near the cloud top are treated. The implementations of this type of scheme vary widely in the degree to which phenomenology is used to simplify the complex physics. Atmospheric models appear to be sensitive to the details of these schemes.

Current state of atmospheric modelling

Among the components of climate models, AGCMs are probably the best verified subsystem models. The AGCM', which produce daily weather forecasts, are subjected to a prediction-analysis-verification cycle with well defined analysis techniques. The weather prediction models have much higher resolution than AGCMs used for climate modelling, because of the requirement for climate models to make much longer simulations. It is known that higher resolution improves the skill of weather predictions. The realism of the precipitation distribution appears to improve markedly with increasing resolution at climate time scales.

The prediction of clouds and the representation of their effect on the short and long wave radiation fields is both empirical and very crude in AGCMs. The cloudiness problem is important for global warming, because the change in the amount of solar radiation reflected to space, due to a few percent change in cloudiness, could compete with the CO_2 induced greenhouse effect. The spatial and temporal inhomogeneity of cloudiness makes the parameterisation of cloud effects for the large scale very difficult.

Dust particles in the atmosphere and their influence on the hydrological cycle, as condensation nuclei and on the radiation budget as reflectors and absorbers of solar radiation, have not been satisfactorily included in atmospheric models.

The representation of processes that maintain the water vapour distribution in the upper troposphere in certain geographical regions is inadequate. Since upper tropospheric water vapour provides the major positive feedback to the greenhouse effect, it is important that the dynamics and physics of the processes that maintain this field be correct. Numerical problems with water vapour are a main reason for the development of a new class of AGCM, just beginning to be used for climate simulations, the "semi-Lagrangian model".

3.3.5 Ocean modelling

Ocean models have varying degrees of complexity, analogous to the climate model hierarchy discussed in Section 3.3.1. The simplest ocean models treat the ocean as a motionless slab of water of fixed depth that stores heat uniformly throughout its depth. This simplest ocean model is known as a *slab mixed layer ocean*. This kind of ocean model, used as the oceanic component of a climate model, can produce a reasonable representation of the amplitude and phase of the annual cycle of SST in much of the extra-tropical ocean, when the depth of the slab is taken to be about 50m. A somewhat more realistic ocean model is the *mixed layer ocean*. This model still takes the ocean to be a motionless slab, but the depth of the slab is calculated internally, rather than specified, and the temperature can be a function of depth in the slab. These additional characteristics of the mixed layer are based on formulae, developed by extrapolation from measurements or deduced from intuitively plausible assumptions. The mixed layer ocean can respond more realistically to atmospheric forcing by storms, for example, than the slab mixed layer ocean.

The mixed layer models address the thermal interchanges between the upper ocean and atmosphere. However, in reality ocean currents transport heat both horizontally and vertically. When these transports are neglected, the climate model, using a mixed layer ocean model, will produce large errors. One enhancement to the mixed layer model that has been used in climate sensitivity experiments (Section 3.3.8) and transient climate experiments (Section 3.3.9) is to include specified heat fluxes by the currents in the mixed layer or slab mixed layer ocean. Inclusion of specified heat fluxes improves simulation of the observed annual mean and annual cycle of SST, but the resulting climate model will not be able to correctly simulate or predict many types of coupled climate variability, such as the large inter-annual variability of SST in the Eastern tropical Pacific (part of the El Niño/Southern Oscillation or ENSO phenomenon) or variability associated with the ocean's thermohaline circulation. When the mixed layer ocean model is used for a transient climate experiment, it should be kept in mind that the experiment is simulating the future climate, assuming that the heat transports by ocean currents do not change in response to the changing forcing.

Ocean models that simulate the time evolution of the ocean currents and the physical properties of the water contained in those currents can be termed *ocean general circulation models (OGCM)*. The OGCM begins with the discretised Newton's laws for fluids (the Navier-Stokes equations), the equation of state, which gives the density as a function of thermodynamic variables, and the mathematical statement of conservation of energy, just as the atmospheric GCM (AGCM) discussed in Section 3.3.4. The OGCM differs from the AGCM in two important respects. First, the density of sea water depends on salinity (the concentration of dissolved salts), as well as temperature, whereas the density of air depends primarily on temperature and water vapour concentration (humidity). Therefore the OGCM contains budget equations that determine salinity changes, while the AGCM calculates humidity changes. The other major difference is that the continents divide the ocean into basins, which communicate through narrow passages, whereas the atmosphere has no such boundaries. Then representation of topography is a more serious issue in the OGCM and has a great influence on the technical aspects of OGCM design.

As noted above, the physical laws have been known for hundreds of years. However, computational resources that allow models of the global ocean circulation to be developed, which can be considered potentially realistic, have only recently become available Current computers allow climate simulations using OGCMs with horizontal resolution of about 4° in the longitudinal and latitudinal directions and less than 20 levels in the vertical direction. Features with horizontal scales less than 1000 km are unresolved with this grid structure and must be parameterised if they contribute significantly to the budget equations on the resolved scales.

The parameterisations commonly used in OGCMs are vertical and horizontal diffusion of momentum, heat, and salt. These parameterisations are based on theories of turbulence that have not been verified over most of the parameter range at which they are applied. In the simplest formulations, vertical diffusion coefficients are specified and sometimes enhanced near the upper surface of the ocean, to represent the turbulent mixed layer forced by air-sea interactions. Horizontal diffusion is used primarily to damp small-scale motions, so that the resolved scale motions look smooth. Horizontal diffusion schemes are employed for numerical reasons and have little physical basis. Another parameterisation is used to obtain the vertical structure of absorption of solar radiation below the sea surface.

The typical horizontal scale of oceanic storms (meso-scale eddies, discussed in Chapter 2) is small compared to the OGCM resolution. Parameterisation of the effect of the oceanic meso-scale eddies on the resolved scales is a serious research question for the ocean, until computers are sufficiently powerful to resolve the eddies in climate simulations, whereas atmospheric storms are resolved by the coarse resolution AGCMs, used for climate simulation and their effects need not be parameterised.

Whereas the AGCM can be viewed as driven from below by SST, the OGCM is driven from its upper boundary by wind stress, which transfers momentum from the atmosphere to the ocean, heat flux, which adds heat, and fresh water flux, which affects the salinity. Conceptually, the response of the OGCM to the wind stress forcing is referred to as *the wind driven circulation* and the collective response to the heat and fresh water flux is referred to as the *thermohaline circulation.* The existence of the thermohaline circulation presents significant technical obstacles to climate simulation, since the time scales for climate variability, associated with the thermohaline circulation, can be thousands of years. The OGCM may require thousands of years of simulation to achieve equilibrium to a change in the surface forcing, whereas the time scale for the AGCM to approach equilibrium is on the order of months.

Developmental research for OGCMs is active in the areas of improvement of numerical methods for the resolved scales and improvement of the parameterisation schemes. Advances in computing have allowed the resolution of OGCMs to be increased to the point of resolving the oceanic meso-scale eddies *(eddy-resolving OGCM)*. However, the eddy resolving OGCM is not practical for climate modelling, because of the limitations of current computing technology.

3.3.6 Modelling other subsystems.

While the models of the atmosphere and ocean are the basic building blocks of the climate model, and the climate GCM in particular, there are several other subsystems with potentially important roles in determining climate. Models of various levels of sophistication have been developed for these subsystems. These subsystem models are described below.

- **Land:**

  A model of the land surface is required to calculate land surface temperature, evaporation, snow cover, and rainfall runoff, quantities of obvious importance for human society. Until recently, land surface models used in GCM climate modelling have been simple but crude "bucket models." As the name implies, the bucket model has a certain capacity for storing water. When precipitation exceeds evaporation sufficiently, so that water accumulates beyond that capacity, the bucket overflows into runoff. Evaporation is taken to be some fraction of potential evaporation, where the fraction is proportional to the percentage of capacity that the bucket is filled.

Recently, more sophisticated land surface biosphere models have been developed that include the influence of local soil and plant properties on the evaporation and water holding capacity. While constructed more realistically, these biosphere models introduce a very large number of new parameters into the model that either have not or cannot be measured. An additional issue with the biosphere models is the manner in which parameterisation of land surface heterogeneity on the climate model scale is handled. If climate changes, the vegetation distribution will of course also change, which may in turn feed back on the climate. Empirical models that predict vegetation distribution as a function of climate parameters are being developed to study this issue.

- **Hydrology:**

  More rain falls on land than re-evaporates to the atmosphere. The excess is either stored in lakes or underground, or flows to the ocean in rivers. The fresh water river flow that reaches the ocean affects the ocean salinity, and hence the oceanic thermohaline circulation. Little is known about the influence of this process on climate variability. A hydrology model is needed to partition the runoff from the land surface into the various components and can be extremely complex. The hydrology models used in global coupled GCMs are very crude, basically assigning the runoff at each point over land to a river outlet into the ocean, instantaneously transporting the runoff to that point, and freshening the sea water at that point.

- **Sea ice:**

  Energy balance climate models demonstrate the potential importance of sea ice feedbacks to climate change. Formation and melting of sea ice is in large part a thermodynamic process. However, simulation of the motion of the ice under the joint influence of the ocean currents and the wind has been found to be a necessary ingredient for realistic simulation of the seasonal variations of the sea ice extent. Some first generation thermodynamic/dynamic models of sea ice exist and are in the process of being verified and included in climate GCMs.

- **Atmospheric chemistry:**

  Realistic simulation of the coupled interactions of atmospheric dynamics, photochemistry, and transport of trace species is important for understanding phenomena such as the "ozone hole" and "acid rain." The chemistry model locally calculates sources and sinks of various species from a system of chemical and photochemical reactions.

The number of reactions considered can be very large. The atmospheric winds carry these species from place to place, so that an additional budget equation must be added to the AGCM for each species for which the sources and sinks are not in local equilibrium. The coupled AGCM/chemistry model can be many times more expensive than the AGCM by itself. Coupled AGCM/chemistry models exist for study of stratospheric ozone, but are not yet used in coupled climate GCMs.

- **Aerosols:**

  Both naturally and anthropogenically produced aerosols have potentially important effects on climate and climate change. Depending on particle size, these aerosols can scatter or absorb both solar and long wave radiation, altering the radiation balance of the climate system. Volcanically produced particles appear to have a substantial effect on global climate for several seasons after they are produced, by reducing solar radiation reaching the ground. Dust raised by Saharan dust storms can be carried for thousands of miles. There is indirect evidence that anthropogenically produced sulphate-aerosols are reducing global warming from increasing CO_2-concentrations. Simulation of these effects requires determination of the sources and radiative properties of the particles, and calculation of the transport by winds, fallout, and removal by precipitation processes. Some early experiments have been conducted with GCMs, simulating the climatic consequences of volcanic eruptions and "nuclear winter."

- **Glaciers:**

  Formation and melting of glaciers can have catastrophic climatic consequences. Ice sheets, several kilometres thick, covered much of North America and Eurasia during the last Ice Age. Melting of the Antarctic ice sheet could inundate many of the world's major cities. Some coupled climate GCMs incorporate simple models for glacial accumulation, flow and melting.

3.3.7 Choices in the philosophy and design of GCMs.

An enormous range of phenomena are encompassed under the umbrella of the "general circulation", which can be defined as the set of circulations involving time scales of a few days and longer, and spatial scales of order about a thousand kilometres and longer. If we think of a hierarchy of time scales, three broad categories emerge. On the seasonal time scale, atmospheric "blocking" (the persistence for many days of anomalous mid-latitude high pressure systems), and the timing and intensity of the

Indian/Asian monsoons are examples of phenomena which affect the climate. On the seasonal-to-inter-annual time scale we consider the tropical oscillation, known as El Niño and the Southern Oscillation (ENSO), in which the entire tropical Pacific atmosphere-ocean system undergoes dramatic shifts, as the intense convection (normally situated in the Western Pacific) moves towards the centre of the tropical Pacific basin (see Section 3.3.5). These "warm episodes" of ENSO occur irregularly with a period of about 2 - 4 years. In addition, the year-to-year changes in the Indian/Asian monsoon are both significant and are thought to be related to ENSO. On decadal to century time scales, the atmospheric response to increases in greenhouse gases and the problem of the formation and circulation of the deep water of the oceans (the thermohaline problem of Section 2.3.2) are of great interest. In addition, the slow increase in the extent of the African desert (in the Sahel region) is a major problem with societal implications, which takes place on these time scales. On even longer time scales (centuries), the glacial cycles manifest themselves, as discussed in Sections 2.4 and 2.5.

Each of these categories has very different modelling requirements. The atmospheric behaviour on the seasonal time scale can be understood in terms of fixed oceanic conditions (mainly sea surface temperature), while it is necessary to include a fairly complete spectrum of atmospheric motions. It is also necessary to include the interactions of the land surface with the atmosphere, particularly for the summer season. These interactions include the storage of precipitated water in the soil and the subsequent evaporation of this water back into the atmosphere, and play a key role in the hydrological cycle in summer over land. Since they are mediated to a large extent by vegetation (biosphere), the GCM used to study these problems should have a biosphere component. An atmosphere-biosphere GCM, utilising a moderate horizontal resolution with fixed oceanic boundary conditions, is appropriate here (a moderate horizontal resolution in this context consist of 42 or more global wave-numbers retained, corresponding roughly to a minimum grid resolution of 3 degrees in both latitude and longitude).

The simulation of the seasonal to inter-annual time scale (ENSO) must involve the upper layers of the ocean, in which the temperature and salinity are fairly well mixed. This is because the changes in the atmospheric circulation in the tropical Pacific that characterise ENSO are coupled to changes in the mixed layer of the ocean. The movement of the intense convection from Western to Central Pacific is in the short term caused by the extension of the very warm tropical sea surface temperatures normally in the Western Pacific to the east (where the sea surface temperature is normally much colder). However, considering the entirety of the ENSO oscillation, neither the atmosphere nor the ocean are the causative factors - they are coupled together. Thus what is needed is an atmosphere - ocean coupled GCM (CGCM), which includes at least the mixed layer of the oceans. If only the tropical atmospheric component of ENSO is to be simulated, the atmospheric component of the model can be quite simple and does not need to include many of the refinements of a full GCM. On the other hand, the simulation of the relationships between the Indian monsoon and ENSO, or the mid-latitude response to the warm episodes require a full atmosphere-biosphere GCM coupled to a mixed layer ocean model.

As the time scale of interest gets longer, increasingly deep layers of the ocean come into play. This is because the thermohaline (deep ocean) circulation becomes important on time scales of decades or longer, and affects the sea surface temperature on these time scales (Section 2.3.2). Since the problem of desertification (increase in desert extent) is thought to be critically linked to sea surface temperature and local land interactions on decadal time scales, the full ocean circulation should be taken into account. The problem of the response of the atmosphere to increases in greenhouse gases also cannot be studied without reference to the deep ocean, since it can store vast amounts of these gases as well as heat. Thus, the simulation of these time scales requires full, coupled oceanic and atmospheric GCMs.

Finally, on the time scales of centuries, in which the glacial cycles dominate, the dynamics of ice in its various forms (that is, the "cryosphere") become as important as the changes in state of the ocean and atmosphere, and coupled atmosphere-ocean-cryosphere models must be used (Section 2.5).

3.3.8 Equilibrium experiments

When the atmospheric composition is fixed and the incoming solar radiation is constant, climate model simulations eventually approach an *equilibrium*, in the sense that the annual mean surface air temperature does not systematically warm or cool, and similarly the precipitation does not show a systematic change of the same sign from year to year. The researcher can perform experiments changing the solar forcing, atmospheric composition, or some other aspect of the model such as resolution or a parameterisation scheme, and calculate the model equilibrium for each case. Then the change in the equilibrium climate gives the model *sensitivity* with respect to the change in the model.

There are some complications to the concept of equilibrium climate. Many simple climate models possess *multiple equilibria* for the same solar forcing, atmospheric composition, and model parameters. Typically, one equilibrium is close to the current climate, and the earth is ice covered in another. The ice covered earth reflects much more of the incident solar radiation to space, leading to the cold surface temperatures and allowing the ice to persist. The climate predicted by the model that possesses multiple equilibria can then be very different for different choices of initial conditions. If the initial state is chosen to be ice covered, it will remain ice covered in equilibrium, while an ice free initial state may lead to the warm climate state. The existence of multiple equilibria is intuitively plausible. One coupled GCM is known to possess two equilibrium states (Manabe and Stouffer, 1988). These two states are both close to the current climate. One has a cold North Atlantic SST and weak thermohaline circulation, while the other has a more realistic warm North Atlantic and stronger thermohaline circulation.

An important class of experiments, the *greenhouse sensitivity* experiment, investigates the sensitivity to changing the concentration of CO_2 in the atmosphere. The typical greenhouse sensitivity experiment compares the equilibrium climate with current CO_2-concentration to the equilibrium climate with double the current CO_2-concentration. There are many possible variations on this theme. The greenhouse sensitivity experiments have been performed with all classes of climate models, from the simplest to the GCM. Most of the GCM greenhouse sensitivity experiments have used a slab mixed layer ocean (Section 3.3.5). In this case, the equilibrium climate is achieved in a few decades of model simulated time. When the oceanic component of the climate model is an OGCM, achieving equilibrium can take thousands of years of model simulated time. The slab mixed layer ocean consequently allows a much more comprehensive investigation of the model climate sensitivity to various changes, than does the OGCM. Of course, the results using the slab mixed layer ocean may differ from those obtained with the OGCM, but this is the sort of trade-off between realism and expediency that must be made because of the cumbersomeness of GCMs.

A detailed discussion of the results of greenhouse sensitivity experiments, as carried out with twenty different models, is given by Mitchell *et al.*, 1990. In all of the experiments the doubled CO_2 climate was warmer than the *control* (current CO_2-concentration) climate, as measured by the global mean surface temperature. The sensitivity of global mean surface temperature to a doubling of CO_2 ranges from 1.9°C to 5.2°C. All of the experiments also found an increase in the global mean precipitation, with warmer climates receiving more precipitation. These results are summarised in Figure 3.3 (after Mitchell *et al.*, 1990).

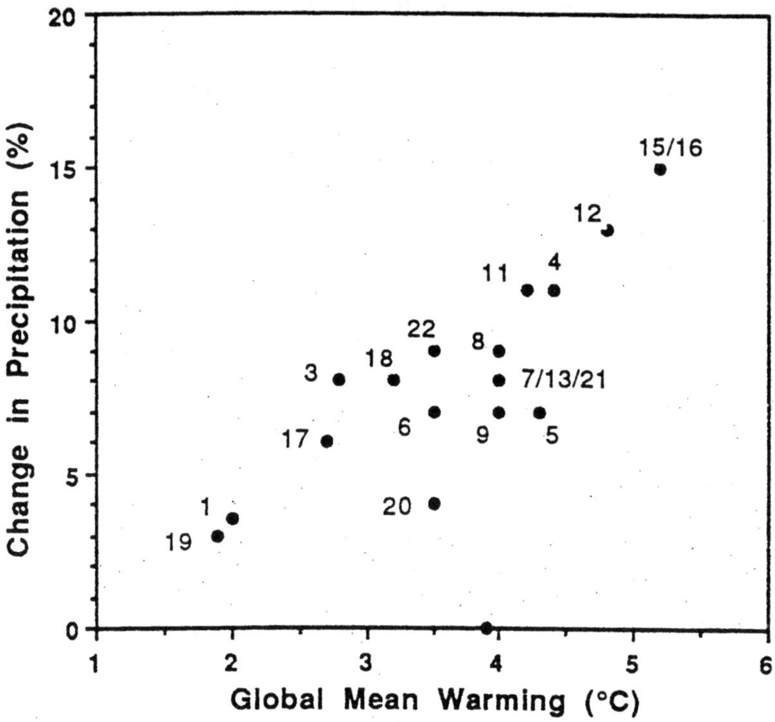

Figure 3.3. Percentage change in globally and annually averaged precipitation as a function of global mean warming from seventeen models (After Figure 5.1 in Mitchell *et al.*, 1990)

The warming directly attributable to the increased CO_2-concentration has been estimated to be about 1°C. The warming above that value is due to *positive feedback* in the climate models. The most important positive feedback involves water vapour. An increased CO_2-concentration increases downward thermal radiation at the surface. The surface and the atmosphere warm in response to this increase. The capacity of the atmosphere to store water vapour (the *saturation humidity*) increases strongly as temperature increases and the amount of water vapour in the atmosphere also increases. Water vapour is an even stronger greenhouse gas than CO_2 , so the downward thermal radiation at the surface is increased, and the surface warms much more than it would if the water vapour amount did not increase. The water vapour feedback is so powerful that a simple climate model suggests that if the Earth were moved to the orbit of Venus, where solar radiation is about 30% higher, then a *runaway greenhouse* might occur, in which the oceans would completely evaporate (Ingersoll, 1969).

An increase in cloudiness with a doubled CO_2-concentration could either enhance or reduce the climate sensitivity. Increased cloudiness can act as a positive feedback by increasing the downward thermal radiation at the ground, but could also produce a *negative feedback* by reducing the solar radiation reaching the surface. Sea ice can also produce a positive feedback. In this scenario, the sea ice amount decreases as the climate warms, leading to a decrease in the amount of solar radiation reflected to space.

It is important to understand the causes of the wide range of results obtained by the different models. All of the models cannot be correct. Some of the models may contain coding errors, since GCM computer codes are extremely complex. Coding errors are almost unavoidable, but the hope is that they do not affect the results. Assuming that there are no coding errors, the differences must be explained by differences in the strength of feedbacks related to different model parameterisations. The strength of the water vapour feedback is similar in most GCMs, although this agreement does prove that the models are correct. Some important differences in climate sensitivity have been traced to differences between cloudiness parameterisations.

Equilibrium experiments allow the potential magnitude of climate change to be estimated, and provide a convenient tool for increasing understanding of the feedbacks influencing these changes. The mechanisms that are likely to produce climate change, and equally importantly those, which are likely to be unimportant, can be identified. Equilibrium experiments also are useful for model inter-comparison and eventually will help in understanding and resolving the causes of the differences between the models. Based on the results from equilibrium experiments, potential climate change due anthropogenic emissions of greenhouse gases must be taken seriously.

3.3.9 Transient experiments

The sensitivity experiments described in Section 3.3.8 evaluate the equilibrium response of the climate model to some specified change. For the equilibrium state to be achieved, the solar forcing, and atmospheric composition are held fixed. Following the behaviour of the model, as it adjusts to the equilibrium state, is not the object of the equilibrium experiment. After the model reaches equilibrium, the climate does not change systematically, and further simulation with the model should give the same climate. The initial condition for the simulation is then irrelevant for an equilibrium experiment, with the caveat that if the model has multiple equilibria, the initial condition should be chosen, so that the simulation produces the equilibrium state relevant to the Earth's current climate. This presents little difficulty in practice.

Predicting the time evolution of the climate is important for practical reasons in developing strategies for dealing with climate change. In the case of simulation of the greenhouse effect, due to an increasing CO_2-concentration, the atmospheric composition may be changing so rapidly that the response of the model surface climate will lag behind the equilibrium climate change, found by equilibrium experiments with the relevant CO_2-concentrations, by decades. The time lag could be due to heat storage in the ocean, for example. If the CO_2-concentration is continually changing, the equilibrium state will never be reached, and the equilibrium experiment may not be useful for quantitative climate prediction. In this case a *transient experiment* will be more relevant. The transient experiment follows the evolution of the climate response to the variations in the CO_2-concentration or other time dependent specified quantity in the model. The transient experiment simulates the time scales for the climate change and the sequence of events in the climate change process. The choice of the initial condition will influence the results from the transient experiment for some time after the beginning of the integration. In contrast to the equilibrium experiment, the choice of the initial condition for the transient experiment, particularly for the initial state of the ocean, may be important.

A common transient experiment is to specify a time dependent scenario for the increase of atmospheric CO_2-concentration, such as a 1% increase per year. An initial condition, representative of the current climate, is used to initiate the experiment. The results for the 1% scenario at the time of CO_2-doubling (about 70 simulated years) will differ from the results of an experiment with a 2% increase of CO_2 per year after the same length of simulation, since the change of the CO_2-concentration will be twice as large as in the latter case. Some of this difference may be explained by appealing to results from equilibrium experiments. However, the difference between the two transient experiments will not necessarily be the same as that obtained from doubled and quadrupled CO_2 equilibrium experiments, due to the time lag effect. Similarly, due to the time lag effect, the results for the two scenario's could differ significantly if compared at the time of CO_2-doubling, which will occur after 35 simulated years in the 2% case, and 70 years in the 1% case.

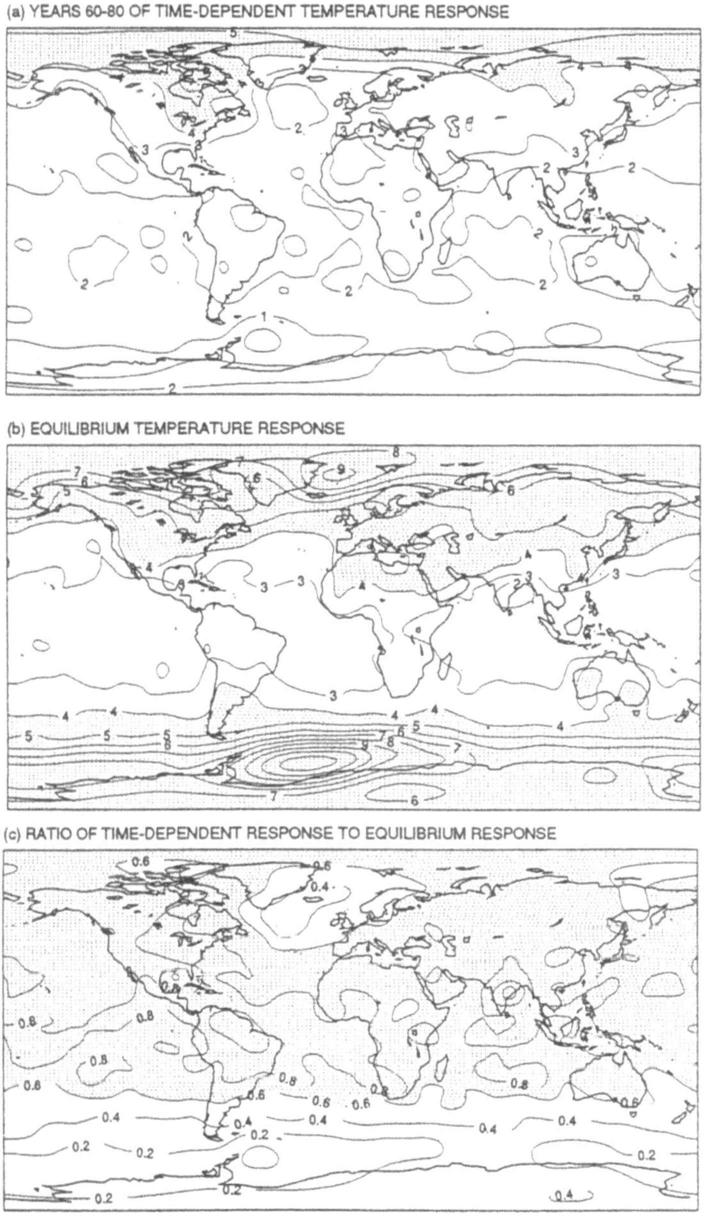

Figure 3.4. (a) The time-dependent response of surface air temperature (°C) in a coupled ocean-atmosphere model to a 1%/yr increase of atmospheric CO_2. The difference between 1%/yr perturbation run and years 60-80 of the control run, when the atmospheric CO_2-concentration approximately doubles, is shown. (b) The equilibrium response of surface air temperature (°C) in the atmosphere-mixed-layer ocean model to a doubling of atmospheric CO_2. (c) The ratio of the time-dependent to equilibrium responses shown above. (After Figure 6.5 in Bretherton *et al.*, 1990)

In a transient experiment the slowly varying elements of the climate system (e.g. the deep ocean) are not in equilibrium and this disequilibrium leads to the time lag effect. In the case of the response to increasing atmospheric CO_2-concentrations, the global warming at any time will be less than the equilibrium response evaluated at the relevant CO_2 value, with the magnitude of the difference being a measure of the imbalance in the climate system. This characteristic relationship between transient and equilibrium experiments is shown in Figure 3.4 (after Bretherton *et al.*, 1990). This figure also shows that transients experiments can give very different results for the spatial distribution of the climate change than equilibrium experiments, due to the interactions between components of the climate system with different intrinsic time scales.

The transient experiment can be viewed as a step towards prediction of climate evolution in the near and long term. A model, which realistically represents the evolution of the components of the climate system that vary on the long time scales of interest, in particular the oceans and sea ice, as well at the behaviour of the atmosphere with its much shorter intrinsic time scale, is necessary for predictive purposes. Therefore, a full dynamical GCM and some type of interactive sea ice are desirable features for the transient experiment model.

The coupled AGCM/OGCM, at its current state of development, generally does not reproduce the current climate when run to equilibrium with current atmospheric CO_2-concentrations in *control simulations*. This phenomenon is known as *climate drift*. The climate drift can be large and is thought to indicate deficiencies in the GCM parameterisations. Since excessive climate drift may lead to unrealistic sea ice distribution and distort the model sensitivity, specified corrections have been added at the atmosphere-ocean interface to force the equilibrium climate to remain close to the observed climate. This procedure is known as *flux correction* or *flux adjustment*. Then, the same flux correction is applied to the model in the transient experiment as in the control. Flux correction is controversial (see Section 3.7). An alternative procedure is to remove the effect of the climate drift *a posteriori* by taking the difference between the transient experiment and the drifting uncorrected control.

Figure 3.5 shows the results from transient experiments carried out with increasing CO_2 scenarios in a number of global coupled AGCM/OGCMs. The curve labelled "IPCC A" shows the results from the IPCC's "business-as-usual" CO_2 scenario (close to the 1% per year case discussed above) using a simplified upwelling-diffusion ocean model coupled to a one dimensional energy balance atmospheric model.

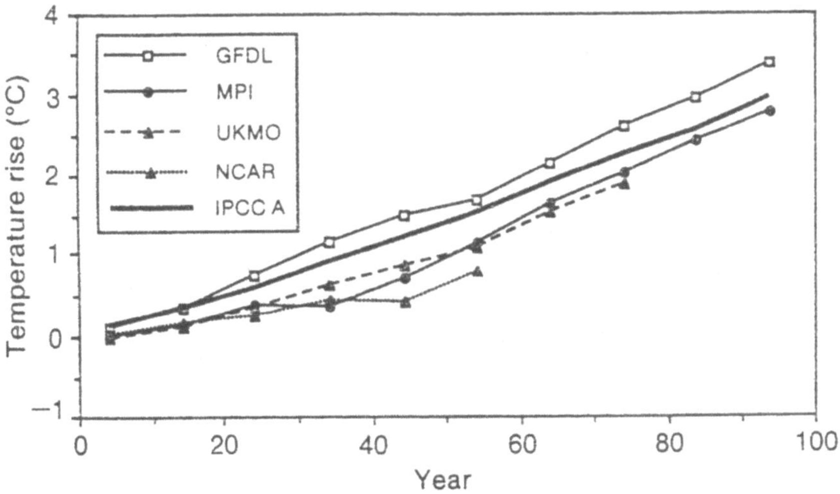

Figure 3.5 Decadal mean changes in globally averaged surface temperature (°C) in various transient coupled ocean-atmosphere experiments. Note that the scenarios employed differ from model to model, and the effect of temperature drift in the control simulation has been removed. The different models are denoted GFDL = Geophysical Fluid Dynamics Laboratory, MPI = Max Plank Institut für Meteorologie, UKMO = United Kingdom Met. Office, NCAR = National Center for Atmospheric Research, IPCC = IPCC 1990 Scenario A "best estimate". (After Gates *et al.*, 1992)

The term "transient experiment" will eventually be replaced by "climate prediction" when sufficient progress is made in the areas of initialisation, model error, and verification for coupled climate models.

3.4 Model calibration

In an effort to ensure that GCMs are as faithful as possible to the observations of the ocean and atmosphere, a process of calibration is required. As the models evolve, they are continually scrutinised in the light of observations to determine if significant physical processes are missing, and to refine those that are present. The latter may involve a reworking of the existing physical parameterisations or simply a new choice for the adjustable parameters that inevitably are part of these parameterisations. Changing the latter is also called "tuning".

The calibration of GCMs to the present climate is a complex and subtle process, for the optimal configuration of one parameterisation depends both on the overall characteristics of the model's resolved flow, and on the behaviour of all the other parameterisations. One important example of this arises when the model's resolution is changed. This will lead to a systematic alteration in the resolved flow, which serves as the input to all the parameterisations, necessitating a re-calibration. Since all the parameterisations simultaneously affect the predicted fields of momentum, temperature and moisture in the atmosphere (momentum, temperature and salinity in the ocean), and hence effectively communicate to each other through these fields, a change in one parameterisation often requires changes in others.

In this section we present three specific examples of model calibration. The first example concerns the relationship of the cumulus convection in the atmosphere to the large scale tropical 30-60 day oscillation, also referred to as the Madden-Julian Oscillation. It consists of a very large scale (zonal wave-number one[1]) tropical circulation which moves eastward around the globe with a variable amplitude and a period of roughly 30 to 60 days. This circulation is most strikingly seen in the divergent component of the circulation. This is the component involving convergence (coming together) and divergence (drawing apart) of air. Low level convergence leads to rising motion, while low level divergence is associated with sinking motion. In the phase of the Madden-Julian oscillation involving low level convergence, there is a corresponding enhancement of the precipitation, while there is a suppression of precipitation in the divergent phase. The oscillation is geographically variable in the sense that when the convergent portion moves over areas of relatively cold tropical sea surface temperatures (such as the Eastern Pacific) it is attenuated, but it redevelops over those areas more favourable for tropical cumulus convection.

Since this oscillation dominates tropical intra-seasonal variability in the atmosphere, much effort has gone into studying the ability of GCMs to simulate it. Models employing a mass flux scheme for cumulus convection originally failed to simulate the correct period for the oscillation, simulating instead a much faster (15 day) eastward travelling circulation (mass flux

[1]A zonal wave-number-one-circulation has a sinusoidal variation around a latitude circle with just one complete sine wave.

schemes of cumulus convection are discussed in Section 3.3.4). Theoretical studies of the oscillation indicated that the lower the level at which atmospheric diabatic heat release occurred in conjunction with cumulus convection, the slower the oscillation ("diabatic heating" is heating, which occurs not simply due to compression of dry air, but in this case due to the condensation of water and release of latent heat). Since the mass flux schemes were predicting almost uniform latent heat release from lower levels to upper levels, a re-calibration of the scheme was suggested. An adjustment in the value of a single parameter, which controls the entrainment of environmental air into the cumulus clouds, led to a profile of diabatic heating more peaked at mid-levels, and consequently a more realistic 30-60 day oscillation.

A good example of a problem in parameterisation, occurring in ocean models, is the effect of the vertical diffusion scheme on the depth of the oceanic mixed layer. Correctly simulating changes in the mixed layer depth is important for modelling ocean climate. The sub-grid scale motions (turbulence), which carry out the vertical diffusion of temperature and salinity necessary to keep the mixed layer homogenous must be parameterised, and there are several methods of doing this. The simplest is to employ classical diffusion with a constant diffusion coefficient determined empirically. A more sophisticated version of this allows the diffusion coefficient to be dependent upon a measure of the local vertical stability of the ocean, with less stable areas implying larger mixing. Finally, a full turbulence closure scheme (similar to that discussed in the context of the atmospheric boundary layer in Section 3.3.4) represents the most complex treatment.

Unfortunately, the choice of mixing scheme has a large influence on the predicted depth of the mixed layer in the ocean, and the nature of this influence is highly dependent upon other factors, such as simulated flux of heat between the ocean and atmosphere and the number of vertical layers used in the ocean model. Since the physical nature of this influence is not completely understood, each ocean model must be tuned individually with respect to the treatment of vertical mixing.

The final example in this section illustrates how a missing physical process can be identified. Higher resolution GCMs tended to generate mid- to high latitude upper level zonal (i.e. west to east) winds, which were too large (too much eastward motion) (see Section 2.3.1 for background on the upper level zonal winds). This phenomenon was far more noticeable in the Northern Hemisphere than in the Southern Hemisphere. Simultaneously the wave motions in the higher resolution GCMs were diagnosed as systematically transporting more eastward momentum towards the poles than the lower resolution GCMs. The poleward transport into this region was consistent in the sense of maintaining the east-west momentum balance. Yet the increase in the model error in the upper level zonal winds, as the resolution was increased, suggested that a physical process had been missed, one that would operate more strongly at high resolution and in the Northern Hemisphere.

The hypothesis of "Gravity Wave Drag" was successful as an explanation on both counts. The basic notion is that the impinging of the wind on rugged terrain (mountains) generates vertically propagating gravity waves, which break at the level of maximum vertical derivative of the zonal wind (known as wind shear). This occurs near the jet level at 200 hPa in the upper troposphere ("Gravity Waves" are waves involving periodic displacements of parcels due only to density differences with their surroundings). Thus, their relatively small eastward momentum is deposited in the jet, providing an effective drag on the very strong winds (with large eastward momentum) at this level. When implemented in GCMs, this mechanism was seen to alleviate the excessive upper level winds.

The problem noted above (excessive upper level winds) was never noted in low resolution models (even without gravity wave drag), because the wave motions in these GCMs fail to transport sufficient momentum towards the pole. Thus, the momentum budget in mid and high latitudes could be balanced without the need for additional drag, and the lack of a gravity wave drag mechanism was never noticed. In effect, the two errors cancelled each other. In current practice a parameterisation of gravity wave drag is present in most models.

These examples suggest some general limitations to the class of models currently used to simulate the ocean and atmosphere. While there are cases, in which physical reasoning leads directly to improvement in the simulation, there are many counter-examples, in which alternative physically based parameterisations lead to very different results, with no clear physical explanation available. In addition, the sensitivity of these parameterisations to a few tuneable parameters suggests that some underlying physics have been missed.

3.5 Model validation

Simplification is the essence of modelling. Moreover, relatively simple parameterisations are usually both easier to understand and to formulate in terms that a computer can use, than are more complicated or physically more complete descriptions. On the other hand, it is clearly important to check, whether the simplified formulation in existing models of the ocean and atmosphere seriously undermine their simulation skills - a step referred to as "model validation."

In order to determine how well GCM simulations capture the behaviour of the Earth's atmosphere and ocean, we compare these simulations with data sets constructed from observations (this is also referred to as "verification"). We compare simulations of different GCMs in order to discover both the problems and successes common to various models, and to learn how the distinct modelling philosophies (Section 3.3.6) compare in their realism. In Section 3.5.1 we discuss some of the many types of observations against which GCMs are compared, and in Section 3.5.2 we introduce a few of the more recent inter-model comparison studies.

3.5.1 Comparison with observational datasets

The process of comparing GCM simulations with datasets constructed from observations of the atmosphere and ocean, a process often referred to as verification, is as important as the modelling itself. Broadly speaking, there are two classes of observational datasets. Primary observational datasets come more or less directly from the observing instruments themselves. Good atmospheric examples are the routine measurements made by radiosondes (instruments carried into the atmosphere by balloons) and aircraft, and the more intensive measurements made by instrument towers during special field experiments.

Another class of primary observations comes from satellite ("remotely sensed") data. The satellites measure only radiation in various wavelength bands, and this information can be used to verify the outgoing short wave radiation (solar radiation reflected from the earth-atmosphere system) and the outgoing long wave radiation (due to emission from clouds and the ground) that are given by the GCMs radiative calculations. Since these quantities play a pivotal role in defining the overall energy balance of the Earth-atmosphere system, they are of vital importance in modelling the climate. See the discussion in Section 2.2.

Satellite measurements of radiance can be used to estimate time mean atmospheric precipitation, although this requires an intermediate physical model. Tropical rainfall can be estimated from the outgoing long wave radiation (OLR), as follows: deep convection leads to very high cloud tops, which are quite cold and hence emit relatively less upward long wave radiation. Since the satellite measures the OLR from the top of the clouds, there will be a correlation between small tropical values of OLR and convective precipitation. Recent microwave measurements can be used to estimate precipitation globally, since these measurements are sensitive to the total amount of liquid water in the atmosphere. These can be combined with the primary measurements of rainfall and snowfall over land to give a global picture of the total precipitation on a season by season basis. This is of great relevance to atmospheric dynamics, for the precipitation gives a measure of the vertically integrated heating, due to latent heat release. While the above discussion has given examples from the atmosphere, similar types of primary data are available for the oceans.

Primary measurements alone are not adequate, however, to define the full four dimensional (three spatial dimensions plus time) structure of either the atmosphere or ocean. This can only be done by using derived observational datasets, in which the primary datasets are merged, using the GCM itself, to produce uniform, gridded four-dimensional datasets called "analyses". The primary data are combined in a process called "data assimilation", which produces values for all the basic variables (horizontal momentum, temperature and moisture for the atmosphere; horizontal momentum, temperature and salinity for the ocean) at regular time intervals. A short-range forecast made by the GCM from the previous analyses is used to provide information in regions where no data are available. Data assimilation is critical for obtaining realistic "initial states" from which to run forecasts with GCMs, whether for the atmosphere alone, the ocean alone, or the coupled system.

The most basic set of statistics of the general circulation that is verified from analyses consists of the (three dimensional) time mean fields, where the averaging period is usually a month or a season, or even a year. Seasonal time means form part of the annual cycle, the regular, smooth component of the atmospheric circulation which is due to the annual cycle of the solar heating. Because of the rotation of the earth there is a very approximate longitudinal independence of the large scale fields such as temperature and zonal wind, particularly in the nearly all ocean covered Southern Hemisphere. It has thus proven useful to separate out the "zonal mean" (average around a latitude circle), with the remaining part of the field

$$[A] = \frac{1}{2\pi}\int_0^{2\pi} A d\lambda \qquad (3.2)$$

referred to as the "eddy component". The zonal mean is denoted by square brackets, where A refers to a two-dimensional field and l is longitude. The eddy component of A is denoted by an asterisk and is defined by A* = A - [A].

The time mean of the field A over a period of time T is denoted by an overbar, as in:

$$\bar{A} = \frac{1}{T}\int_0^T A dt \tag{3.3}$$

and the departure from the time mean (known as the "transient" field) is denoted by a prime, so that

$$A' = A - \bar{A} \tag{3.4}$$

The time average of the zonal (west to east) wind u can be written as

$$\bar{u} = [u] + \bar{u}^* \tag{3.5}$$

The first term is the zonal and time average, which is given as a function of latitude and pressure in Figure 3.6. This represents a basic state of the atmosphere (or ocean) in terms of which many wave quantities can be computed, using linearised versions of the basic equations of motion. These wave quantities include the "stationary eddy field" denoted by the second term in the above equation. The stationary eddy field plays a large role in defining regional climate. In the atmosphere it is forced both by the presence of mountains and by latent and radiative heating, and it can be thought of as consisting of waves, which vertical and horizontal propagation is controlled by the mean zonal wind (first term above).

In the context of the overall balance of momentum, heat and moisture in the general circulation of the atmosphere (or ocean), both the stationary eddies and the transient field introduced above, play a large role. We have for example the following expression for the northward transport of sensible heat in the atmosphere:

$$[\overline{VT}] = [\bar{v}][\bar{T}] + [\bar{v}^* \bar{u}^*] + [\overline{v'T'}] \tag{3.6}$$

in which T is the temperature, v is the meridional (south to north) wind and all the remaining notation has been explained above. In words, the total northward transport of sensible heat is due to the "time mean meridional averaged circulation" (first term), the stationary eddies (second term), and the transients (third term). Together they make up the atmospheric heat

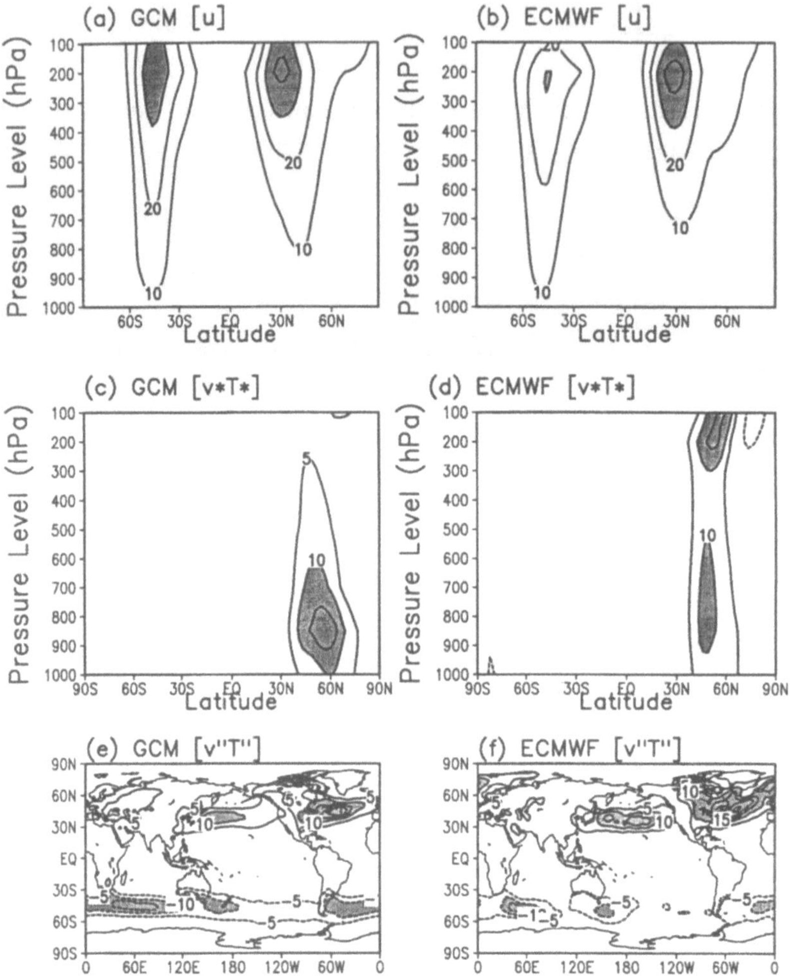

Figure 3.6 The *top panels* show the time mean and zonally averaged u-wind for the Center of Ocean-Land-Atmosphere Studies (COLA) atmospheric GCM (left) and the European Centre for Medium-Range Weather Forecasts (ECMWF) analyses (right panel). The time period is Jan. through Mar. 1983. Units are in meters/sec, the contour interval is 10 m/s and values greater than 30 m/s are shaded. The *middle panels* show the stationary eddy heat flux, time averaged (over the same period as above) and zonally averaged, with the GCM again on the left and the analyses on the right. The units are (°K)(m/s), the contour interval is 5 and values greater than 10 are shaded. The *bottom panels* show maps of the transient heat flux at the 850 hPA level, filtered to retain periods of less than 10 days. The time period is the same as above, and the units are (°K)(m/s). Values grater than 10 in the Northern Hemisphere (less than -10 in the Southern Hemisphere) are shaded, and the contour interval is 10. The GCM is on the left, the ECMWF analyses on the right. In all figures dotted lines denote negative values.

transport discussed in Section 2.3.1. In the tropics the first term is associated with the Hadley cell, which refers to the near-equatorial zonally averaged cell of upward and poleward motion (see Section 2.3.1). These are shown for both the COLA atmospheric GCM and analyses of analyses of the ECMWF in Figure 3.6, where the time mean used is the winter season.

The transient component in general measures the intensity of both the day-to-day weather and the less rapid varying components. The day-to-day weather circulations are in large part due to instabilities, in which small perturbations grow rapidly (see Figure 2.6). The mechanisms involved in this rapid growth can be studied both by controlled experiments with GCMs and by detailed examination of analyses. The instabilities tend to be associated with the upper level jets (strong maxima in the u-wind), which are centred over the east coasts of Asia and North America in the Northern Hemisphere, and over the Indian and Pacific Oceans in the Southern Hemisphere (the signature of the jets in the zonal mean is seen in panels (a) and (b) of Figure 3.6). The regions of instability associated with these jets are known as "storm tracks", and can be diagnosed from GCM output or analyses by time filtering the transient data to retain only periods shorter than about 10 days. The geographical distribution of transient heat transport associated with these short time scales can be seen in panels (e) and (f) of Figure 3.6, which indicates that the GCM simulates the systematic heat transport associated with these transients in a reasonable manner.

3.5.2 Inter-model comparison

The compelling scientific rationales for comparing the results of different GCMs are firstly to isolate the effects that specific physical processes or interactions have on the general circulation, and secondly to document whether the current state of the art in general circulation modelling is adequate to simulate particular aspects of the general circulation. Are important physical processes being entirely ignored or misrepresented? (see, for example, Section 3.4). Are common assumptions made in the modelling process faulty? The first goal is addressed by controlled comparisons, in which the models entering the comparison differ from each other only in a few aspects, such as a parameterisation, or in having different resolution. Unrestricted model comparisons, on the other hand, compare the performance of a set of widely varying GCMs.

The clearest and most extensive controlled comparisons among atmospheric GCMs are those that have examined the effects of varying the horizontal resolution. Integrations of the same GCM at differing resolutions are run from the same initial conditions, under the influence of the same boundary conditions (most notably sea surface temperature). We must recognise that changing the resolution implies more than increasing the number of spectral components retained (or decreasing the grid size). The complex topography (mountains) of the Earth's surface and the gradients of sea surface temperature are much more sharply defined at a higher resolution. The improved forcing of the stationary waves by the mountains, and the added atmospheric sensitivity to sharp anomalies of sea surface temperature are strong arguments for using a higher resolution. Results from such comparisons indicate a consistent improvement in the realism of the stationary wave simulation with increasing resolution only up to a point, beyond which little improvement is seen in the stationary waves. This saturation point occurs at about T42 spectral resolution, corresponding to a $2.8^{\circ} \times 2.8^{\circ}$ grid spacing (T stands for triangular truncation in a spectral discretisation technique; 42 for the maximum global wave-number retained). The level of transient activity (measured for instance by the temporal variance of the basic fields) also tends to increase with resolution, but this effect does not have such a well defined saturation point.

A good example of an unrestricted model comparison is that carried out by the Monsoon Numerical Experimentation Group (MONEG), who compared the simulation of the summer Indian Monsoon in a large number of GCMs and in two different sets of analyses. The period is the summer (June through August) of 1988, a year in which the Indian Monsoon was very good in the sense of having significantly higher than normal rainfall. Very important for the Indian Monsoon is the wind at lower levels (850 hPa) in the Indian Ocean.. Here the flow goes almost due eastward in the Arabian Sea, transporting the moisture, needed for the Monsoon rainfall to India. Thus, the amount of rainfall simulated is sensitive to the precise configuration of this current, as it approaches and crosses the west coast of India. Two simple parameters, which describe this flow, are the maximum wind speed over India itself and the latitude at which this maximum is attained. Each model simulation is represented by its own acronym in Figure 3.7, and the two analyses (indicated by the letters "ECMWF" and "NMC") by dark squares. Not only is there a great deal of variation between the GCMs in representing the Monsoon flow, but the two sets of analyses (which each represent a valid set of observations) disagree. Research, seeking to relate the Monsoon circulation to the parameterisations of the boundary layer and cumulus convection as well as the treatment of orography, is ongoing.

The most extensive unrestricted comparison between models that has occurred to date is the Atmospheric Modeling Intercomparison Project (AMIP), being sponsored by the World Climate Research Program (WCRP). A large number (30) of modelling groups throughout the world integrated their GCMs for 10 years (starting from 1 January, 1979), using a common set of sea surface temperature and sea-ice fields as boundary conditions. The range of models used was very wide, both in terms of horizontal and vertical resolution and in terms of the philosophy behind the parameterisations of physical processes (see Section 3.3.6). The basic outcome of this comparison is that although variations between models of course exist, there are a number of common errors made by many of the models.

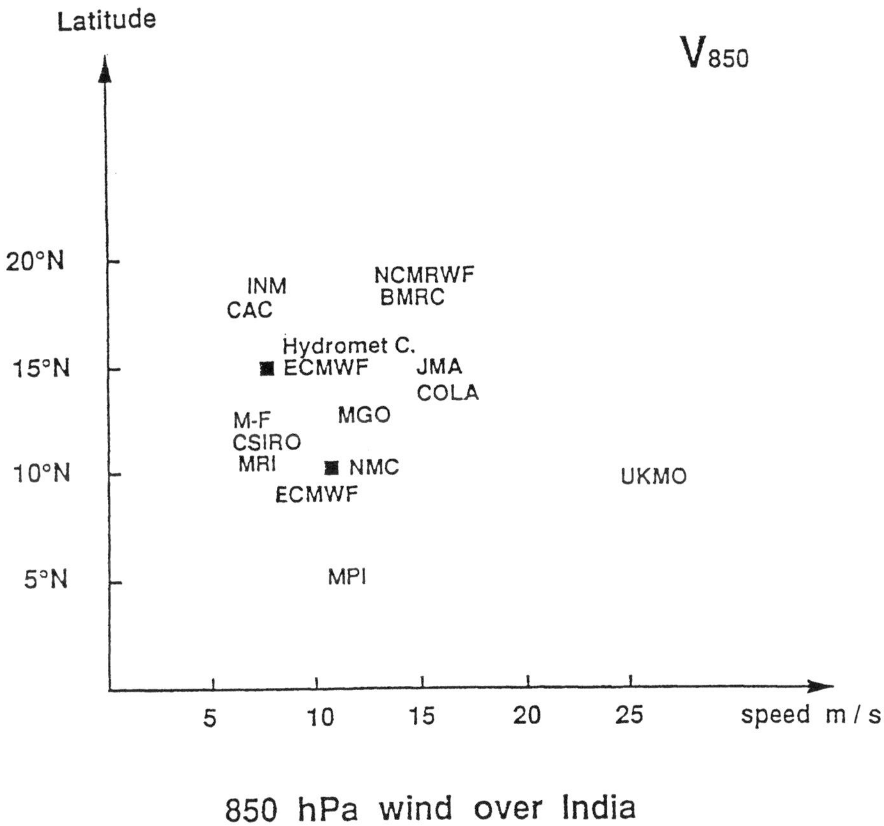

Figure 3.7 A comparison between two different analyses (ECMWF and NMC, labelled by squares) and many GCM simulations of the low-level (850 hPa) winds over India during the period of June through August of 1988. The abscissa shows the maximum wind speed over India and the ordinate the latitude at which this wind speed is reached.

One of the more interesting AMIP comparisons involves the meridional energy transports by the atmosphere-ocean system, by the atmosphere alone and the implied oceanic transports. The latter is seen to vary dramatically from model to model and in some cases lies outside the range estimated from the all available observations. It turns out that the treatment of the cloud-radiation interaction is the key player here and that it is apparently not being well handled by many models.

The annual mean meridional transport of energy by the ocean can be calculated from net ocean surface energy flux, which is the sum of the net downward flux of radiation at the ocean surface and the fluxes of sensible heat (internal energy) and latent heat (energy tied up in water vapour) from the atmosphere to the ocean. All are quantities predicted by the GCMs, so that the implied oceanic transport can be computed. In the Northern Hemisphere all the GCMs give northward transport, although the magnitude varies by as much as a factor of 10. In the Southern Hemisphere even the sign is in doubt, with most models yielding northward (that is, equatorward) transport, which is in disagreement with the estimates from observations, which show southward (poleward) heat transport in this region.

Another estimate of this transport may be obtained by using satellite data to estimate the total (atmosphere plus ocean) annual mean heat transport, which can be calculated from the net downward radiation at the top of the atmosphere (see Figure 2.7). Subtracting from this the atmospheric energy transport calculated directly from the simulated atmospheric fields of temperature, pressure and specific humidity give a hybrid implied ocean heat transport. This quantity shows much greater agreement between GCMs, in particular giving southward (poleward) transport everywhere in the Southern Hemisphere. Since the only term involving the radiative heating is taken from observations in this method, it is free of the model errors in cloud-radiation interaction. Clouds influence atmospheric radiative heating primarily by trapping longwave energy within and beneath the cloud layer. See the discussion in Section 2.4.4.

The errors in the net ocean surface energy flux due to the errors in cloud radiative forcing do not affect the atmospheric simulations in these atmospheric GCMs, because the sea-surface temperature (SST) is prescribed. However, when atmospheric and oceanic GCMs are coupled and the SST is predicted, these errors become significant, causing the coupled models to "drift" away from conditions typical of the current climate.

3.6 Climate predictions

Although it is well known that the day to day changes of weather cannot be predicted for periods beyond 1-2 weeks, it has been suggested that the climate variations (climate being defined as the space-time average of weather) can be predicted if averaged over certain space and time scales.

How well we can predict climate variations or if we can predict them at all, depends upon our ability to understand and model the mechanisms that produce climate variations. For brevity, the mechanisms that produce climate variations can be described in two categories.

a) Internal:
These are climate variations that are produced by the internal dynamics of, and interactions among the atmosphere, biosphere, cryosphere and hydrosphere components of the coupled climate system. This category will include climate variations due to tropical ocean-atmosphere interactions (El Niño-Southern Oscillation or ENSO), land atmosphere interactions (droughts, desertification, deforestation), deep ocean-atmosphere interactions (thermohaline circulation), and ocean-land-atmosphere interactions (monsoon floods and droughts, heat waves and cold spells).

b) External:
These are climate variations that are caused by factors which are external to the climate system itself, and are not initially caused by the internal dynamical and physical processes. It should be recognised that this classification is only for convenience and even if the primary causes are external, the details of the regional climate variations are determined by interactions and feedbacks among the external and internal processes. As an example, this category will include climate change and climate variations due to changes in solar forcing (either due to changes in the solar constant or changes in Earth's orbit around the sun or Earth's axis of rotation), and changes in the chemical composition of the Earth's atmosphere (either due to human induced changes in the concentration of greenhouse gases or due to volcanoes).

It should also be noted that the predictability of climate variations, due to the *internal* mechanism, sensitively depends upon the initial state of the climate system, and therefore there is a finite limit of predictability of these variations. However, it may be possible to predict a new equilibrium climate for a different external forcing, which is independent of the initial state of the climate system. The above two types of predictions were earlier suggested by E. Lorenz, the originator of the ideas referred to as the "Butterfly Effect" and Chaos, as the climate prediction of the first kind and the second kind, respectively.

In the following two subsections (3.6.1 and 3.6.2) we describe the state of our current knowledge about our ability to predict seasonal to inter-annual variations and decadal variations respectively. It is generally agreed that seasonal to inter-annual variations are predominantly due to *internal* mechanisms and, in fact, successful predictions of certain aspects of seasonal to inter-annual variations have been made without any consideration of external forcing, due to greenhouse gases or volcanoes. The mechanisms for decadal to century variations, however, are not well understood and the relative roles of *internal* and *external* mechanisms in producing decadal to century scale variations and their predictability is a topic of current research.

3.6.1 Prediction of seasonal to inter-annual variations

As discussed before, seasonal to inter-annual variations are caused by the internal dynamical processes of the coupled climate system. However, it is convenient to further subdivide these processes into two categories: the fast atmospheric variations associated with the day to day weather; and slowly varying changes in sea surface temperature (SST), soil wetness, snow cover and sea ice at the Earth's surface. The latter act as slowly varying boundary conditions for the fast weather variations. It is now well known that seasonal mean atmospheric circulation and rainfall are strongly influenced by changes in the boundary conditions at the Earth's surface, and this has provided a scientific basis for dynamical prediction of seasonal and inter-annual variations.

Thus, if it were possible to predict the boundary conditions themselves, it would be possible to predict the atmospheric circulation and rainfall. The extent, to which slowly varying boundary conditions influence atmospheric circulation and rainfall, strongly depends upon the latitudinal position of the region under consideration. Therefore, we will describe the predictability of seasonal to inter-annual variations separately for the tropical and the extra-tropical regions. The predictability of the seasonal mean circulation and rainfall for the current season and for the same season one year in advance is also very different, depending upon the region and season under consideration. Therefore, we will describe this subsection under four separate items:

a) Prediction of seasonal variations (tropics)

The tropical circulation is dominated by large-scale east-west (Walker circulation) and north-south (Hadley circulation) overturnings. These large-scale features have a well defined annual cycle, associated with the annual cycle of SST and the solar heating of land masses. Weak tropical disturbances (for example, easterly waves, lows and depressions) are superimposed on these large-scale features. Changes in the location and intensity of these large-scale overturnings are caused by changes in the boundary conditions at the Earth's surface. For example, when the Central Pacific ocean is warmer than normal, the climatological mean rainfall maximum shifts eastward and produces floods over the Central Pacific Islands and drought over Australia and India. When the Northern tropical Atlantic, in the months of March, April and May, is warmer than normal, the inter-tropical convergence zone does not move as far south as normal and gives rise to severe droughts over Northeast Brazil. Likewise, seasonal rainfall over sub-Sahara Africa is strongly influenced by the location of the inter-tropical convergence zone, which in turn is influenced by the boundary conditions over the global tropical oceans and local land conditions. A simple conceptual model to understand the causes of tropical floods and droughts is to consider space and time shifts of the climatological annual cycle of rainfall. These shifts are caused by anomalous boundary conditions at the Earth's surface.

A large number of climate model simulations have well established the validity of the conceptual model described above. It is now well accepted that the potential for dynamical prediction of the seasonal mean circulation and rainfall in the tropics is quite high. The most important dynamical reason for high seasonal predictability in the tropics is the absence of strong dynamical instabilities, which produce large amplitude weather fluctuations. Day to day weather fluctuations are relatively weak in the tropics, and therefore, it is possible for changes in boundary conditions to exert a large influence on the seasonal mean circulation and rainfall.

b) Prediction of seasonal variations (extra-tropics)

Extra-tropical weather fluctuations, especially during the winter season, are caused by strong dynamical instabilities, which produce very large day to day changes. The seasonal mean circulation in the extra-tropics, therefore, is not influenced by slowly varying boundary conditions to the same extent as it does in the tropics. The inherent chaotic nature of the extra-tropical circulation makes it less likely that useful seasonal predictions can be made. In some special cases, especially during the winter season, when tropical SST anomalies are large in amplitude and in spatial scale, and there is a significant change in the dominant tropical heat sources, it has been found that these tropical changes also affect the extra-tropical circulation. For example, it has been shown that during the years of strong El Niño events in the tropical Pacific, there is a well defined predictable pattern of winter season climate anomalies over the Pacific North America region.

The potential for prediction of the seasonal mean circulation over the extra-tropics is higher during spring and summer season, because, not unlike the tropics, the day to day weather changes are not strong and therefore it is possible that seasonal variations are controlled by changes in the boundary conditions. It are the local land boundary conditions, which are more important during the spring and summer seasons, because the solar forcing is large, and the large scale dynamical environment is not favourable for propagation of remote influences from tropical SST and heating changes.

c) Prediction of inter-annual variations (tropics)

The El Niño-Southern oscillation (ENSO) phenomenon is the most outstanding example of tropical **inter-annual** variability, for which there are sufficient oceanic and atmospheric observations to describe, and which has been successfully predicted by several dynamical models of the coupled tropical ocean-atmosphere system. ENSO is an a-periodic (with quasi-periodicity of 3-5 years) phenomenon characterised by alternating episodes of warmer than average and colder than average SST in the Central and Eastern Pacific. When the SST is warmer than average, the surface pressure is higher than average in the Central Pacific ocean and lower than average in the Eastern Indian ocean, droughts occur over Australia and India, and floods occur over the west coast of South America and Central Pacific Islands. ENSO is produced by an interaction between the upper ocean and the overlying atmosphere. Several dynamical models of the coupled ocean-atmosphere system have successfully simulated the ENSO related **inter-annual** variations and it has been demonstrated that the range of predictability of the coupled tropical ocean-atmosphere system is about 1-2 years. It has been further recognised that, just as the memory for predictability of weather resides in the (initial) structure of atmosphere, and boundary conditions are crucial for predictability of seasonal averages, the memory for predictability of ENSO primarily resides in the (initial) structure of the upper layers of the tropical Pacific ocean.

Variations in the Indian monsoon rainfall are another example of **inter-annual** variability, which has been successfully predicted, using empirical techniques. It has been found that winter seasons with excessive snowfall over Eurasia are followed by below average monsoon rainfall over India and vice versa. This relationship, along with a strong association between warmer than normal equatorial Pacific SST and deficient monsoon rainfall over India, is routinely used to predict summer monsoon rainfall over India.

d) Prediction of inter-annual variations (extra-tropics)

With the exception of the influence of tropical Pacific SST anomalies on winter season circulation over North America, there are no well recognised and generally accepted mechanisms that can be invoked to predict inter-annual variations over the extra-tropics. There are strong correlations between the extra-tropical SST/sea ice and extra-tropical circulation. However, these correlations occur either for simultaneous variations or for atmospheric anomalies, forcing (i.e. ahead of) SST and sea ice anomalies. It is likely, although it can not be proven, due to lack of appropriate land-surface datasets, that anomalous soil wetness, albedo and vegetation cover during spring and summer season can produce significant anomalies of circulation and rainfall over land.

3.6.2 Prediction of decadal variations

As described before, decadal variations can occur either due to internal dynamical mechanisms of the coupled climate system, or due to external forcing. For either case, it is reasonable to state that our understanding of the physical mechanisms is insufficient, and our ability to model the decadal variations is inadequate, and therefore, at present, a scientific basis to make decadal predictions does not exist. However, it is instructive to review our current knowledge of the evidence for and understanding of the nature of decadal variations and the potential for their predictability. We present this discussion separately for *internal* and *external* decadal variations.

Internally forced decadal variations

Analysis of past observations in the atmosphere and oceans has revealed several examples of decadal variability, which can be attributed to the internal dynamics and interactions among the atmosphere, ocean and land processes. The examples include fluctuations of the thermohaline circulation (especially in the Atlantic); persistent droughts in sub-Sahara Africa; multi-decadal variability of the Indian monsoon rainfall; changes in the level of the Great Salt Lake, Utah and SST anomalies in the North Atlantic and the North Pacific. In addition, there are decadal changes in the intensity and frequency of El Niño events, which can produce, in turn, decadal changes in global circulation. As yet, no systematic study of predictability of decadal variations has been carried out. There have been only a few, if any, model simulations of decadal variability using realistic models of the atmosphere and ocean. Currently, there are several national and international efforts underway to observe, model and, if possible, predict decadal variability.

Externally forced decadal variations

The most extensively discussed example of decadal (and longer) variation of climate is that due to the increase in the concentration of greenhouse gases. Figure 3.8 shows a time series of the observed global mean surface temperature anomaly. It can be seen that there is a well defined long-term trend, as well as decadal variations in the global mean surface temperature. GCM sensitivity experiments (see Section 3.3.8) have been carried out to predict the long-term trend. However, it is unclear at this stage if decadal variations due to greenhouse gases are predictable or not. For one thing, the magnitude of these changes is quite small, especially compared to the error in the model's ability to simulate present climate. While the magnitude of the observed and model simulated regional anomalies is quite large compared to the fluctuations of the global mean, it is far more difficult to predict regional averages, because these fluctuations are largely determined by internal dynamics mechanisms, which are inherently unpredictable.

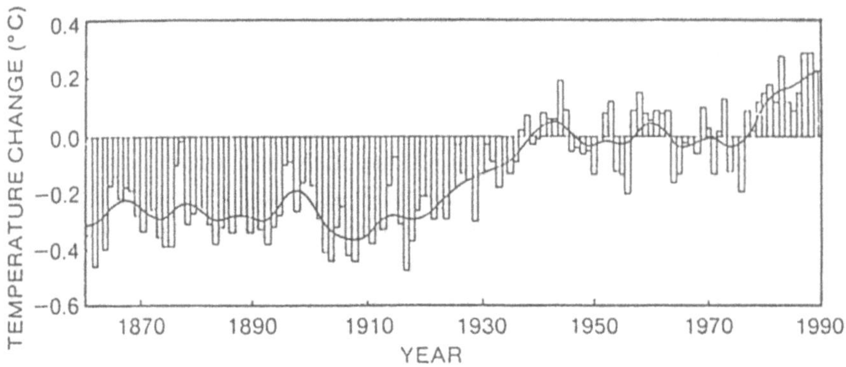

Figure 3.8 Global-mean combined land-air and sea-surface temperatures, 1861-1989, relative to the average for 1951-80 (Source: IPCC, 1990)

3.6.3 Prediction of changes in variability due to climate change

Till now we have discussed the predictability of seasonal, inter-annual and decadal averages. It is also of interest, scientifically as well as from a societal point of view, what, if any, changes in the frequency of extreme events (viz frequency and intensity of hurricanes, severe floods and droughts), changes in the amplitude of diurnal cycle, and changes in the intensity and frequency of El Niño events can be predicted. This question has been addressed mainly in the context of climate change, due to increase in greenhouse gases. There are no conclusive results yet, as some model calculations show some change in hurricane frequency (and amplitude), whereas some other calculations do not show any significant change. Actual observations show a clear tendency for reduced hurricane frequency in the Atlantic during El Niño years.

3.7 Limitations in present climate-modelling

The two major limitations in present-day climate modelling are lack of important ocean data and inadequate models. The subsurface ocean has been observed only sporadically in space and time. This hinders verification of OGCMs and the development of initial conditions for climate prediction.

All types of climate models suffer from both conceptual and practical problems. The simple climate models discussed in Section 3.2 rely for their solution on a small number of assumptions that are known to be violated by the observed climate system. For example, terms in the equations of motion are neglected or approximated. Despite this, simple climate models are useful in helping to understand the results obtained from GCMs. However, GCMs are also based to an uncomfortable extent on assumptions about the behaviour on the unresolved scales, the parameterisations. While there is general agreement on which processes need to be parameterised, there usually exist multiple parameterisation schemes for each process and few compelling reasons for choosing between them. Associated with the parameterisations, the GCM has many more adjustable parameters than the simple climate model. The appropriate range of values for these parameters is in many cases poorly known. Even the number of adjustable parameters in climate GCMs is not well known. Ideally, the values for the parameters would come from measurements. In practice, many parameter values are chosen to reduce errors in the GCM simulation of the current climate. This procedure is known as model tuning. The parameters that are considered tuneable are a matter of taste, sometimes including the solar constant. The parameters of simple climate models are also tuned, but the range of options is much more limited, and the cause and effect relationships better understood than in GCMs. Despite extensive tuning, simulations of the current climate with GCMs have large errors.

A climate GCM contains many complex interactions between the component models. The coupled model may exhibit large errors that are not apparent from inspection of the results from the component models, and which are resistant to tuning. For example, the uncoupled AGCM is integrated by specifying observed SST. The results may have important errors in the heat flux or wind stress at the atmospheric ocean interface. However, the consequences of these errors may not be obvious, until the AGCM is coupled to an OGCM. The coupled model simulation of SST will then have large errors that are immediately apparent. The SST errors can adversely affect the simulation of the sea ice, and lead to potentially incorrect estimates of climate sensitivity. This has in fact been the experience with many coupled GCMs. The process of coupling highlights those processes, which are important for the interactions between the subsystems.

When tuning fails, flux adjustment (Section 3.3.9) is sometimes used to correct errors in the mean quantities of coupled GCMs, such as annual mean SST and annual cycle of SST. With flux correction, specified fluxes are added at the atmosphere-ocean interface to those calculated internally by the model. These fluxes are chosen to minimise the error in the coupled simulation, usually of SST, and the requirement that physical laws be locally or globally satisfied is suspended. Flux correction is obviously a questionable procedure, and the results from flux corrected models should be viewed with suspicion. Many objections have been raised to flux correction, based on results from simpler climate models. On the other hand, there are strong arguments that climate projections with GCMs will be wrong if the model simulation of current climate is too far from the actual climate. These conflicting points of view will be reconciled when climate GCMs are developed that produce good simulations of the current climate without flux correction.

The ultimate test of the climate GCM will be the verification of a large enough number of predictions on the time scale of interest. This may be possible in a few decades for prediction of inter-annual variability of SST in the tropical Pacific associated with ENSO. However, action taken to deal with inter-decadal climate change, produced by anthropogenic greenhouse emissions, will have to rely on projections from models that have not been extensively verified on that time scale.

3.7.1 The different subsystems

The uncoupled component models have been developed independently from each other. Each component model is generally the product of a single scientific discipline. The effect of the limitations of the different subsystems of climate simulation will not be known, until these limitations have been superseded - the limitations introduce uncertainty, but not necessarily error into climate model projections.

a) Ocean

The major limitation to ocean models is currently poor resolution. Ocean models, with the resolution currently used for climate studies, have difficulty in simulating the steep temperature gradients in the tropical upper oceans, called the thermocline, and do not resolve motions on the scale of oceanic "storms", which is known as the Rossby radius. In the ocean the Rossby radius is on the order of 50 km, an order of magnitude smaller than the corresponding scale for the atmosphere. Ocean models that can resolve the Rossby radius are known as "eddy-resolving". The ocean models used in climate modelling are not eddy resolving, and require much stronger parameterised horizontal mixing to obtain realistic-looking large scale results than the eddy-resolving models. The magnitude and structure of the errors introduced by the low resolution is not known. Computational limitations have prevented the use of eddy-resolving OGCMs in climate modelling. Ocean modelling also suffers from a shortage of data, with which to construct realistic oceanic initial conditions and for use in model verification.

b) Land

Land surface models that have been used in climate modelling have been very crude. Models that include the important effects of plants on heat transfer between the land and atmosphere, essentially by representing the plant and soil properties in each atmospheric grid cell by a single huge plant, are now being implemented for study of climate change. Obviously it would be more realistic to model a collection of many different plant species. The plants in the current land surface models are specified in their distribution and physical properties as a function of time of year. If the climate is affected by the vegetation type, it is important to allow the model climate to influence the distribution of the different types of vegetation. Realistic models, in which the vegetation type is determined by the local climate, do not yet exist.

The surface hydrological models, currently used in climate models, are extremely crude. River flow models, important in both the response of the land surface to climate change and in the fresh water balance of the oceans, are just beginning to be developed for coupling into climate models.

c) Sea ice

Energy balance models demonstrate the potential importance of sea ice feedbacks to climate change. Formation and melting of sea ice is in large part a thermodynamic process. However, simulation of the motion of the ice under the joint influence of the ocean currents and the wind has been found to be a necessary ingredient for realistic simulation of the seasonal variations of the sea ice extent. Some probably oversimplified thermodynamic/dynamic models of sea ice exist and are in the process of being verified and included in climate models.

d) Atmospheric chemistry

Current climate models do not include representations of the transport and chemistry of the important naturally occurring and anthropogenically produced trace species.

3.7.2 The complex interaction

The separate component models are constructed and verified with specified realistic external conditions, those conditions that are the outputs from the other component models. The verification quantities are usually variables internal to the component model, such as the mid-tropospheric circulation field in the case of the AGCM, and not those quantities that are important when the component models are coupled together, such as surface wind stress and heat flux over the oceans, produced by the AGCM. When the component models are coupled together and the coupling quantities are determined internally, the coupled models can and do develop unrealistic climates. The process of coupling highlights those processes, which are important for the interactions between the subsystems. The identification of the important coupling processes and the improvement of their representation can only be accomplished within the framework of the coupled model.

The development of the current coupled atmosphere-ocean-land-sea ice climate models has created a number of problems that have to be overcome. Climate adjustment due to poor initialisation and climate drift due to errors in the model physics have been discussed in Section 3.3. Serious errors occur in all AGCMs in areas like the stratospheric circulation and mean precipitation distribution. When mixed layer oceans are included at the lower boundary, the surface temperature begins to differ from the observed present climate, especially with warmer and more zonally homogenous tropical oceans. These differences are removed by specifying a heat flux "below" the mixed layer ocean. When a fully interactive OGCM is used, the surface temperature simulation can become radically different from the present climate. This could be indicative of serious errors in the model formulations, but the most severe effects can be apparently removed by applying specified fluxes at the air-sea interface. These flux adjustments can correct errors in the mean quantities, such as monthly mean surface temperature and its annual cycle. However, the effect of flux adjustment on the climate anomalies, the prediction of which is the object of climate modelling, are not known.

An active area of research, employing coupled atmosphere-ocean GCMs, is the simulation and prediction of the El Niño Southern Oscillation (ENSO) phenomenon, which dominates the inter-annual variability of tropical climate in the Pacific (see Section 3.6.1). A good simulation of both the annual cycle of SST and its inter-annual variability in the near-equatorial Pacific has proven elusive. Models, which simulate the annual cycle well, do poorly at simulating inter-annual variability and vice versa. The errors are so severe that models, which have mediocre simulations of both phenomena, can perhaps claim to be the best overall. Models, which use flux adjustment to assure the verisimilitude of the annual cycle, have been as successful or more successful than models, which do not use flux adjustment in ENSO prediction.

Coupled model sea ice simulations without flux adjustment are not yet satisfactory. The AGCMs and OGCMs have particular mathematical problems near the poles, and the simulations by the component models near the North Pole are usually poor by themselves, which leads to poor simulations of the sea ice extent and duration.

3.8 Discussion

The current climate models do not take into account several feedbacks. This is partly because many of the feedbacks are not well understood and therefore difficult to model, and partly because the computational needs of such models can be prohibitive. The biogeochemical cycles in particular have a very long time scale and it is not clear if a complex model of weather and climate needs to be integrated for hundreds of thousands of years to investigate the possible interaction between biogeochemical cycles and cycles of water and energy (see Chapter 4). The current models also do not have an adequate treatment of solar cycles and glacial cycles. In order to investigate the role of long-period global cycles, it may be necessary to develop simpler models of the fast components of climate i.e. atmosphere, land and upper oceans, as discussed in the next chapters.

References

Arakawa, A., Schubert, W.H. Interaction of a cumulus cloud ensemble with the large scale environment. Part I. *J. Atmos. Sci.*, 31, 674-701, 1974.

Betts, A. K., Miller, M.J. A new convective adjustment scheme. Part II: single column tests using GATE, BOMEX, ATEX, and Arctic air-mass data sets. *Quart. J. Roy. Meteor. Soc.*, 112, 693-709, 1986.

Bretherton, F. P., K. Bryan and J. D. Woods. *Time-Dependent Greenhouse-Gas-Induced Climate Change*. Climate Change: The IPCC Scientific Assessment (eds. J. T. Houghton, G. J. Jenkins, and J. J. Ephraums). Cambridge University Press, Cambridge, 173-193, 1990.

Bryan, K. A numerical investigation of a nonlinear model of a wind-driven ocean. *J. Atmos. Sci.*, 20, 594-606, 1963.

Budyko, M. I. The effect of solar radiation variations on the climate of the earth. *Tellus*, 21, 611-619, 1969.

Charney, J. G., R. Fjortoft, Neumann, J. von. Numerical integration of the barotropic vorticity equation. *Tellus*, 2, 237-254, 1950.

Deardorff, J. W. Parameterization of the planetary boundary layer for use in general circulation models. *Mon Wea. Rev.*, 100, 93-106, 1972.

Emanuel, K. A. A scheme for representing cumulus convection in large scale models. *J. Atmos. Sci.*, 48, 2313-2335, 1991.

Gates, W. L., J. F. B. Mitchell, G. J. Boer, U. Cubasch and V. P. Meleshko. *Climate Modelling, Climate Prediction and Model Validation*. Climate Change 1992: The Supplementary Report to the IPCC Scientific Assessment (eds. J. T. Houghton, B. A. Callander and S. K. Varney). Cambridge University Press, Cambridge, 97-134, 1992.

Goody, R. M. and Y. L. Yung. *Atmospheric Radiation:Theoretical Basis*, 2nd ed. Oxford Univ. Press, 528 pp, 1989.

Goody, R. M. and Y. L. Yung. *Atmospheric Radiation: Theoretical Basis*, 2nd ed. Oxford Univ. Press, 528 pp, 1989.

Hays, J.D., J. Imbrie, and N.J. Shackleton. Variations in the Earth's orbit: Pacemaker of the ice ages. *Science*, 194, 1121-1132, 1976.

Ingersoll, A. P. The runaway greenhouse: a history of water on Venus. *J. Atmos. Sci.*, 26, 1191-1198, 1969.

IPCC. Climate Change. *The IPCC Scientific Assessment*. Editors: J.T. Houghton, G.J. Jenkins, J.J. Ephraums, Cambridge Univ. Press, 365 pp, 1990.

Kuo, H. L. Further studies of the parameterization of the influence of cumulus convection on large scale flow. *J. Atmos. Sci.*, 31, 1232-1240, 1974.

Leith. C. E. Numerical simulation of the earth's atmosphere. In *Methods in Computational Physics*, 4, B. Adler, S. Ferenbach, and M. Rotenberg (eds.), Academic Press, 385 pp, 1965.

Manabe, S., J. Smagorinsky, and R. F. Strickler. Simulated climatology of a general circulation model with a hydrologic cycle. *Mon Wea. Rev.*, 93, 769-798, 1965.

Manabe, S. and R. J. Stouffer. Two stable equilibria of a coupled ocean-atmosphere model. *J. Climate*, 1, 841-866, 1988.

Mellor, G. L. and T. Yamada. Development of a turbulence closure model for geophysical fluid problems. *Rev. Geophys. Space Phys.*, 20, 851-875, 1982.

Mesinger F. and A. Arakawa. Numerical Methods Used in Atmospheric Models, 1. *GARP Pub. Ser.*, 17, WMO, Geneva, 64 pp, 1976.

Mitchell, J. F. B., S. Manabe, T. Tokioka and V. Meleshko. *Equilibrium Climate Change*. Climate Change: The IPCC Scientific Assessment (eds. J. T. Houghton, G. J. Jenkins, and J. J. Ephraums). Cambridge University Press, Cambridge, 131-172, 1990.

Miyakoda, K. and J. Sirutis. Comparative integrations of global spectral models with various parameterized processes of sub-grid scale vertical transports - descriptions of the parameterizations. *Beitr. Phys. Atmos.*, 50, 445-487, 1977.

North, G. R. Theory of energy-balance climate models. *J. Atmos. Sci.*, 32, 2033-2043, 1975.

North, G. R., R. F. Cahalan and J. A. Coakley. Energy balance climate models. *Rev. Geophys. Space Phys.*, 19, 91-121, 1981.

Orszag, S. A. Transform methods for calculation of vector coupled sums: application to the spectral form of the vorticity equation. *J. Atmos. Sci.*, 27, 890-895, 1970.

Peixoto, P.J. and A.H. Oort. Physics of Climate, *American Institute of Physics*, NY, 1992.

Phillips, N. A. The general circulation of the atmosphere: a numerical experiment. *Quart. J. Roy. Meteor. Soc.*, 82, 123-164, 1956.

Phillips, N. A. A coordinate system having some special advantages for numerical forecasting. *J. Meteor.*, 14, 184-185, 1957.

Phillips, N. A. Numerical integration of the primitive equations on the hemisphere. *Mon. Wea. Rev.*, 87, 333-345, 1959.

Ramanathan V. and J. A. Coakley. Climate modeling through radiative-convective models. *Rev. Geophys. Space Phys.*, 16, 465-489, 1978.

Randall, D.A., Harchvardham, D.A. Dazlich, and T.G. Corsetti. Interactions among radiation, convection, and large scale dynamics in a general circulation model. *J. Atmos. Sci.*, 46, 1943-1970, 1989.

Robert, A. The behavior of planetary waves in an atmosphere model based on spherical harmonics. Arctic Meteor. Research Group, *McGill Univ. Pub. Meteor.*, 77, 59-62, 1965.

Schneider, S. H. and R. E. Dickinson. Climate modeling. *Rev. Geophys.* 2, 447-493, 1974.

Sellers, W. D. A climate model based on the energy balance of the earth-atmosphere system. *J. Appl. Meteor.*, 8, 392-400, 1969.

Slingo, J.M. The development and verification of a cloud prediction scheme for the ECMWF model. *Quart. J. Roy. Meteor. Soc.*, 13, 899-927, 1987.

Smagorinsky, J. On the numerical integration of the primitive equations of motion for baroclinic flow in a closed region. *Mon. Wea. Rev.*, 86, 457-466, 1958.

Smagorinsky, J. On the numerical prediction of large scale condensation by numerical models. *Geophys. Monogr.*, 5, 71-78, 1960.

Smagorinsky, J. General circulation experiments with the primitive equations: I. The basic experiment. *Mon. Wea. Rev.*, 91, 99-164, 1965.

Smagorinsky, J., S. Manabe, and J. L. Holloway. Numerical results from a nine-level general circulation model of the atmosphere. *Mon Wea. Rev.*, 93, 727-768, 1965.

Staniforth, A.N. The application of the finite element method to meteorological simulations - a review. *Int. J. Num. Methods*, 4, 1-12, 1984.

Chapter 4

GLOBAL BIOGEOCHEMICAL CYCLES

J. Rotmans and M. den Elzen

4.1 Introduction

The elements of carbon, hydrogen and oxygen, and the basic nutrient elements nitrogen, phosphorus and sulphur, are essential for life on earth. The term 'global biogeochemical cycles' is used for describing the transport and transformation of these substances in the global environment. In recent decades detailed studies have been carried out on the global biogeochemical cycles of the basic elements, in particular carbon (C), nitrogen (N), phosphorus (P), and sulphur (S) (Bolin *et al.,* 1979; Bolin and Cook, 1983; Bolin *et al.*, 1983; Schlesinger, 1991; Butcher *et al.*, 1992; Wollast *et al.,* 1993; Den Elzen *et al.*, 1995).

Through an intricate web of biogeochemical processes the cycles of elements C, N, P, S are always in a state of disturbance. These changes could be natural, as part of a constant disturbance or variability (like the constant change in solar external forcing, temperature and sea level changes, changes in atmospheric composition over the different glacial-interglacial scales, etc.). However, these changes could also be human-induced, leading to new, extremely fast global environmental changes, as observed over the past decades. The main causes of these changes are the changes in the rates of fluxes and flows of chemicals, gases and other compounds to the compartments water, air and soil. Human activities, like agricultural, industrial, transport and urbanisation activities, but also biomass burning (deforestation), lead directly to changes in the biogeochemical cycles, thereby changing the composition of the atmosphere (e.g. greenhouse gases, stratospheric ozone depleting gases, and toxic pollutants), and the chemistry of aquatic systems and soils (SO_4, NO_3, PO_4, heavy metals, pesticides and organic micro-pollutants). Evidently all these changes differ with respect to scales in time (seasonal, yearly, decades till centurial) and space (local, regional, continental till global).

Research has hitherto mainly been focused on separate global cycles rather than on interactions between the various global cycles. It is only in the last decade that a start has been made on quantitative studies of the interactions between the global cycles, using sophisticated compartment models (e.g. Keller and Goldstein, 1994; Hudson *et al.*, 1994; MacKenzie *et al.*, 1992). What is lacking so far, however, is an integrated framework which describes the global cycles of carbon, nitrogen, phosphorus and sulphur: where each originates, where it remains and how the various global budgets can be balanced.

In order to demonstrate how current global environmental problems, and the global climate change problem in particular, could be reconsidered in terms of human disturbances of the various global element cycles, this Chapter discusses the current and future state of the global element cycles. It will be discussed to what extent these global cycles are being perturbed by human activities, in a direct and indirect way, and what the possible future changes could mean in terms of global climate change. Finally, various of the most important interactions between the global element cycles of C, N, P and S are treated.

4.2 The global carbon cycle

4.2.1 Introduction

Among the group of elements essential to life on Earth carbon is the most important one. All life forms on Earth are primarily composed of carbon, so that studying the global carbon cycle in the past and present gives an indication of the comparative state of the biosphere. The natural global carbon cycle encompasses exchanges of CO_2 , carbonates, organic carbon, etc., between three reservoirs: the atmosphere, the hydrosphere, and the terrestrial biosphere, of several billions of tonnes of carbon per year (see Figure 4.1).

The anthropogenic increment of carbon due to the burning of fossil fuels and changing land use is disturbing the balance of the global carbon cycle, which will ultimately change the Earth's climate (Intergovernmental Panel on Climate Change (IPCC), 1990, 1992, 1994, 1995).

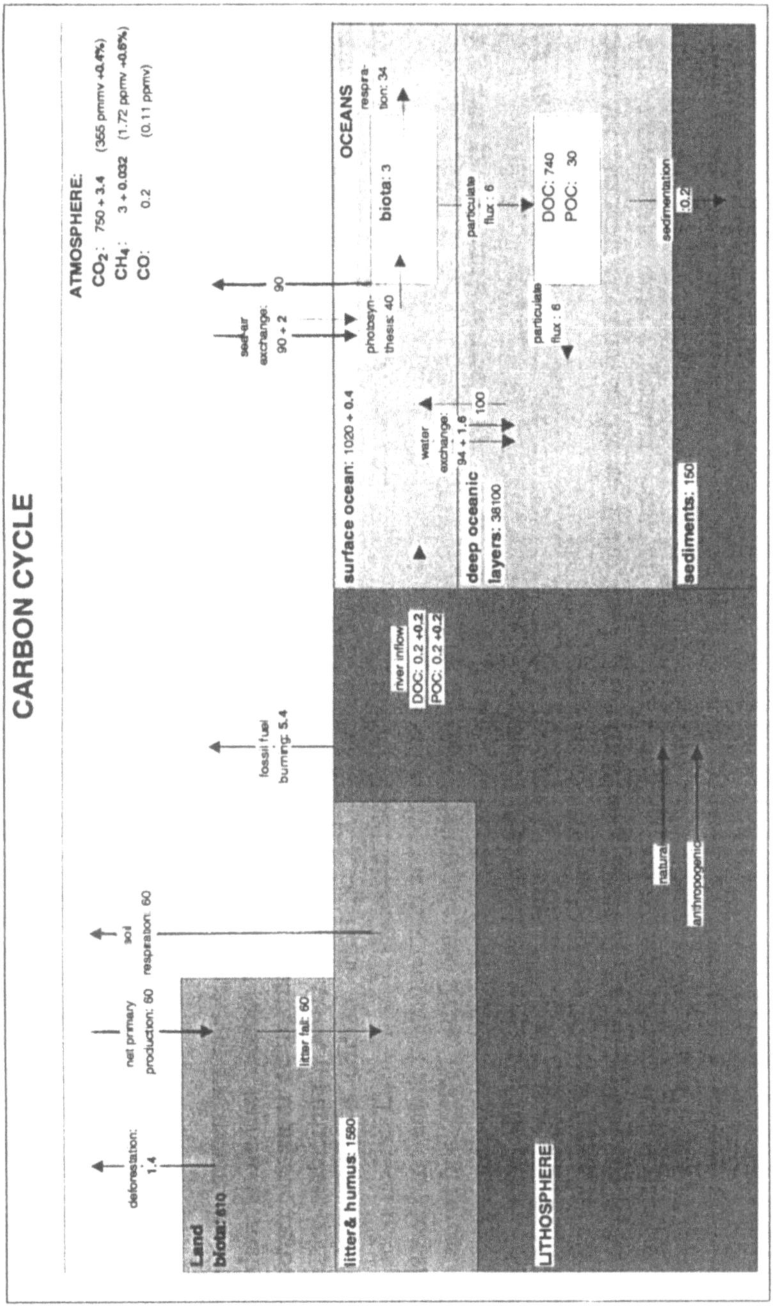

Figure 4.1 The global carbon cycle (IPCC, 1990;1992;1994). The numbers in boxes indicate the size in GtC of each reservoir. On each arrow is indicated the magnitude of the flux in GtC/yr (DOC = dissolved organic carbon, POC = particulate organic carbon)

4.2.2 The present global carbon cycle

Carbon is present in the atmosphere mainly as CO_2, with minor amounts present as methane (CH_4), carbon monoxide (CO) and other gases. In the ocean, carbon is mainly present in four forms: particulate inorganic carbon, dissolved organic carbon (DOC), particulate organic carbon (POC), and the marine biota itself. The marine biota, although it is a small carbon pool, is importance for the distribution of numerous elements in the sea (Broecker and Peng, 1992), and forms the major input of organic carbon in the oceans. The ocean itself is by far the largest of the carbon reservoirs, with a continuous exchange of CO_2 between the atmosphere and oceans. The net flux is driven by the difference between the atmospheric partial pressure of CO_2 and the equilibrium partial pressure of CO_2 in surface waters. Rapid exchanges of carbon occur between the terrestrial biosphere, soils and atmosphere, through the processes of photosynthesis and respiration. These natural exchanges are influenced by the impact of a whole range of human activities as listed above.

The global carbon cycle can also be described in terms of chemical transformations (Figure 4.2).

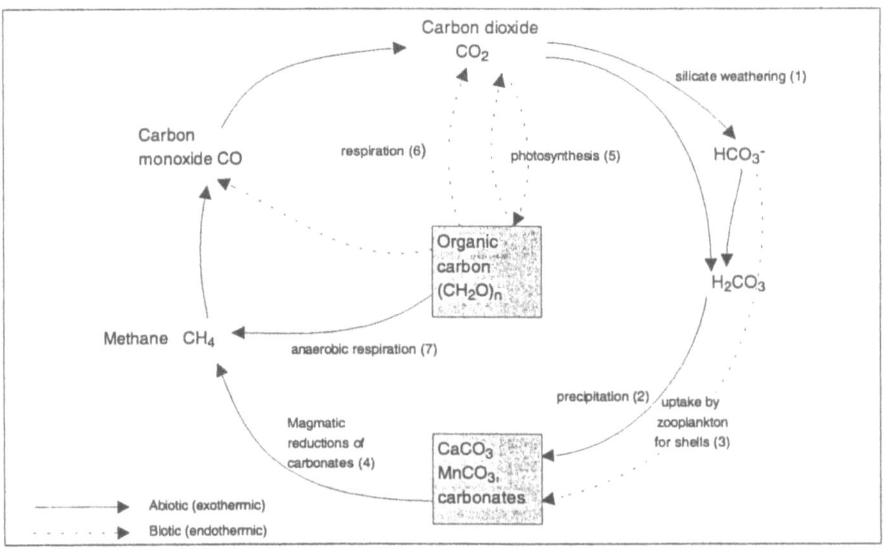

Figure 4.2 Carbon cycle: chemical transformations (source: Ayres, 1992)

This cycle involves a slow carbon cycle, in which the inorganic reservoir of carbon (sedimentary carbonate rocks, such as calcite ($CaCO_3$) and dolomite ($CaMg(CO_3)_2$) plays a major role. The main processes in this slow carbon cycle are: (i) the uptake of carbon dioxide from the atmosphere by weathering of silicate rocks, where the calcium, magnesium and bicarbonate ions, as well as the dissolved silica in the surface waters are carried to the oceans(1); (ii) inorganic precipitation of the dissolved compounds (2); and (iii) sedimentation of calcium carbonate ($CaCO_3$) to the ocean floor (3). The net reaction of the above described mechanisms is the following:

$$CaSiO_3 + CO_2 \text{ ----> } CaCO_3 + SiO_2 \qquad (4.1)$$

In itself, this would lead to the uptake of all atmospheric carbon dioxide within 400,000 years. However, there is also a reverse reaction to (4.1) which reconverts sedimentary calcium and/or magnesium carbonate rocks into calcium or magnesium silicate, releasing gaseous CO_2 and CH_4 (4). The eventual release is due to volcanic eruptions or hot springs (Ayres, 1992). All the above processes are related to climate conditions, and explain the major features of paleoclimatic history. However, these extremely long-term exchange mechanisms cannot explain the short-term correlations between the changes in climate and atmospheric CO_2 levels. These are part of the rapid carbon cycle, which is biologically controlled by the following main processes: photosynthesis (5), whereby CO_2 is biologically transformed into sugars and cellulose, and the complementary process, photorespiration (6). The net product of these two processes is referred to as: net photosynthesis. Another biotic respiration process is the so-called 'dark' respiration, and this includes all plant respiration processes minus photorespiration. The net biological process resulting from these three processes is referred to as: net primary production (net photosynthesis minus respiration of non-photosynthetic plant parts).

A sub-cycle, which may be considered as part of the global carbon cycle is the CH_4-CO-OH cycle, which involves the atmospheric reservoirs of methane (CH_4) and carbon monoxide (CO) (see Figure 4.2). The methane concentration is determined by the sources and sinks of methane. The major sink processes are the removal of methane by (i) oxidation by hydroxyl radicals to carbon monoxide which in turn is oxidised to CO_2 (see equation 4.2), and (ii) soil uptake.

$$CH_4 + 2\,OH \text{ <----> } CO2 + 3\,H2 \qquad (4.2)$$

Biogenic methane is produced during microbial decomposition of organic material in an anaerobic environment. This process, methanogenesis, is highly complicated and temperature dependent. Biogenic methane sources are: natural wetlands, wetland rice cultivation, landfills and deposits of solid waste, ruminants (enteric fermentation and methane from animal waste), biomass burning, termites, oceans and fresh water systems. Fossil sources and non-biogenic sources are: natural gas vents, coal mining and methane hydrate destabilisation. The concentration of carbon monoxide is determined on the one hand by the sources (direct emissions from e.g. fossil fuel combustion and biomass burning, as well as emissions through oxidation of non-methane hydrocarbons (NMHC) and oxidation of methane), and on the other hand by the sinks: principally the oxidation by hydroxyl radicals and the uptake by soils.

4.2.3 Anthropogenic perturbation of the global carbon cycle

As denoted in Figure 4.1, the natural global carbon cycle encompasses exchanges of CO_2, carbonates, organic compounds, etc. between the atmosphere, hydrosphere and terrestrial biosphere, of several billions of tonnes of carbon per year. The anthropogenic increment due to the burning of fossil fuels and changing land use is relatively small compared with most of the natural exchanges of carbon between the reservoirs. However, there is strong evidence that the anthropogenic increment of carbon is disturbing the balance of the global carbon cycle, leading to an increase in the atmospheric CO_2 concentration. Such evidence includes the isotopic dilution of the atmosphere and surface oceans with respect to radiocarbon and stable carbon isotopes. The measured increase which, as the subject of numerous studies during the past decades indicate, will lead to a change in the Earth's climate (IPCC 1990; 1992; 1994 and 1995).

However, there are considerable uncertainties in our knowledge of the present sources of and sinks for, the anthropogenically produced CO_2. In fact, the only well understood source is fossil fuel combustion, while in contrast, the source associated with land-use changes is less understood. The amount of carbon remaining in the atmosphere is the only well-known component of the budget. With respect to the oceanic and terrestrial sinks, the uncertainties are likely to be in the order of $\pm$ 25% and $\pm$ 100%, respectively, mainly resulting from the lack of adequate data and from the deficient knowledge of the key physiological processes within the global carbon cycle (IPCC, 1994).

The carbon balance concept

The uncertainties concerned can be expressed explicitly by the formulation of a basic mass conservation equation, which reflects the global carbon balance, as defined in:

$$dC_{CO2}/dt = E_{fos} + E_{land} - S_{oc} - E_{for} + I \quad (4.3)$$

where dC_{CO2}/dt is the change in atmospheric CO_2 (in GtC/yr), E_{fos} denotes the CO_2-emission from fossil fuel burning, E_{land} the CO_2-emission from changing land use, S_{OC} the CO_2-uptake by the oceans, and E_{for} the CO_2 uptake through forest re-growth (all in GtC/yr). The best available current knowledge on the sources and sinks of CO_2, which comprises a mixture of observations and model-based estimates, does not permit us to obtain a balanced carbon budget (Den Elzen *et al.*, 1997). To balance the carbon budget, another term, *I*, is introduced, which represents the missing sources and sinks. *I* might therefore be considered as an apparent net imbalance between the sources and sinks ('missing carbon sink'). The analysis of the net imbalance in the global carbon cycle has become a major issue in the last decade (Tans *et al.*, 1990), particularly the question of how to account for very large carbon sinks in the terrestrial biosphere in the northern temperature latitudes. Table 4.1. presents the global carbon cycle over the 1980s in terms of anthropogenically induced perturbations to the natural carbon cycle, as is given by the IPCC in 1994 (Schimel *et al.*, 1995). Schimel *et al.* stated that the imbalance of 1.4 ± 1.6 GtC/yr may be inappropriate, since sink mechanisms (CO_2 fertilisation [0.5-2.0 GtC/yr for the 1980s], N fertilisation [0.2-1.0 GtC/yr] and climatic effects [0-1.0 GtC/yr]) would account for it. Lately, the role of nitrogen fertilisation as a major sink mechanism has been questioned by Nadelhoffer *et al.* (1999), so that so that there is still no conclusive answer to the puzzle of the missing carbon sink (Schindler, 1999).

Table 4.1 Components of the carbon dioxide mass balance (Schimel *et al.*, 1995)

Component (in GtC/yr)	**1980-1989**
Change in atmospheric mass of CO_2 (dC_{CO2}/dt)	3.2 ± 0.2
Uptake by the oceans (S_{OC})	2.0 ± 0.8
Emissions from fossil fuel burning (E_{fos})	5.5 ± 0.5
Emissions from land use change (E_{land})	1.6 ± 1.0
Uptake by northern hemisphere forest re-growth	0.5 ± 0.5
Net imbalance ($I = dC_{CO2}/dt + S_{OC} - E_{fos} - E_{land}$)	1.4 ± 1.6

In order to test various hypotheses regarding how to balance the global carbon budget, models can be used. In particular integrated assessment models are useful, which attempt to simulate the cause-effect chains of the global carbon cycle (in terms of emissions, biogeophysical processes, interactions and feedbacks, and temperature effects, see also Chapter 7). The latest models also include interactions between the global carbon and nitrogen cycles. Experiments with these integrated assessment models underline the importance of negative feedback processes in the global carbon cycle. In these models feedback mechanisms such as the CO_2-and N-fertilisation effect, and temperature feedbacks on net primary production and respiration, can be used to balance the global carbon budget.

4.2.4 Conclusions

Although the anthropogenic perturbation of the global carbon cycle is small compared to the huge amounts of carbon forms exchanged between the various reservoirs in a natural way, there is strong evidence that the global carbon cycle is significantly disturbed. This human-induced disturbance will result in a global climate change, but the rate of change, the magnitude and the regional distribution of an anticipated climate change is uncertain (see also Chapters 2 and 3). There are still considerable uncertainties with regard to the sources and sinks of the global carbon cycle, and the 'missing carbon sink' has become a major issue during recent years. A possible (partial) explanation for this imbalance could be the extra carbon sink due to terrestrial feedbacks as CO_2-and N-fertilisation.

4.3 The global nitrogen cycle

4.3.1 Introduction

Although the atmosphere consists of 78% nitrogen (N), most biological systems are N-limited, because these systems are unable to use unreactive N_2. Therefore N must first be converted to reactive forms of N by the process of N-fixation by bacteria so that it can be used by biological systems until it is converted back to N_2 by denitrification. Through anthropogenic perturbation of the cycle, whether associated with agricultural, industrial and household activities or with the burning of fossil fuels and biomass, the N-compounds play an important role in a wide range of environmental issues, such as climate change, soil acidification and nitrate pollution of groundwater and surface water.

4.3.2 The present nitrogen cycle

The global nitrogen cycle comprises a rich variety of important biological and abiotic processes that involve numerous compounds in the gas, liquid and solid phases, which are described in Table 4.2. In Figure 4.3 the global nitrogen cycle is portrayed in terms of a complex set of transfers between air,

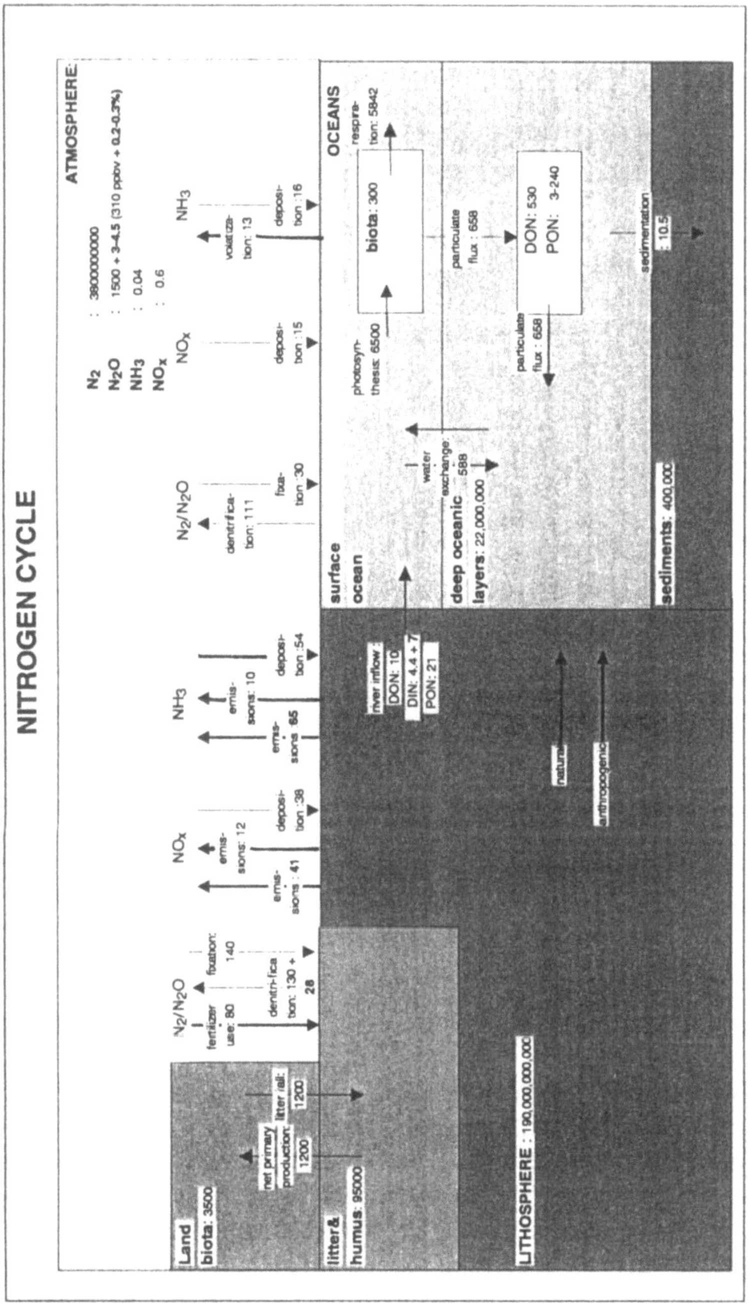

Figure 4.3 The global nitrogen cycle (based on Schlesinger, 1991; MacKenzie *et al.*, 1992). The numbers in boxes indicate the size in TgN of each reservoir. On each arrow is indicated the magnitude of the flux in TgN/yr (DON = dissolved organic nitrogen, PON = particulate organic nitrogen, DIN = dissolved inorganic nitrogen)

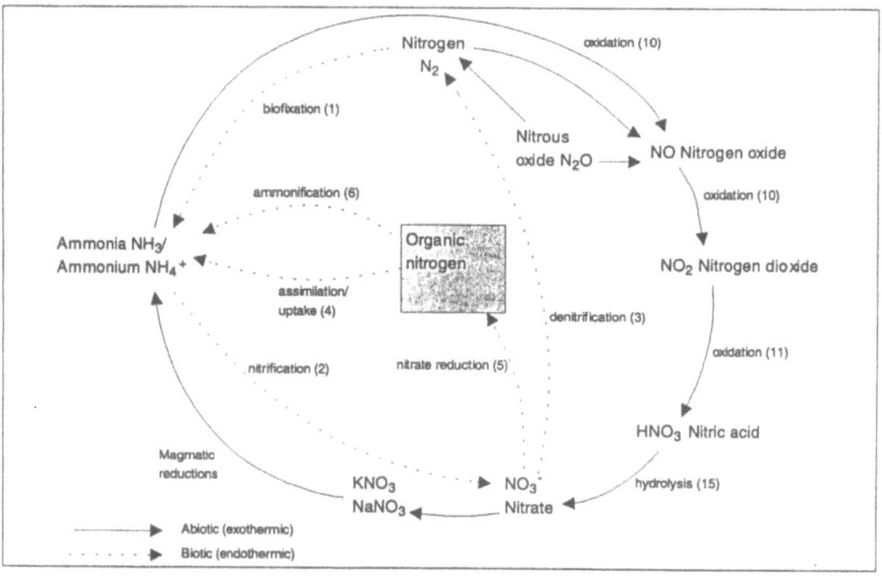

Figure 4.4 The nitrogen cycle: chemical transformations (adapted from Ayres, 1992)

land and sea, while Figure 4.4 depicts the global nitrogen cycle in terms of chemical transformations, and of biological and abiotic processes.

The biological processes are the primary mover in the terrestrial and oceanic nitrogen cycles, and represent about 95% of all nitrogen fluxes. The major process in nitrogen cycling and the most important source of nitrogen for plant growth is the process of nitrogen fixation (1), whereby atmospheric molecular nitrogen (N_2) is converted to ammonia (NH_3) by bacteria. This process is followed by bacterial nitrification (2), which converts ammonium (NH_4^+) to nitrites (NO_2) and nitrates (NO_3). Through the process of denitrification (3), nitrogen is recycled to the atmosphere.

Another route is through the process of inorganic N-uptake by plants through the processes of ammonia assimilation or ammonification (4) (the process by which NH_3 or NH_4^+ is taken up by an organism to become part of its biomass in the form of organic nitrogen compounds), and assimilatory nitrate reduction (5), followed by the nitrogen uptake by the organisms as biomass. This uptake process is followed by a process of decay of plant organic matter, contributing to the soil organic nitrogen pool. Then, in the process of volatisation (6) (the breaking down of organic nitrogen compounds into NH_3 or NH_4^+), inorganic N is produced, which is partly immobilised by decomposer organisms to become part of its biomass.

Table 4.2 Major chemical forms and processes of the nitrogen cycle

Major chemical forms of nitrogen		
Ammonia/Ammonium	NH_3/NH_4^+	A major nutrient, ammonia is a gas and is highly soluble in water
Nitrogen gas	N_2	Principal constituent of the atmosphere, unusable (molecular nitrogen) except by a few nitrogen-fixing bacteria
Nitrous oxide	N_2O	Combustion product. In the stratosphere formed by N_2O due to photochemical reactions. This catalyst can destroy O_3-molecules via catalytic cycle reactions, and thus NO is a natural regulator of stratospheric ozone
Nitric oxide	NO	Important role in stratospheric ozone depletion and climate change, product of denitrifying bacteria and other anthropogenic and natural sources
Nitrogen dioxide	NO_2	Formed by the reaction of NO with oxygen in the air, NO_2 is an urban air pollutant, a precursor of acidic deposition, and aids the formation of tropospheric ozone (photochemical smog)
Nitrite ion	NO_2^-	Intermediate product in the conversion of ammonia to nitrate
Nitrate ion	NO_3^-	Principal nutrient form, component of acid rain

Major biological processes of nitrogen	
Nitrogen fixation (1)	Conversion of N_2 to NH_3, NH_4^+ and any organic nitrogen compound by nitrogen-fixing bacteria and algae
Nitrification (2)	Subsequent oxidation of ammonia (NH_3/NH_4^+) to nitrate (NO_2^- or NO_3^-). The key early steps are performed only by a few types of bacteria. The reactions involving these nitrogen transformations are the following: $2\ NH_4^+ + 3\ O_2 \dashrightarrow 2\ NO_2^- + 2\ H_2O + 4\ H^+$ $2\ NO_2^- + O_2 \dashrightarrow 2\ NO_3^-$
Denitrification (3)	A chain of reactions by a few types of bacteria that ultimately return nitrogen to the atmosphere as N_2O and Nn4 2, by the reaction: $5\ CH_2O + 4\ H^+ + 4\ NO_3^- \dashrightarrow 2\ N_2 + 5\ CO_2 + 7\ H_2O$
Volatisation (4)	The breaking down of organic nitrogen compounds into NH_3 or NH_4^+
Assimilatory nitrate reduction (5)	The reduction of NO_3^-, followed by the uptake of the nitrogen by the organisms as biomass
Ammonia assimilation (6)	The process by which NH_3 or NH_4^+ is taken up by an organism to become part of its biomass in the form of organic nitrogen compounds

Major chemical processes of nitrogen	
Homogeneous gas phase reactions:	
Oxidation of NH_3 (7)	Ammonia is oxidised by OH radicals, which is a source and a sink of NO_X (Logan, 1983), although a minor sink of NH_3, with the following reactions: $NH_3 + OH \dashrightarrow NH_2 + H_2O$ Then NH_2 can react either with O_3, NO or NO_2: $NH_2 + NO/NO_2 \dashrightarrow N_2/N_2O$ (No_X concentration < 60 pptv) $NH_2 + O_3 \dashrightarrow$ NO/others (NO_X concentration > 60 pptv)

Major chemical processes of nitrogen	
Stratospheric ozone destruction:	
Source of NO_x (8)	Upward diffusion of N_2O in the stratosphere and its subsequent reaction with O atoms, or photolysis: $N_2O + hv \text{ ---> } N_2 + O$ $N_2O + O(^1D) \text{ ---> } 2\ NO$ The first reaction brings a single ozone molecule, and the second reaction leads to the production of catalyst NO, which leads to catalytic ozone destruction.
Catalytic reactions (9)	$O_3 + NO \text{ ---> } NO_2 + On4\ 2$ $NO_2 + O \text{ ---> } NO + O_2$ Net reaction: $O_3 + O \text{ ---> } 2\ O_2$
Acidification:	
Formation of NO_2 (10)	Net reaction: $CH_4 + 2O_2 + 2NO \text{ ---> } CH_2O + 2NO_2 + H_2O$
Formation of HNO_3 (11)	$NO_2 + OH \text{ ---> } HNO_3$
Tropospheric ozone increase/photochemical smog:	
Production of tropospheric ozone (12)	$NO_2 + hv \text{ ---> } NO + O$ $O_2 + O + M \text{ ---> } O_3 + M$
Heterogeneous reactions:	*The major sinks of the compound NH_3, NH_4^+ and HNO_3 are the following heterogeneous reactions and removal by rainfall:*
Removal of NH_3 (13)	$NH_3 + H_2O \text{ <--> ab } NH_4^+ + OH^-$
Removal of NH_4^+ (14)	$2\ NH_3(g) + H_2SO_4(l) \text{ ---> } (NH_4)_2SO_4(s)$
Removal of HNO_3 (15)	$NH_3(g) + HNO_3(g) \text{ ---> } NH_4NO_3(s)$ $NH_3(g) + HNO_3(aq) \text{ ---> } NH_4^+(aq)\ NO_3^-(aq)$

Although the abiotic fluxes are much smaller than the biological fluxes, they are quite important with respect to major environmental issues, as denoted in Table 4.2. The abiotic processes can be categorised into homogeneous and heterogeneous reactions. For brevity's sake, the complex chemistry involved will not be discussed here. For details the reader is referred to: Logan *et al.* (1981), Crutzen (1988), and Butcher *et al.* (1992).

4.3.3 Anthropogenic perturbation of the global nitrogen cycle

Natural balance

In pre-industrial times the only sources of nitrogen entering the terrestrial biosphere must have been from lightning, precipitation (ammonia, nitrate) and biological nitrogen fixation. The total nitrogen fixation flux is estimated at 170 TgN/yr, which supplies about 14% of the nitrogen that is made available for plant use each year (1200 TgN/yr). The remainder is delivered from internal cycling and the decomposition of dead organic matter in soils. Since the process of denitrification is the least understood process, it is normally used for balancing the terrestrial N cycle, resulting in an estimate of the "pre-industrial" denitrification flux of 130 TgN/yr.

The world's oceans receive about 36 TgN/yr from rivers, about 30 TgN/yr by biological fixation, and about 50 TgN/yr in precipitation. Nutrients are continuously removed from the surface water by downward sinking of dead organisms and faecal pellets, and they are regenerated in the deep ocean, where the concentrations are much higher. Permanent burial of organic nitrogen in sediments is insignificant, so most of the input to the oceans must be returned to the atmosphere as N_2 by denitrification.

Human disturbance

The net effect of various anthropogenic inputs on the entire nitrogen cycle certainly appears significant. However, the complexity of the data makes it difficult to provide evidence to support this hypothesis. The most important human impact on the nitrogen cycle is fertiliser use. We assume that the fertilisers input of 80 TgN/yr results in: a total direct loss of NH_3/NO_3^- of 13 TgN/yr comprising 9 TgN/yr by direct NH_3 release (volatisation), and 4 TgN/yr by NO_3^- from leaching. The remainder of the fertiliser input is either taken up by plants, or is released to the atmosphere by denitrification (N_2 and N_2O). The total denitrification flux is estimated at about 28 TgN/yr, a statistic based on 3.5 TgN/yr due to N_2O releases from fertilised soils and a denitrification flux of nitrogen (N_2) from cultivated soils of 24.5 TgN/yr. The uptake by plants can be calculated using yield response curves, and is presumed to be around 40-60% of the total fertiliser input. The remainder of about 39 TgN/yr is taken up by plants, 19 TgN/yr being removed with the harvesting of crops and 20 TgN/yr accumulating in organic soil fractions. Harvesting activity and subsequent consumption, together with waste products in sewage are important mechanisms for transporting fertilisers in the form of organic nitrogen compounds to the surface waters.

The cultivation of crops (e.g. legumes and the use of Azolla in rice paddies) also causes an anthropogenic enhancement of the nitrogen fixation (about 40 TgN/yr).

Other important anthropogenic sources are: the emissions of NH_x, NO_x and N_2O from fossil fuel burning (about 30 TgN/yr), biomass burning (about 14 TgN/yr) and domestic animals (32 TgN/yr), although only the first is an additional source of nitrogen into the global nitrogen cycle (from atmospheric N_2) ,whereas the others are human perturbations within the global nitrogen cycle (from terrestrial organic nitrogen). Owing to the short residence time of NO_x and NH_x, the emissions of these compounds are nearly all once again deposited on land.

Representing the global nitrogen cycle in terms of flows between the different environmental compartments, yields three mutually-dependent sub-cycles: the ammonia cycle, the NO_x cycle, and the nitrogen fixation/denitrification cycle (N_2O and N_2), which are all heavily disturbed, and will be briefly discussed below.

Subcycles

Ammonia cycle - The total global flux of NH_3 into the atmosphere is estimated at 75 TgN (about 65% of which is anthropogenic), and is dominated by the emissions from land (especially from the manure from domestic animals, from soils of ecosystems and fertilised agricultural land (see Table 4.3)). The primary sink of NH_3 from the atmosphere is the deposition of NH_4^+ in rainfall and dry-out, at rates of about 40 TgN/yr on land and 16 TgN/yr in the ocean. The other important sink of NH_3, namely the oxidation of NH_3 by OH, is estimated at less than 1 TgN/yr.

Although a balanced ammonia budget is assumed, there are numerous uncertainties in the estimates of sources and sinks of ammonia. This is illustrated by a decreasing trend in the deposition of NH_4^+ of about 34% during the period 1963 to 1982 in the United States (Likens *et al.*, 1984), which is surprising, given the increased fertiliser use and agricultural practices. Possible explanations are decreasing emissions from the other sources over the same period, or increased emissions of SO_2, which lead to an increase in the residence time and in the potential for long-range transport of NH_3/NH_4^+ in the atmosphere.

The ammonia cycle is disturbed by human activities, whereby emissions from fertilised agriculture and domestic animals are the most important. In many areas, this increased release of ammonia to the atmosphere leads to atmospheric deposition of nitrogen far from the major sources, where forests growth consequently declines.

Table 4.3 The global nitrogen cycle, as subdivided into the three nitrogen cycles: the ammonia cycle, the NO_X cycle and the fixation/denitrification cycle (fluxes in TgN/yr)

Ammonia cycle	**Schlesinger and Hartley (1992)**	**Den Elzen *et al.* (1995)**
Sources		
NH_3 volatisation		
Domestic animals	32 (24-40)	
Sea	13 (8-18)	
Soils	10 (6-45)	
Fertilisers	9 (5-10)	
Biomass burning	5 (1-9)	
Human	4	
Energy	~2	
Total	*75 (50-128)*	
Sinks		
NH_4 deposition		
Deposition on land	40	54
Deposition on oceans	16	16
NH_3 oxidation	1	5
Total	*57*	*75*

NOx-cycle	**Prather *et al.* (1994) Watson *et al.* (1990)**	**Den Elzen *et al.* (1995)**
Sources		
Fossil fuel combustion	24	
Soil release	12 (5-20)	
Biomass burning	8 (2.5-13)	
Lightning	5 (2-20)	
NH_3 oxidation	3	
Transport from stratosphere	0.6 (mostly HNO3)	
Aircraft	0.4	
Total	*53*	
Sinks		
Deposition on land		38
Deposition on oceans		15
Total		*53*

N_2O cycle Nitrogen fixation/denitrification cycle	Leggett *et al.* (1992)	Prather *et al.* (1994)
Natural sources		
Oceans		3 (1.0-5.0)
Tropical soils		3 (2.2-3.7)
Dry savannes		1 (0.5-2.0)
Temperate soils/forests		1 (0.1-2.0)
Grasslands		1 (0.5-2.0)
Total	*8.3*	*9 (6-12)*
Anthropogenic sources		
Cultivated soils	2.2	3.5 (1.8-5.3)
Biomass burning	1.3	0.5 (0.2-1.0)
Industrial sources	1.1	1.3 (0.7-1.8)
Cattle and feed lots		0.4 (0.2-0.5)
Total	*4.6*	*5.7 (2.9-8.6)*
Total sources	**12.9**	**14.7 (10-17)**
Sinks		
Stratospheric photochemistry	9.5	12.3 (9-16)
Removal by soils		?
Atmospheric increase	~3.5	3.9 (3.1-4.7)
Implied total sources		*16.2 (13-20)*

NO_x cycle - This cycle comprises the balance of the sources of NO_x (NO and NO_2) on the one hand, and the sinks, the wet and dry deposition of NO_3^-, NO and NO_2 on the other (see Table 4.3).The main anthropogenic source of NO_x is fossil fuel combustion, which is still increasing globally. Of all the NO_x produced about 60% (30-35 TgN/yr) is anthropogenic. The most important removal mechanisms of NO_x from the atmosphere are the oxidation of NO_2 by reaction with OH to HNO_3 and at night with O_3 to NO_3^- The tropospheric lifetime of NO_x is short: only a few days. The nitrate which is deposited forms a significant contributor to acid deposition. The deposited NO_3^- on the land and ocean system can be incorporated into the biomass, and enters the fixation/denitrification cycle, or accumulate in the ocean.

Nitrogen fixation/denitrification cycle - In the atmosphere this cycle, with its principal components N_2O and N_2, is heavily perturbed by humans, mainly by fertiliser use. For N_2O, the global atmospheric concentration is at present increasing at a rate of about 0.2% per year (about 4 TgN/yr). The sources and sinks of N_2O are summarised in Table 4.3. The uncertainties in the emissions of N_2O from the various sources are quite large. The major sinks of N_2O are stratospheric photolysis, yielding N_2 and O_2, or oxidisation by ozone to NO_x, which contributes to stratospheric ozone depletion. The uptake of N_2O by soils is probably a minor sink. The atmospheric lifetime of N_2O is estimated at about 130 years (110-168) years. The sources and sinks of N_2O clearly show the enormous impact of human activities on the cycling of nitrogen, although the overall effect of this disturbance is only poorly understood. The same holds for the cycling of N_2, which is closely linked to the N_2O cycle, since both substances are produced by denitrification.

4.3.4 Conclusions

Human activities have resulted in a significant disturbance of the global nitrogen cycle. Through the processes of fertiliser production, energy production, and cultivation of crops, an additional amount of about 150 TgN/yr is added to the natural nitrogen cycle. Only for a part of this anthropogenic N it is known where it remains. N_2O is accumulating in the atmosphere at a rate of about 4 TgN/yr, and plays a role in global climate change. Coastal oceans receive another 41 TgN/yr via rivers, much of which is buried or denitrified. Open oceans receive 18 TgN/yr by atmospheric deposition, which is incorporated into oceanic N pools. The remaining 90 TgN/yr are either stored on continents in groundwater, soils, or vegetation, or denitrified to N_2.

4.4 The global phosphorus cycle

4.4.1 Introduction

Phosphorus (P) is one of the most important elements on Earth because of its essential role in many of the biogeochemical processes. Together with nitrogen, phosphorus limits biological productivity on land and in the sea. Whereas on land nitrogen is the primary limiting nutrient, scientific controversy still exists whether phosphorus or nitrogen is the limiting nutrient in the marine environment. However, this depends on the location, in particular on the distance from the source. Furthermore, phosphorus is an essential element of DNA and the phospholipid molecules of cell membranes, and therefore plays an important role in the biological cycle. The main human modification to the global phosphorus cycle is the change in the phosphorus flux carried in rivers to the oceans as a result of fertiliser runoff, pollution and erosion, phenomena associated with the environmental issue of eutrophication.

4.4.2 The present phosphorus cycle

The global phosphorus cycle is relatively simple - and also significantly different – when compared to the other biogeochemical element cycles, since it has no gaseous component. The major reservoirs of the P-cycle are land, oceans and sediments, while the atmosphere plays a subsidiary role, because there is only a small transport of P within the atmosphere on aerosol and dust particles and P dissolved in rain and cloud droplets.

Phosphorus is always found in combinations with oxygen, i.e. phosphate PO_4^{3-}. The proportion of the individual types of ionic forms of phosphorus depends on the following equilibria:

$$H_3PO_4 \;\text{<--->}\; H^+ + H_2PO_4^- \;\text{<---->}\; 2H^+ + HPO_4^{2-} \;\text{<---->}\; 3H^+ + PO_4^{3-}\text{b} \quad (4.4)$$

The position of the equilibria (to right or left) also depends on the pH value in surface water or soil solution. The PO_4-ions form at the presence of cations such as Ca^{2+}, Mg^{2+} and Na^+, PO_4^{3-}- species. The most abundant phosphate mineral is apatite, which occurs for more than 95% of all P in the Earth's crust. Nearly all phosphorus on land is originally derived from mechanical and chemical weathering of these apatites. The largest accumulations of P on the Earth's surface are massive apatite deposits (phosphorites), which mining provides for about 80% of the total phosphate production and 95% of the total remaining reserve.

Phosphorus has also many organic forms and is a constituent of DNA, RNA, and of cell membranes. It therefore has an important role in the biological cycle, every organism contains phosphorus.

The global phosphorus cycle roughly consists of an inorganic and organic P-cycle, and is depicted in Figure 4.5. The inorganic phosphorus cycle can be considered as the overall phosphorus cycle, which starts with the weathering of rock material and the transport of P in both dissolved and particulate form via rivers to the oceans, where it is deposited into the sediments. Occasionally, this movement of P is interrupted by the involvement of biological systems, such as terrestrial ecosystems and aquatic ecosystems in lakes and oceans. Phosphorus is important in the total biological cycle, since it is one of the least available elements (together with nitrogen) required for plant growth, and therefore sets limits to plant growth for some ecosystems, i.e. tropical forests (in general nitrogen is more limiting) (Vitousek and Howarth, 1991). The total net primary production, and thus the magnitude of the flux cycling between the soil and vegetation, is about 200 TgP/yr. The uptake of phosphate by vegetation is in balance with the return of P to soils from the biota due to the decay of dead organic materials. Lakes, other surface waters and coastal waters are also important components of the terrestrial P-system and the P-cycling, especially in view of the role of phosphorus in their eutrophication.

The remainder of the P (not participated in the biological cycle) is eventually transported to the estuaries both in dissolved and particulate form. The latter will rapidly settle to the sea floor and be incorporated into the sediments, while the dissolved P will enter the surface water of the ocean, and participate in the biological cycles. The current river flux of P to the oceans is about 21 TgP/yr, nearly all in particulate form (20 TgP/yr), which is probably 2-3 times greater than in the pre-industrial times due to erosion, pollution, and fertiliser runoff (see Section 4.4.3).

The total P-flux to the oceans primarily occurs via river transport and to a small extent via atmospheric dust and aerosol transport. This flux must be in balance with the annual increase of the permanent burial of phosphorus in ocean sediments, since there are no gaseous losses of P to the sea. However, the total burial of phosphorus is probably not known within a factor of 10, and much of the river input is apparently deposited on the continental shelf without ever interacting with the ocean biogeochemistry. As stated earlier, the net primary productivity in the ocean is limited by nutrients (phosphorus and nitrogen) in the surface waters. The concentration of phosphate in the surface water of the ocean is low, whereas the deep oceans contain a substantial amount of P.

The total P-flux to the oceans primarily occurs via river transport and to a small extent via atmospheric dust and aerosol transport. This flux must be in balance with the annual increase of the permanent burial of phosphorus in ocean sediments, since there are no gaseous losses of P to the sea. However, the total burial of phosphorus is probably not known within a factor of 10, and much of the river input is apparently deposited on the continental shelf without ever interacting with the ocean biogeochemistry. As stated earlier, the net primary productivity in the ocean is limited by nutrients (phosphorus and nitrogen) in the surface waters. The concentration of phosphate in the surface water of the ocean is low, whereas the deep oceans contain a substantial amount of P.

A human-induced addition to the river input of P by fertiliser use would result in an extra accumulation of organic carbon due to increased net primary productivity in the oceans. This is based on the assumption that phosphate limits the productivity in the oceans, which seems likely on the longer time-scales, since there is no alternative source or substitute for phosphate, while in the case of N there are organisms, e.g. algae, capable of fixing N from the large atmospheric N_2 pool (Jahnke, 1992).

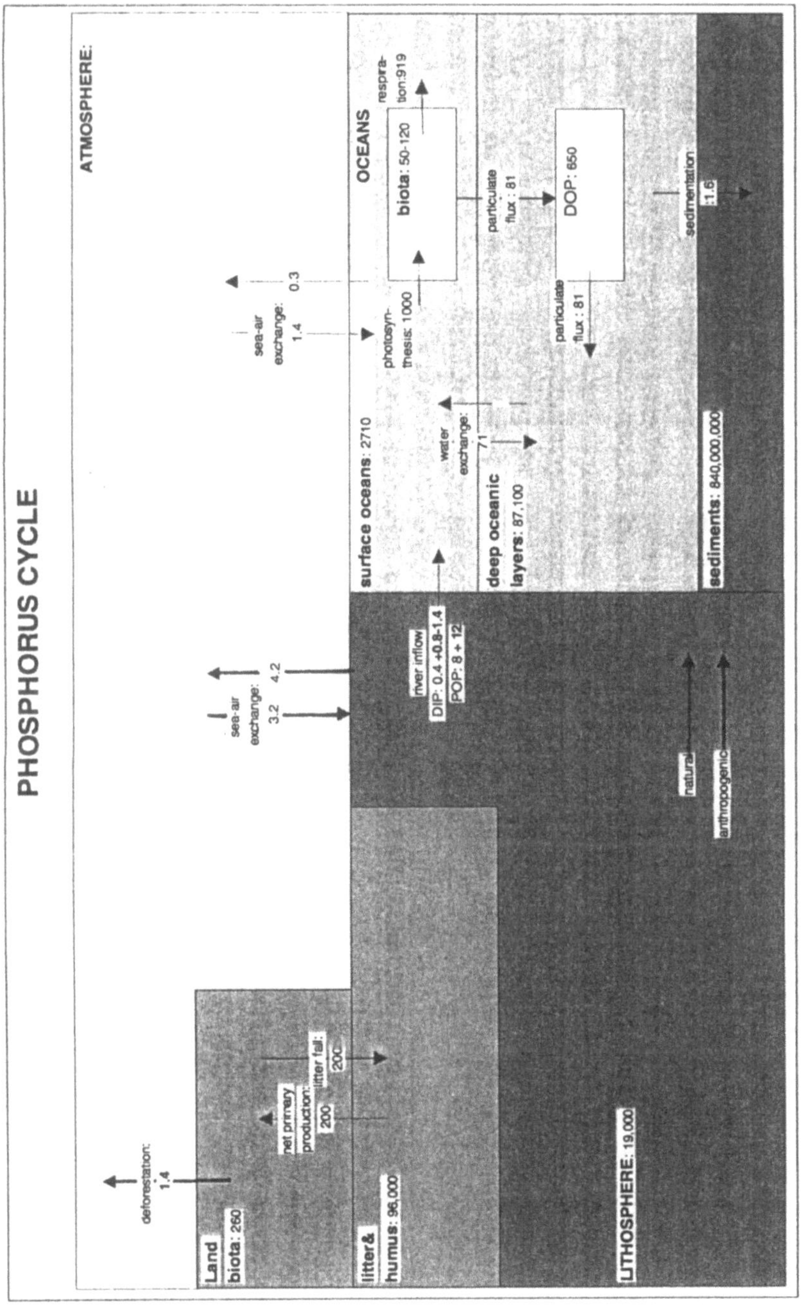

Figure 4.5 The global phosphorus cycle (Schlesinger, 1991). The numbers in the boxes indicate the size in TgP of each reservoir. On each arrow is indicated the magnitude of the flux in TgP/yr (DOP = dissolved organic phosphorus, POP = particulate organic phosphorus, DIP = dissolved inorganic phosphorus)

4.4.3 Anthropogenic perturbation of the global phosphorus cycle

The most visible mechanisms by which humans have altered the P cycle is evidently the mining of phosphorus rock of which a large part is brought into the environment (P-fertilisers, P-mineral supplementation to ruminants), or indirectly via detergents, as well as the increased phosphate flux through waste waters (human sewage and industrial use). The human impact can best be detected by analysing the increase in the total phosphorus flux from the rivers to the oceans. The natural rate of phosphorus migration from land to the hydrosphere by leaching, natural erosion and runoff is estimated to be between 6-13 TgP/yr, whereby recent estimates are closer to 6 than 13 TgP/yr. Current annual river transport to the ocean is estimated as 21 TgP/yr (19-24 TgP), which means a factor 2-3 higher than in pre-industrial times as a result of three determinants: erosion, pollution (municipal wastes) and fertiliser runoff. The key question to be answered is: what is the contribution of each determinant to both the total and natural oceanic transport flux?

Firstly, the total present flux due to fertiliser use (about 15 TgP/yr) arriving in the surface waters is estimated at about 2.3 TgP/yr, of which 90% reaches the oceans. There are three different routes via which phosphate migrates from the terrestrial biosphere to the hydrosphere: leaching from soils; soil erosion; and harvesting of crops (via waste-water). The latter is thought to be most important, as denoted in Table 4.4, while leaching and erosion of the fertilised top layer of soils is currently considered less important because of the strong adsorption of phosphate by soil particles (which could nevertheless change in the near future if phosphate saturation of soils occurs).

Table 4.4 Components of the total transport P-flux to the oceans

Source	
Natural flux	8.0 TgP/yr
Anthropogenic	
-erosion of fertilised top-layer of soils	0.5 TgP/yr
-sewage water (harvesting of crops by fertiliser use)	2.0 TgP/yr
-leaching from agricultural soils	1.0 TgP/yr
-cleaning applications	1.0 TgP/yr
-anthropogenic erosion	8.5 TgP/yr
Total flux	21 TgP/yr

Secondly, the phosphate flux through waste waters reaching the surface waters after industrial and human use (cleaning applications) is about 1 TgP (Table 4.4). Thirdly, the process of erosion due to deforestation and agricultural practices is considered to be by far the main cause of the increase in phosphate transport by the rivers to the oceans, contributing about two-thirds to the total anthropogenic flux (Table 4.4).

The difference between the total phosphate flux to the hydrosphere (20-24 TgP/yr) (rivers, fresh waters, and oceans) and the phosphate flux to the oceans (20 TgP/yr), excluding the flux through the atmosphere, represents the P which is sequestered in rivers and fresh waters. This phosphate flux to fresh waters is currently estimated at about 1-2 TgP/yr, which is likewise twice the natural flux.

4.4.4 Conclusions

Humans have altered the natural phosphorus cycle in a substantial manner, mainly through the mining of phosphorus rock, through detergents, and through waste waters. The best way to measure the human impact is by analysing the increase in the total phosphorus flux from the rivers to the oceans. The current river flux appears to be 2-3 times higher than the natural flux. The human disturbance contributes to global climate change through two biogeochemical feedbacks affecting the marine production: the P-fertilisation effect on marine production, and the eutrophication effect. As described in section 4.6, the net oceanic biological primary productivity is limited by nutrients, i.e. phosphate, and therefore an increase in total phosphate in the sea would raise the primary production by the oceans, and thus lower the atmospheric CO_2 concentration, suppressing the climate change signal.

On the other hand, eutrophication stimulates the growth of some plant and algae species under conditions of abundant nutrients. This can lead to algae blooms and reduction of marine production, especially in coastal waters, which amplifies the climate change problem, as also addressed in section 4.6.

4.5 The global sulphur cycle

4.5.1 Introduction

Sulphur plays various roles in the chemical functioning of the Earth. In its oxidised form, SO_4^{2-}, it is a key nutrient for the sustaining of life, since, when it is assimilated by organisms, it is reduced and converted into organic sulphur, which is an essential component of proteins (of which enzymes, hormones, chlorophyll, and genes are made). The living biosphere contains relatively little sulphur, although it is the 14th most abundant element in the Earth's crust, and sulphate is the second most abundant anion in rivers (after bicarbonate, HCO_3^-) and in seawater (after chloride).

During recent decades, the oxidation of large-scale emissions of sulphur dioxide (SO_2) from fossil fuel combustion, biomass burning and industrial sources has largely contributed to the widespread acidification of soils and fresh water (lakes). A further intriguing aspect of the global sulphur cycle is that sulphate in the atmosphere is thought to play a major role in the Earth's climate system, although scientific knowledge about climate regulation by sulphate aerosols is rather limited. Finally, of all global element cycles (C, N, O, P and S) the sulphur cycle is one of the most heavily disturbed by human activities. Anthropogenic sulphur emissions are nowadays twice as high as natural sulphur emissions (Den Elzen *et al.*, 1995).

4.5.2 The present sulphur cycle

The global sulphur cycle features a rich variety of important biological and abiotic processes that involve many important compounds in their gaseous, liquid and solid phases. These are set out in Figures 4.6 and 4.7. Figure 4.6 traces the various flows of the different sulphur- containing compounds in the atmosphere, terrestrial biosphere, soils, and oceans due to natural and anthropogenic emissions on land and in the oceans.

Figure 4.7 charts the sulphur cycle in terms of the various biological and abiotic processes and the main compounds involved. In the global sulphur cycle, the reduced forms of sulphur (H_2S, CS_2, COS and DMS) are gradually oxidised by atmospheric oxygen (mainly as OH) to sulphur oxides (SO_2, SO_3) and finally via oxidation to sulphuric acid (H_2SO_4), or via heterogeneous reactions in the aerosol phase to sulphate aerosols (SO_4^{2-}). These particles are deposited in wet or dry form, thus creating the phenomenon of acidification.

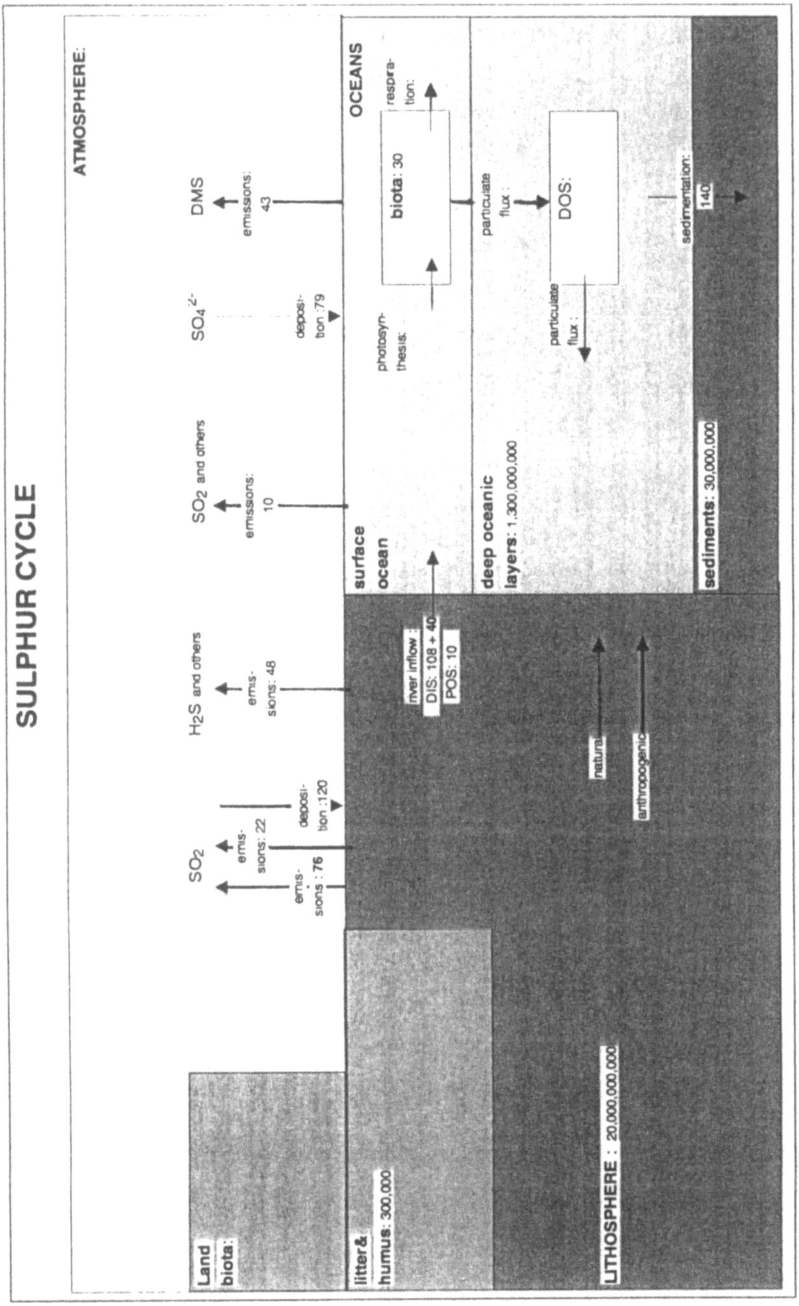

Figure 4.6 The global sulphur cycle (based on Schlesinger, 1991; Charlson *et al.*, 1992; Jonas *et al.*, 1994). The numbers in boxes indicate the size in TgS of each reservoir. On each arrow is indicated the magnitude of the flux in TgS/yr (DOS = dissolved organic sulphur, POS = particulate organic sulphur, DIS = dissolved inorganic sulphur)

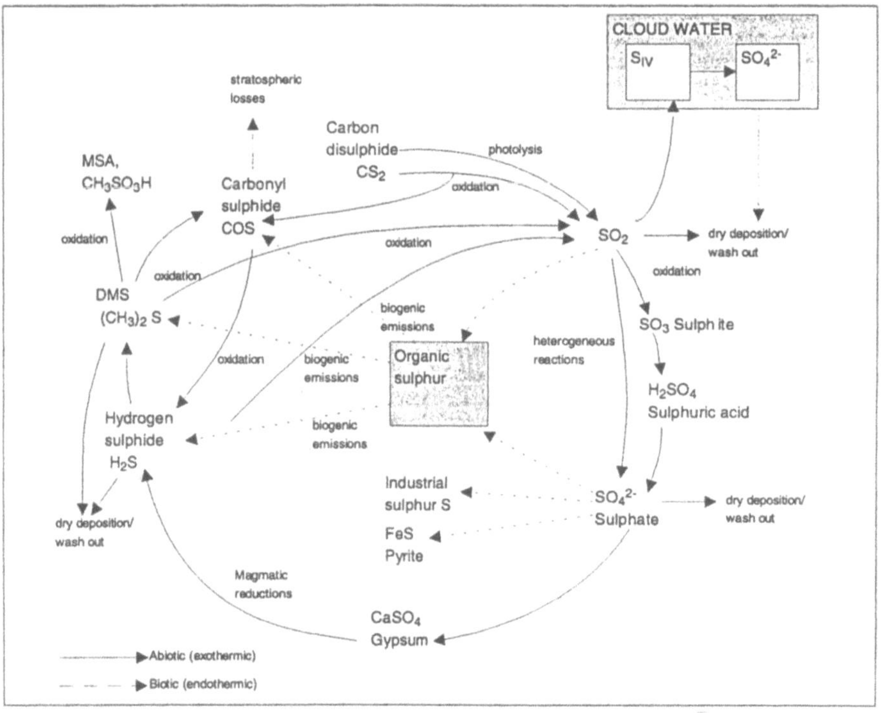

Figure 4.7 The sulphur cycle: chemical transformations (based on Ayres, 1992; Charlson *et al.*, 1992)

Since the major pathway within the sulphur cycle – the emissions of sulphur compounds up to deposition as sulphate (via the gas-phase, or the cloud-liquid phase) – involves a change in chemical oxidation state and physical phase, the lifetime of sulphur in the atmosphere is governed by both the kinetics of the oxidation reactions and the frequency of cloud and rain (Charlson *et al.*, 1992). The overall process is rapid, on the order of days, which means that the atmospheric sulphur cycle is a regional phenomenon, distributing nearly all forms of sulphur relatively close to the sources. The reverse conversion, i.e. from sulphur back to states of higher thermodynamic potential, is accomplished either by biological activity or by magmatic reactions in the Earth's mantle (Ayres, 1992).

The Earth's surface-atmosphere exchange processes are particularly important in the global sulphur cycle. In the oceans, biological processes (phytoplankton production) result in the reduction of seawater sulphate to reduced sulphur gases to the atmosphere, and mainly to dimethylsulfide (DMS) (about 40 TgS/yr, with an uncertainty range of 10-50 TgS/yr). The mean lifetime of DMS is about 1 day as a result of the oxidation to SO_4, in which form it is mostly re-deposited to the oceans. The major biogenic emissions from land are the H_2S emissions, which are estimated at 32 ± 27 TgS/yr. This makes the current estimate of the total biogenic emissions from land and sea about 90 TgS/yr (with an uncertainty range of 60-110 TgS/yr). Further sources of sulphur (mainly SO_2) emissions in the atmosphere are formed by volcanic eruptions, which, in case of large eruptions, are capable of emitting around 17-30 TgS/yr of H_2SO_4. This is further discussed in Den Elzen *et al.* (1995), and also in Chapter 2.

4.5.3 Anthropogenic perturbation of the global sulphur cycle

In spite of the fact that our knowledge of the global sulphur cycle is primitive, it is known that current human perturbation of the sulphur cycle is extreme. Roughly-speaking, the pre-industrial inputs to the land surface have more than quadrupled, from about 26 TgS/yr to the present value of about 112 TgS/yr as a result of atmospheric deposition and fertiliser use.

The major human impact is the anthropogenic release of SO_2 from fossil fuel combustion, estimated at 70-80 TgS/yr, and from biomass burning, estimated at 10 TgS/yr. Approximately 90% of these anthropogenic emissions of sulphur occur in the Northern Hemisphere. From 1860 to 1980 the globally averaged anthropogenic emissions of SO_2 increased at a rate of 2.9% per year, from about 2 TgS/yr in 1860 to about 70 TgS/yr, representing a 35-fold increase. The historical anthropogenic SO_2 emissions are represented in Figure 4.8. The anthropogenic emissions of sulphur into the atmosphere (largely from coal combustion) now exceed the natural sulphur emissions by a factor of about 2 (Charlson *et al.*, 1992). Human activities also affect the transport of S in rivers. The current inputs to the oceans indicate that these inputs from river transport have roughly doubled since pre-industrial times, from about 72 Tg/yr to about 200 Tg/yr currently, due to air pollution, runoff of fertilisers, mining and erosion and other human activities. These inputs to oceans appear to exceed deposition to the ocean bottom by as much as 100 TgS/yr. This extra man-made input of about 100 TgS/yr to the oceans is still difficult to record, since the sulphur content of the ocean is about 12 10^8 TgS.

4.5.4 Conclusions

It is obvious that the current global sulphur cycle is extremely unbalanced. Due to human activities such as fossil fuel combustion, fertiliser use, and mining, human sulphur inputs are twice as large as the natural inputs. The global sulphur cycle plays an important role in the issues of acidification and climate change. SO_2 emissions from fossil fuel and biomass burning and industrial activities lead to acidity of rainwater, and the deposition on land leads to soil acidification, and acidification of fresh water (mainly lakes). Sulphate in the atmosphere, in the marine regions mainly coming from DMS, and in urban areas from anthropogenic SO_2 emissions, has a major role in Earth's climate. As treated in Chapter 2, sulphate aerosols have a direct and indirect influence on the Earth's radiative forcing.

4.6 Interactions between the global element cycles and climate change

On a global scale the biogeochemical cycles not only have common sources, but are also interlinked through mutual physical, chemical and biological interactions and feedbacks and through common environmental impacts. To illustrate this, in this section we will seek to demonstrate how the problem of global climate change can be described in terms of interlinkages and disturbances between the various disturbed global element cycles.

Generally, the main interactions between the global element cycles manifest themselves at three different levels: at the level of common sources, the level of physical interrelations, and at the level of feedbacks. Because the first two levels are discussed in other chapters (Chapters 2 and 5), in this chapter we will focus on the major climate-related feedbacks between the global element cycles.

Feedbacks

The various global element cycles are interlinked through a whole range of climate-related feedback mechanisms. Such mechanisms can amplify (positive feedback) or dampen (negative feedback) the response of the climate system, due to enhanced concentrations of the elements of C, N, P and S resulting from anthropogenic perturbations. Our understanding of these feedbacks is surrounded by considerable uncertainties, chiefly because of the lack of knowledge of their characteristics under both present and future circumstances.

In general, two classes of climate feedbacks can be discerned: geophysical and biogeochemical feedbacks. Geophysical climate feedbacks are caused by physical processes directly affecting the response to radiative forcing. The most important ones are the water vapour and cloud feedbacks, and snow-ice albedo. These radiative feedbacks are incorporated in the present climate models, and the climate sensitivity of these models is highly dependent on the representation of these physical feedback processes. The wide uncertainty range of this climate sensitivity is mostly due to the lack of knowledge about the characteristics of these feedbacks under the present and new circumstances.

Here we will further concentrate on the main biogeochemical climate feedbacks, which are those related to the response of the marine and terrestrial biosphere and components of the geosphere, thereby indirectly altering the radiative forcing. These feedbacks, which are summarised in Table 4.5, will be discussed below.

Carbon cycle

The major biogeochemical feedbacks within the global carbon cycle are the terrestrial feedbacks (Schimel *et al.*, 1995): CO_2- and N-fertilisation and temperature feedbacks on the net primary production (net photosynthesis minus respiration of non-photosynthetic plant parts) and on soil respiration.

Short-term experiments under optimal conditions of water supply, temperature and nutrients (non-stressed conditions) show that most plants achieve an increase in photosynthetic rate and plant growth at enhanced CO_2 levels: up to 20 to 40% when CO_2 is doubled. This process is called the CO_2 fertilisation effect. However, the final overall impact of this feedback on the net primary production depends very much on the availability of water and nutrients in these natural systems (Schimel *et al.*, 1995). So far, model studies show that nutrients limit the response of terrestrial ecosystems to CO_2 fertilisation to 20-40% of the potential with unlimited nutrients (Comins and McMurtrie, 1993).

Temperature feedbacks involve the combined effects of the stimulation of photosynthesis and respiration at increased temperatures (e.g. Harvey, 1989; Schimel *et al.*, 1995). The effect of rising temperature on photosynthesis will, up to an optimum level, be to stimulate plant growth and thus carbon storage by vegetation (negative feedback). Another feedback involves the temperature effect on soil respiration, resulting in an increase in the release of CO_2 to the atmosphere (positive feedback). The overall impact of such temperature-related feedbacks may become a significant positive climate feedback in the future (Schimel *et al.*, 1995).

Table 4.5 Principal feedbacks in the enhanced greenhouse effect

Feedback	Sign of the feedback [1)]
Geophysical feedbacks	
Water vapour	++
Snow-ice albedo	+
Clouds	?
Sulphate aerosols	-
Indirect effects of CFCs	-
Biogeochemical feedbacks	
Carbon cycle:	
Ocean temperature	+
Ocean eddy diffusion	+
Vegetation respiration	+
Carbon dioxide fertilisation	-
Soil moisture	-/+
Methane sub-cycle:	
Methane hydrates	+
Methane from wetlands	+
Methane from rice paddies	+
Methane from permafrost	+
Nitrogen cycle:	
N-fertilisation	-
Soil acidification	-
Phosphorus cycle:	
P-fertilisation	-
Eutrophication effect	+
Sulphur cycle:	
DMS-feedback	-
Stratospheric ozone depletion	
UV-B feedback on plankton	+

[1)] Sign of the feedbacks: ++ = strongly amplifying global warming; + = amplifying global warming; - = damping global warming; and ? = effect of feedback is unknown or uncertain.

Nitrogen cycle

The major biogeochemical feedbacks within the nitrogen cycle are nitrogen fertilisation and temperature feedbacks on nitrogen mineralisation. N-fertilisation is related to anthropogenic releases of NO_X and NH_3 The consequent depositions of their oxidation products increase the level of nutrients in the soils of natural ecosystems, which increases the productivity of the ecosystems. This increases the CO_2 uptake by vegetation by about 0.2-1.0 GtC/yr over the 1980s, with a proposed upper bound of about 2 GtC/yr

(Peterson and Melillo, 1985; Schlesinger, 1993; Schindler and Baley, 1993). However, high levels of nitrogen addition are also associated with acidification, which in the long term may damage ecosystems and reduce carbon storage.

Global warming could increase decomposition and nitrogen mineralisation, and thereby increase the net primary production, especially in the northern latitudes (Rastetter *et al.*, 1991; Hudson *et al.*, 1994). The temperature feedback could possibly offset the increased CO_2 releases from the northern soils (temperature feedback on soil respiration).

Phosphorus cycle

The net primary production in marine ecosystems is mainly limited by nutrients, in general: phosphorus. In the phosphorus cycle, there exist two main biogeochemical feedbacks, affecting the marine production, namely: the P-fertilization effect on marine production and the eutrophication effect.

The main interaction of the phosphorus cycle and the other biogeochemical C-, N- and S-cycles is via biological processes that occur in the oceans. The C:N:P ratios for the production of oceanic plankton remain constant for biological processes (like an increase in the respiration rate, or weathering rates). However, this ratio can also change due to an adjustment of the system itself, for example: a shift in the algae community in a lake towards blue-green species (efficient atmospheric N-fixing species) can bring the availability of N up to the level of P. In general, it is believed that there is a shortage of phosphorus in aquatic ecosystems (in case of nitrogen the feedback process is analogous), and that an increase in phosphorus (by fertiliser use for example) can therefore lead to an increase in the biological production (eutrophication). This increases the CO_2-uptake in the oceans, thus decreasing the atmospheric carbon dioxide levels. On the other hand, an increase in the marine production also leads to an increased production of DMS, and amplifies the DMS feedback (see next paragraph).

Sulphur cycle

In the tropospheric sulphur cycle, there is a linkage between the marine production and climate, the so-called DMS-feedback. Shaw (1983) and Charlson *et al.* (1987) were among the first who argued that the biogeochemical sulphur cycle could act as a "planetary thermostat", due to a negative feedback mechanism that would tend to stabilise the climate system against perturbations (cooling and warming). This mechanism consists of the production of dimethylsulfide (DSM) by oceanic phytoplankton production and its subsequent transport to the marine boundary layer. There the DMS is oxidised to sulphate particles, which act as cloud condensation nuclei (CCN). An increase in CCN is believed to cause an increase in cloud droplet concentration and cloud albedo, and increased albedo leads to less radiation and lower surface temperatures, which might be expected to affect phytoplankton productivity. In case of global climate change, this feedback mechanism could significantly reduce the warming effect. Lawrence (1993) estimated that the feedback strength is about 20% of that which would be necessary to completely counteract a perturbation to global climate change.

4.7 Discussion

Human activities have significantly affected the functioning of the Earth's system, and its related global biogeochemical cycles. These large-scale human-induced changes of the biogeochemical cycles, in particular the cycles of carbon (C), nitrogen (N), phosphorus (P), and sulphur (S), lead to a whole range of mutually-dependent environmental issues.

An integrated assessment of these global environmental issues, denoted as global environmental change, requires an analysis of the common causes (human activities), common impacts (on ecosystems and on the economy), and of the many interactions and feedback mechanisms between the global biogeochemical cycles. Analysing the global biogeochemical cycles helps us to better understand a variety of global environmental problems. Stratospheric ozone depletion, acidification and human-induced global climate change are just visible symptoms of significant disturbances of the global biogeochemical cycles. Current global environmental problems are often nothing but accumulations of human-produced substances in one compartment (the atmosphere, hydrosphere, or the terrestrial biosphere). These accumulations, however, move in time from one compartment to the other, thereby enhancing the interaction between the various global cycles. This dynamic process of moving accumulated substances between the various compartments, makes that in the future we will again certainly be surprised by emerging new global environmental problems.

References

Ayres, R.U. (1992), Industrial metabolism and the grand nutrient cycles, 92/6/EPS, Centre for the Management of Environmental Resources, INSEAD, Fontainebleau, France.

Bacastow, R., and Keeling, C.D. 1973. Atmospheric carbon dioxide and radiocarbon in the natural carbon cycle: II. Changes from A.D. 1700 to 2070 as -deduced from a geochemical model, in Carbon and the Biosphere, edited by G.M. Woodwell and E.V. Pecan, 86-135. U.S. Atomic Energy Comission, Washington, D.C

Bolin, B. (1970a), Changes of land biota and their importance for the carbon cycle, Science 196, 613-615.

Bolin, B. (1970b), The carbon cycle, Scientific American 223 (3), 124-132

Bolin, B., Degens, E.T., Kempe, S. and Ketner, P. (eds.) (1979), The global carbon cycle, John Wiley, New York.

Bolin, B. and Cook, R.B. (eds.) (1983), The major biogeochemical cycles and their interactions, SCOPE 21, John Wiley & Sons, New York, U.S.A.

Bolin, B., Crutzen, P.J., Vitousek, P.M., Woodmansee, R.G., Goldberg, E.D., and Cook, R.B. (1983), Interactions of biogeochemical cycles, in The major biogeochemical cycles and their interactions, B. Bolin and C.B. Cook (eds.), SCOPE 21, John Wiley & Sons, New York, U.S.A.

Broecker, W.S., Peng, T.H. 1992. Interhemispheric transport of carbon dioxide by ocean circulation. Nature 356, 587-589.

Butcher, S.S., Charlson, R.J., Orians, G.H. and Wolfe, G.V. (1992), Global biogeochemical cycles, Academic Press, London.

Butcher, S.S. (1992), Equilibrium, rate and natural systems, in Global biogeochemical cycles, Butcher, S.S., Charlson, R.J., Orians, G.H. and Wolfe, G.V. (eds.), Academic Press, London.

Charlson, R.J., Lovelock, J.E., Andrea, M.O., Warren, S.G. (1987). Oceanic phytoplankton, atmospheric sulphur, cloud albedo and climate. Nature, 326, 655-661.

Charlson, R.J., Schwartz, S.E., Hales, J.M., Cess, R.D., Coakley, J.A., Hansen, J.E. and Hofmann, D.J. (1992), Climate forcing by anthropogenic aerosols, Science 255, 423-429

Comins, H.N. and McMurtie, R.E. (1993). Long-term response of nutrient limited forests to CO_2 enrichment; equilibrium behaviour of plant-soil models. Ecological Applications, 3(4), 666-681.

Crutzen, P.J. (1988), Tropospheric ozone: an overview, In: Tropospheric Ozone-Regional and Global Scale Interactions, I.S.A. Isaksen (ed.), 3-32, NATO ASI Series, Vol. 227, Reidel, Boston, Mass.

Den Elzen, M.G.J., Beusen, A.H.W. and Rotmans, J. (1995). Modelling biogeochemical cycles: an integrated assessment approach. RIVM Report No. 461502007, Bilthoven, the Netherlands.

Goudriaan, J. and Ketner, P. (1984), A simulation study for the global carbon cycle including mans impact on the biosphere, Climatic Change 6, 167-192.

Gugten, van der D. (1988), The global phosphorus cycle: anthropogenic disturbtion and the effects (in Dutch), TNO SCMO report R 88/08, Delft, the Netherlands.

Harvey, L.D. (1989). Effect of model structure on the response of terrestrial biosphere models to CO_2 and temperature increases. Global Biogeochemical Cycles, 3(2), 137-153.

Houghton, J.T., Jenkins, G.J. and Ephraums, J.J. (eds.) 1990. Climate Change: The IPCC Scientific Assessment. Intergovernmental Panel on Climate Change. Cambridge University Press, Cambridge.

Houghton, J.T., Callander, B.A. and Varney, S.K. (1992), Climate Change 1992: The Supplementary Report to the IPCC Scientific Assessment. Intergovernmental Panel on Climate Change. Cambridge University Press, Cambridge.

Hudson, R.J.M., Gherini, S.A., and Goldstein, R.A. (1994). Modelling the global carbon cycle: nitrogen fertilization of the terrestrial biosphere and the missing CO_2 sink. Global Biogeochemical Cycles, 8(3), 307-333.

International Geosphere-Biosphere Programme (IGBP) (1992), Global change: reducing uncertainties, The Royal Swedish Academy of Sciences, Stockholm, Sweden.

Intergovernmental Panel on Climate Change (IPCC) (1990a), Climate Change: The IPCC Scientific Assessment, Houghton, J.T., Jenkins, G.J.. and Ephraums, J.J. (eds.), Cambridge University Press, Cambridge, U.K.

Intergovernmental Panel on Climate Change (IPCC) (1990b), Climate Change: The IPCC Impacts Assessment, Australian Government Publishing Service, Australia.

Intergovernmental Panel on Climate Change (IPCC) (1991), Climate Change: the IPCC Response Strategies, Island Press.

Intergovernmental Panel on Climate Change (IPCC) (1992), Climate Change 1992: the Supplementary Report to the IPCC Scientific Assessment, Houghton, J.T., Callander, B.A. and Varney, S.K. (eds.), Cambridge University Press, Cambridge, U.K

Intergovernmental Panel on Climate Change (IPCC) (1994), Climate Change 1994: radiative forcing of climate change and an evaluation of the IPCC 1992 emission scenarios, Cambridge University Press, Cambridge, U.K.

Intergovernmental Panel on Climate Change (IPCC) (1995), The Science of Climate Change, Cambridge University Press, Cambridge, U.K.

Jahnke, R.A. (1992), The Phosphorus Cycle, in Global Biogeochemical Cycles, eds. S.S. Butcher, R.J. Charlson, G.H. Orians, G.V. Wolfe, Academic Press, London.

Keeling, C.D. and Bacastow, R.B. (1977): Impact of industrial gases on climate. In: Energy and Climate. Stud. Geophysical, 72-95, National Academy of Sciences, Washington, D.C.

Keller, A.A. and Goldstein, R.A. (1994). The human effect on the global carbon cycle: response functions to analyse management strategies. World Resource Review, 6(1), 63-87.

Lawrence, M.G. (1993), An empirical analysis of the strength of the phytoplankton-dimethylsulfide-cloud-climate feedback cycle, Journal of Geophysical Research 98, 20663-20673.

Legget, J.A., Pepper, W.J., and Swart, R.J. (1992). Emissions scenario for the IPCC: an update. In: Climate Change 1992: the Supplementary Report to the IPCC Scientific Assessment, Houghton, J.T., Callander, B.A. and Varney, S.K. (eds.), Cambridge University Press, Cambridge, U.K, 69-95.

Likens, G.E., Bormann, F.H., Pierce, R.S., Eaton, J.S. and Munn, R.E. (1984), Long-term trends in precipitation chemistry at Hubbord Brook, Atmospheric Environment 18, 2641-2647.

Logan, J.M., Prather, M.J., Wofsy, S.C., and McElroy, M.B. (1981), Tropospheric chemistry: a global perspective, Journal of Geophysical Research 88, 10785-10807.

Logan, J. (1983), Nitrous oxides in the troposphere: global and regional budgets, Journal of Geophysical Research 88, 10785-10807.

MacKenzie, F.T., Ver, L.M., Sabine, C., Lane, M. and Lerman, A. (1992). C, N, P, S global biogeochemical cycles and modeling of global change. In: Wollast, R., MacKenzie, F,T and Chou, L. (eds.) Interactions of C, N, P, and S biogeochemical cycles and global change, 4, 1-64, Springer Verlag.

Nadelhoffer, K.J., Emmett, B.A., Gundersen, P., Kjonaas, O.J., Koopmans, C.J., Schleppi, P., Tietema, A., and Wright, R.F. (1999), Nitrogen deposition makes a minor contribution to carbon sequestration in temperate forests, Nature 398, 145-148.

Peterson, B.J. and Melillo, J.M. (1985). The potential storage of carbon caused by eutrophication of the biosphere. Tellus, 37B, 117-127.

Prather, M., Derwent, R., Ehhalt, D., Fraser, P.., Sanhueza, E. and Zhou, X. (1994), Other trace gases and atmospheric chemistry, In Houghton, J.T. *et al.* (eds.), Radiative forcing of climate, Scientific Assessment, Intergovernmental Panel on Climate Change (IPCC). Cambridge University Press.

Quay, P.D., Tilbrook, B. and Wong, C.S. (1992), Oceanic uptake of fossil fuel CO_2: carbon-13 evidence, Science 256, 74-79.

Rastetter, E.B., Ryan, M.G., Shaver, G.R., Melillo, J.M., Nadelhoffer, K.J., Hobbie, J.E. and Aber, J.D. (1991). A general biogeochemical model describing the responses of the C and N cycles in terrestrial ecosystems to changes in CO_2, climate and N deposition. Tree Physiology, 9, 101-126.

Rotmans, J. and Elzen, M.G.J. den (1992), A model-based approach to the calculation of Global Warming Potentials (GWPs), The International Journal of Climatology, Vol. 12, 865-874.

Rotmans, J., and Elzen, M.G.J. den (1993), Modelling feedback mechanisms in the carbon cycle, Tellus 45B.

Rotmans, J., Hulme, M. and Downing, T. (1994), Climate change implications for Europe: an application of the ESCAPE model, Global Environmental Change, in press.

Rotmans, J. and de Vries, H.J.M. (1997), Perspectives on Global Change: the TARGETS approach, Cambridge University Press, Cambridge, U.K.

Sarmiento, J.L., Siegenthaler, U. and Orr, J.C. (1992), A perturbation simulation of CO_2-uptake in an ocean General Circulation Model, Journal of Geophysical Research 97, 3621-3645.

Schimel, D., Enting, I.G., Heimann, M., Wigley, T.M.L., Raynaud, D., Alves, D and Siegenthaler, U. (1995), CO_2 and the carbon cycle, In Houghton, J.T., . (eds.), Radiative forcing of climate, Scientific Assessment, Intergovernmental Panel on Climate Change (IPCC). Cambridge University Press.

Schindler, D.W. (1999), The mysterious missing sink, Nature 398, 105-107.

Schindler, D.W. and Baley, S.E. (1993). The biosphere as an increasing sink for atmospheric carbon: estimates from increased nitrogen deposition. Global Biogeochemical Cycles, 7(4), 717-733.

Schlesinger, W.H. (1991). Biogeochemistry: an analysis of global change. Academic Press, London.

Schlesinger, W.H. (1993). Reponse of the terrestrial biosphere to global climate change and human perturbation. Vegetation, 104/105, 295-305.

Schlesinger, W.H. and Hartley, A.E. (1992), A global budget for atmospheric NH_3, Biogeochemistry 15, 191-211.

Shaw, G.E. (1983). Climatic Change, 5, 297.

Tans, P.P., Fung, I.Y. and Takahashi, T. (1990), Observational constraints on the global atmospheric CO_2 budget, Science 247, 1431-1438.

Taylor, K.E., and Penner, J.E. (1994), Anthropogenic aerosols and climate change, Nature 369, 734-737.

Vitousek, P.M. and Howarth, R.W. (1991), Nitrogen limitation on land and in the sea: how can it occur?, Biogeochemistry 2, 86-115.

Watson, R.T., Rodhe, H., Oeschger, H. and Siegenthaler, U. (1990), Greenhouse gases and aerosols, In Houghton, J.T., Jenkins, G.J. and Ephraums, J.J. (eds.), Climate Change: the IPCC Scientific Assessment, Intergovernmental Panel on Climate Change (IPCC), Cambridge University Press.

Wigley, T.M.L. (1991), Could reducing fossil-fuel emissions cause global warming?, Nature 349, 503-506.

Wigley, T.M.L. (1994), Outlook become hazier, Nature 369, 709-710.

WMO - World Meteorological Organization (1992): Scientific Assessment of Ozone Depletion: 1991. WMO/UNEP, WMO Global Ozone Research and Monitoring Project, Report no. 25.

Wollast, R., Mackenzie, F.T. and Chou, L. (eds) (1993), Interactions of C, N, P and S biogeochemical cycles and global change, NATO ASI Series I, vol.. 4, Springer Verlag, Berlin, Germany.

Chapter 5

CAUSES OF GREENHOUSE GAS EMISSIONS

K. Chatterjee

5.1 Introduction

Anthropogenic emissions, a result of human activities, play a crucial role in disturbing the radiative balance of the atmosphere, as described in the previous chapters, giving rise to global warming. Human activities are not only increasing the concentrations of existing greenhouse gases, but are also adding new infrared absorbing gases such as halons and chlorofluorocarbons, resulting in the enhancement of the greenhouse effect. For a thousand years prior to industrial revolution, abundances of the greenhouse gases were relatively constant. However as the world's population increased, as the world became more industrialised and as agriculture developed, the abundances of greenhouse gases increased markedly (Box 5.1). The biggest single contributor is carbon dioxide, but chlorofluorocarbons, methane, ozone and nitrous oxide are also important. Figure 5.1 shows the contribution of various greenhouse gases to radiative change in 1980's. Carbon dioxide emission accounted for about 55% of the total (IPCC 1990; UNCTD 1992).

The cumulative effect of these greenhouse gases is to increase average global temperature, which alters the associated atmospheric circulation and wind patterns. Global precipitation patterns are also likely to undergo changes, influencing agriculture and water resource regimes. Another major impact of global warming will be a rise in sea-level due to oceanic thermal expansion and melting of land based ice masses, posing a serious threat of inundation to low lying coastal areas. There is a possibility that global warming would increase the frequency and intensity of tropical storms. These storms at present develop only over seas warmer than about 26°C.

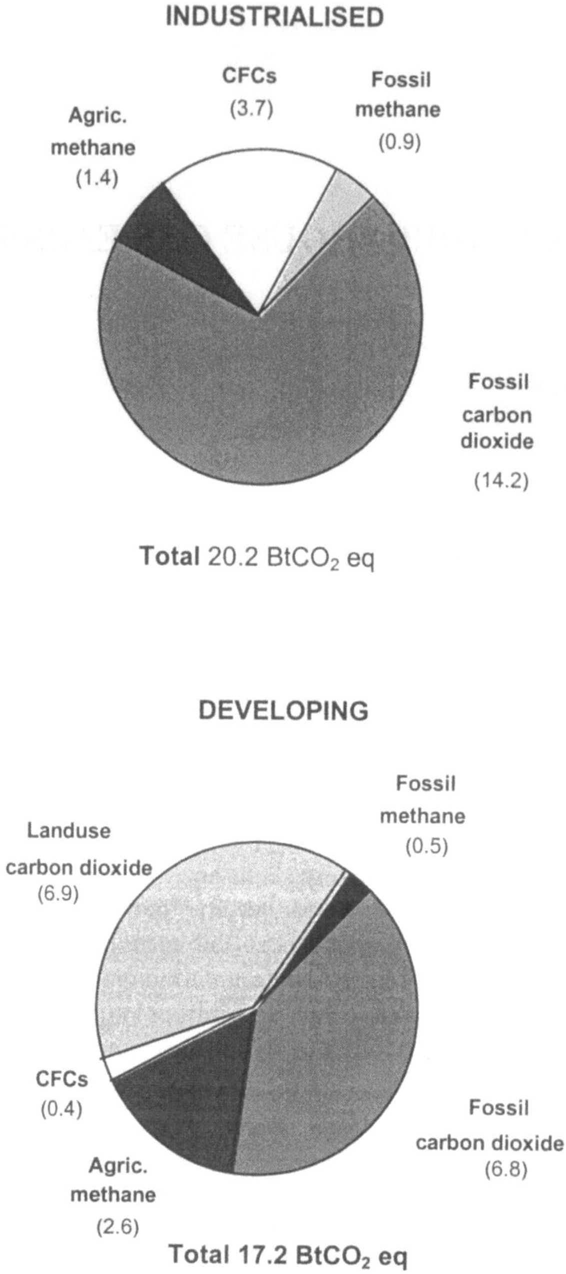

Figure 5.1 The contribution of different gases to radiative change during the 1980s in industrialised and developing countries

With global warming, the size of ocean areas warmer than this temperature will increase. Flora and fauna will also be threatened by changes in rainfall and temperature.

It is extremely difficult to assess the contribution of human induced greenhouse gases on climate change because of the variety of anthropogenic and natural sources of greenhouse gas emissions emitting one or several of the greenhouse gases. Moreover, the atmospheric cycle of some greenhouse gases including both sources and sinks is still subject to scientific uncertainty because of the non-linear relationships between atmospheric physics, dynamics, chemistry and radiation processes. Nevertheless, it is becoming clearer that rapid population growth, increasing human activities including industry, energy production, transport and construction play a major role in climate change (Box 5.2)

Box 5.1
Greenhouse gases and world population

a) Human induced Greenhouse Gases:

Carbon Dioxide (CO_2): *is responsible for more than half the human contribution to the greenhouse effect. Every year human activities, like burning of fossil fuels and cement production, are responsible for putting 6 billion metric tonnes of this fossil carbon from the earth into the air in the form of 22 billion tonnes of carbon dioxide. The destruction of forests and the degradation of soils add an estimated 5.9 billion tonnes of carbon dioxide to the atmosphere.*

Methane (CH_4): *is released as a result of combination of carbon and microbial decay in the absence of oxygen. This occurs from rice paddy cultivation (in wet land), burning biomass, landfills and digestive system of cattle and termites.*

Nitrous Oxide (N_2O): *is produced from soil cultivation, biomass burning, and fossil fuel combustion. Other major sources are the production of nylon and nitrogen fertilisers.*

Ozone (O_3): *is a powerful greenhouse gas, particularly formed in the lower atmosphere (near ground) from vehicular pollution in the sunlit atmosphere.*

Chlorofluorocarbons (CFCs): *are responsible for depletion of stratospheric ozone and greenhouse effect. Chlorofluorocarbons are to be phased out totally by 2010 by all the countries. Industrialised countries have already phased out the chlorofluorocarbons.*

Box 5.1 Continued

b) Concentrations of atmospheric greenhouse gases:

Atmospheric greenhouse gases	***Range of atmospheric gas concentrations from the dawn of the human species to 1750***	***Atmospheric greenhouse gas concentrations in 1994***	***Rates of increase in 1990***	***Percentage increase from pre-industrial levels***
Carbon dioxide	*180 - 295 ppm*	*353 ppm*	*25%*	*0.5%*
Methane	*0.3 - 0.8 ppm*	*1.72 ppm*	*115%*	*0.5 - 1%*
Nitrous oxide	*275 - 295 ppm*	*310 ppm*	*9%*	*0.25%*
CFC-11	*(compound not yet invented)*	*280 ppt*	-	*4%*
CFC-12	*(compound not yet invented)*	*484 ppt*	-	*4%*

Source: Second Assessment Report, IPCC (1996)

c) World population:

World population	***Number of people***
World population, 1750	*760 million*
World population, 1990	*5.3 billion (annual growth rate is 1.6%)*
World population, 2025 (projected)	*7.8 - 12.5 billion*

Box 5.2
Some of the key findings of Second Assessment Report
(IPCC, 1996)

- *Atmospheric concentrations of greenhouse gases are still rising and thus increasing their contribution to global warming. This warming of the surface of the earth tends to bring about other associated changes in the earth's climate.*
- *The enhanced concentration of aerosols (dust and other small particles) in the lower atmosphere resulting from the combustion of fossil fuels, biomass burning and other sources cause some cooling effect reducing the net effect of warming due to greenhouse gases by about one third (ranging between 20 and 40%). This cooling is not uniform world-wide as aerosols have a very short atmospheric life time in comparison to greenhouse gases (e.g. carbon dioxide has an atmospheric life time of 200-500 years). At the local level it can be large enough to more than offset the greenhouse gases induced warming. On the continental to hemispheric scale it affects climate change patterns.*
- *Climate models project an increase in global mean surface air temperature of about 2 °C by the year 2100, in addition to what has been induced by human emissions so far; the uncertainty of this projection is 1-3.5 °C.*
- *Because of the ocean's thermal inertia, only 50-90% of the eventual equilibrium temperature would have been realised by 2100. Temperature would continue to increase beyond 2100, even after the stabilisation of global climate.*
- *Projected sea level rise as a result of thermal expansion of the oceans and melting of glaciers and ice sheets with a best estimate of 50 cm by the year 2100. Sea level would continue to rise even after the global climate and mean temperature would be stabilised.*
- *Regional temperature changes may differ substantially from the change in global mean temperature.*
- *Uncertainties still remain concerning some key factors. Nevertheless the balance of evidence now suggests that there is a discernible human influence on climate.*

5.2 Industry

Industry plays a central role in furthering development goals. It has both positive and negative impacts upon the environment. Industry makes equipment and instruments to monitor our environment. Industry also helps us to adopt new technologies to preserve our environment for the benefit of the present as well as future generations. Some of the negative impacts are industrial waste and exploitation of natural resources beyond sustainability.

In the early stages of industrialisation, local or global environment issues were not a primary consideration and were often ignored in the face of the benefits of industrial development. There was a growing concern when environment began to show signs of stress due to the intensity of industrial processes and their impacts particularly from energy generation, mineral extraction, chemical, iron and steel, cement and non-ferrous metal industries.

Industrial activities including manufacturing, energy production and supply, transport and construction are responsible for emissions of greenhouse gases. The industrial sector contributes towards 20% of the global warming (EPA, 1989), playing a major role in climate change.

Industries are wholly responsible for the emissions of chlorofluorocarbons. Chlorofluorocarbons, apart from being greenhouse gases are also responsible for depletion of the stratospheric ozone layer. Industrial processes such as nylon manufacturing also emit nitrous oxide. One third of human induced methane emissions comes from gas drilling and transmission, waste landfills, coal mining and leaks from the production and distribution of petroleum and natural gas. Use of solvents in industries leads to anthropogenic emissions of non-methane hydrocarbons often called volatile organic compounds or VOCs. Industry is also responsible for emission of carbon monoxide as a result of incomplete combustion of fossil fuels and biomass.

All these pollutants undergo photochemical reactions in the sunlit troposphere producing tropospheric ozone that is another important greenhouse gas.

5.2.1 Main developments in developed countries

The main contributor to human induced global climate change is the emission of carbon dioxide from the burning of fossil fuels, which increased exponentially since 1860 at the rate of 4% per year, with major interruptions during the two world wars. This has resulted in the increase of the atmospheric concentrations of carbon dioxide from 280 parts per million by volume (ppmv) in the pre-industrial period to the current atmospheric concentrations of 353 ppmv. Developed countries account for 95% of industrial carbon dioxide emission. Their annual carbon dioxide release reaches up to 5 metric tonnes of carbon per capita per year.

Industry is now faced with the need to find ways and means to reduce its environmental effects. Curative approaches are costly and do not eradicate the sources of danger. Preventive strategies are required. The present trend in developed countries is to shift away from a "pollute-and-cure" to an "anticipate-and-prevent" approach. This trend has resulted largely from a consideration of economic trade-offs.

5.2.2 Main developments in developing countries

In developing countries particularly in Asia (excluding Japan), Latin America and Africa, the industrial revolution began by the middle of the twentieth century (post-war), but proceeded at a slower rate. In Africa, the process of industrialisation remains nascent. However, the rate of increase of carbon dioxide emission is much larger in the developing countries (about 6% per year) as compared to Western Europe.

Economic growth will continue to be the primary objective of the world development strategy for many decades, which would need systematic technology applications to an ever-widening range of productive activities. Although other sources of energy undoubtedly have a role to play, a large proportion of this energy must come, in the foreseeable future, from fossil fuels. However, the stress on personal consumption which has been such a major factor in OECD (Organisation for Economic Co-operation and Development) industrialised societies for example, is unlikely to be seen as a possible option for many developing countries in reaching more general development goals. For those developing countries that decide to follow the well-trodden path to growth via industrialisation, management of resources to attain socio-economic goals becomes imperative. Other developing countries may decide to follow a pattern of development that is not directly linked to industrialisation (agrarian economic development).

The current technological revolution, well underway, leads to "dematerialisation" of the economy and society-in the sense that less and less resources, energy and environment-content are required for each unit gross domestic product (GDP). Developing countries cannot ignore the implication of this revolution on their own economic growth. They now have greater opportunities to profit from the technological improvements now achieved by developed countries and hence avoid the need for the expensive clean up of mistakes made during the process of industrialisation. Use of best available technology should be the development choice of the developing countries, provided the developed nations come forward to transfer additional funds and state of the art technologies to the developing nations. Otherwise, the growth of industrial sector comprising of chemical and non-chemical industries will severely affect the climate by emitting greenhouse gases and ozone depleting substances.

5.2.3 Chemical industry

Though the chemical industry on one hand proved to be a boon for improving the quality of life, on the other hand it has proved disastrous to the environment. Chemical industry embraces sectors like fertilisers, pesticides, pharmaceuticals, detergents, plastics, petrochemicals and dyes, each of which has become vital for man's well being.

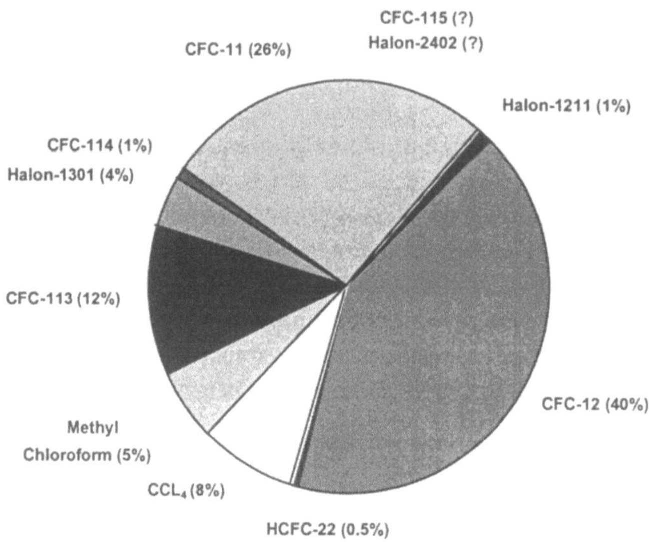

Figure 5.2 Share of contribution to ozone depletion of CFCs and Halons (controlled items)

In the context of climate change the chlorofluorocarbon and halon industry needs a special mention as it contributes significantly to greenhouse warming and is primarily responsible for the depletion of the stratospheric ozone layer. Chlorofluorocarbons emerged as safe chemicals during the1930s for use as refrigerants and later found use in foam making, as an aerosol propellant, a cleansing agent in electronic industry and a fire extinguisher in many critical defence and computer applications. In the 1980's they were found to be primarily responsible for the appearance of the Antarctic hole during September-October every year and for the general depletion of the ozone layer on a global scale. Each of the chlorofluorocarbons (and halons) can be characterised in terms of its ability to deplete ozone by the index known as ozone depletion potential (ODP). Figure 5.2 shows the relative contribution of different chlorofluorocarbons in depleting the ozone layer.

Considering the harmful effects of the chlorofluorocarbons, the world community through international agreements, the Vienna Convention and the Montreal Protocol on substances that deplete the ozone layer has decided to completely phase out chlorofluorocarbons production and use, latest by 2010. Developed countries have already phased out the chlorofluorocarbons.

5.2.4 Non-chemical industrial sectors

The non chemical industrial sectors such as ferrous and non ferrous metallurgy, paper and pulp, cement, glass making and burnt bricks are responsible for considerable waste, atmospheric pollution and carbon dioxide emission to the atmosphere. The pollutants discharged by these industries damage the health of local people, reduce output from local agriculture and industry, and damage infrastructure and buildings. The carbon dioxide emissions from the non-chemical industrial sector add to greenhouse warming and global climate change. Some of the non-chemical industries such as iron and steel, cement and bricks that emit greenhouse gases such as carbon dioxide are discussed here.

Iron and steel industries

Japan, United States, China, Russian Federation, Ukraine, Korean Republic, Italy, Brazil and India are the ten major producers and consumers of iron and steel. These ten countries total production in 1992 amounts to about 506 million tonnes of crude steel against world's total production of approximately 721 million tonnes.

In the process of iron making, limestone ($CaCO_3$) and metallurgical coal is used in the blast furnace resulting in the emission of greenhouse gases. Most of the iron produced in the world is used in the production of steel. The remainder is converted to iron castings, ferroalloys and iron powder. Iron and steel sector contributes 9.8% of the total carbon dioxide emission to the atmosphere.

Cement

Cement is a major industrial commodity, manufactured commercially in at least 120 countries globally. Cement is essential for the development of a country as it provides the foundation for the infrastructure required for economic growth.

Production of cement contributes to atmospheric pollution in the form of particulates that affect the environment in the vicinity of cement plants. The manufacturing process of cement also contributes to the emissions of carbon dioxide to the atmosphere. Carbon dioxide emission from cement industry comes from both the energy consumed under the manufacturing and calcination processes. Some researchers have calculated emission from the calcination process alone. Figure 5.3 depicts regional emissions of carbon dioxide from cement production, (1989) and global carbon dioxide emissions from cement production since 1950.These figures do not include carbon dioxide emissions from fuels used in the manufacturing process (Maryland *et. al.*, 1989). This estimate of emission is based on the amount of cement that is produced, multiplied by an average emission factor and assuming that an average lime content of cement is 63.5%. The emission factor calculated for the purpose comes to 0.498 tonnes of carbon dioxide to one tonne of cement.

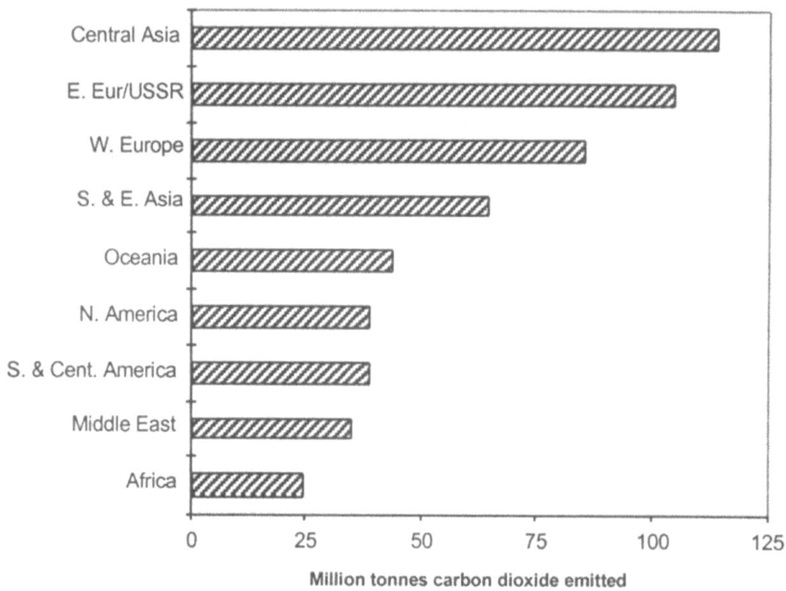

Figure 5.3a Regional emissions from cement production, 1989

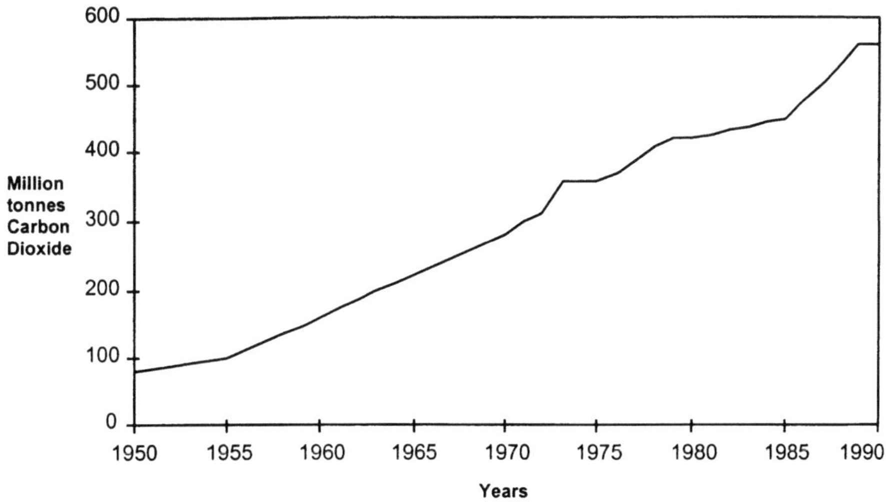

Figure 5.3b Global carbon dioxide emissions from cement production

The current global emissions of carbon dioxide from cement production during 1991 are estimated at 593.57 million tonnes (World Resource, 1994). Cement production and related emissions of carbon dioxide have risen at roughly three times the rate of population growth over the entire period. Regional carbon dioxide emission data show that carbon dioxide emissions from cement production from all regions are significant, reflecting the global nature of cement production.

Burnt bricks

Burnt clay bricks are manufactured in kilns and clamps. The production of brick requires mechanical energy in the processing of wet clay and thermal energy in burning the bricks. Mostly coal or wood is used as fuel for burning bricks. In the informal sector, rice husks and agricultural wastes are also used. Brick burning operations release considerable greenhouse gases including carbon dioxide.

In addition to contributing to greenhouse warming, the brick making industry also damages soil in the vicinity and pollutes the air. The major hazard, is of course, damage to the soil fertility resulting from indiscriminate exploitation of agricultural land, both for seasonal installation of kilns and using clay for brick making.

5.2.5 Future projections

Economic and population growth will continue in the future. This in turn will increase energy use, agricultural and industrial activities that will result in net increase of greenhouse gas emissions. This increase in greenhouse gas emissions will severely affect the climate in future. Apart from increasing ozone depleting substances and greenhouse gases in the atmosphere several major trends will be affected by the world's industrial growth and its environmental implications. They arise as a result of the growing interdependence of national economics in terms of trade and investment; research, development and transfer of technology; financial flows; access to raw materials, energy sources, markets, technology and labour. These major trends are:

- The progress of industrialisation of the developing countries and relocation of some basic and consumer industries from the developed to developing countries;
- The ever accelerating pace of urbanisation in developing countries, which has implications on food production, housing and city planning, transport, energy and water supplies, health and sanitation services, education, tourism, amenities and recreation facilities;
- The emergence of new models of agriculture;
- The increasing role of multinationals;
- The increasing importance of small and medium size industries indigenous to the countries concerned;
- Greater societal awareness of environmental problems at a global, national and local level;
- Growing importance of cleaner production, waste minimisation, disposal and recycling;
- The role of industry in reducing hazards arising from the transport, storage and disposal of wastes resulting from its own activities and from those of other sectors of the economy.

5.3 Energy resources

Energy resources are central to the world's economic development, yet unsustainable use of these resources are harming both the local and global environments. World-wide use of fossil fuels (coal, oil and gas) accounts for about 50% of man-made greenhouse emissions (Engelman, 1994). 80% of greenhouse gas emission in the energy sector is due to the emission of carbon dioxide. The increasing use of fossil fuels due to increase in population, industrialisation and transportation has resulted in an exponential increase in carbon dioxide emission (Figure 5.4).

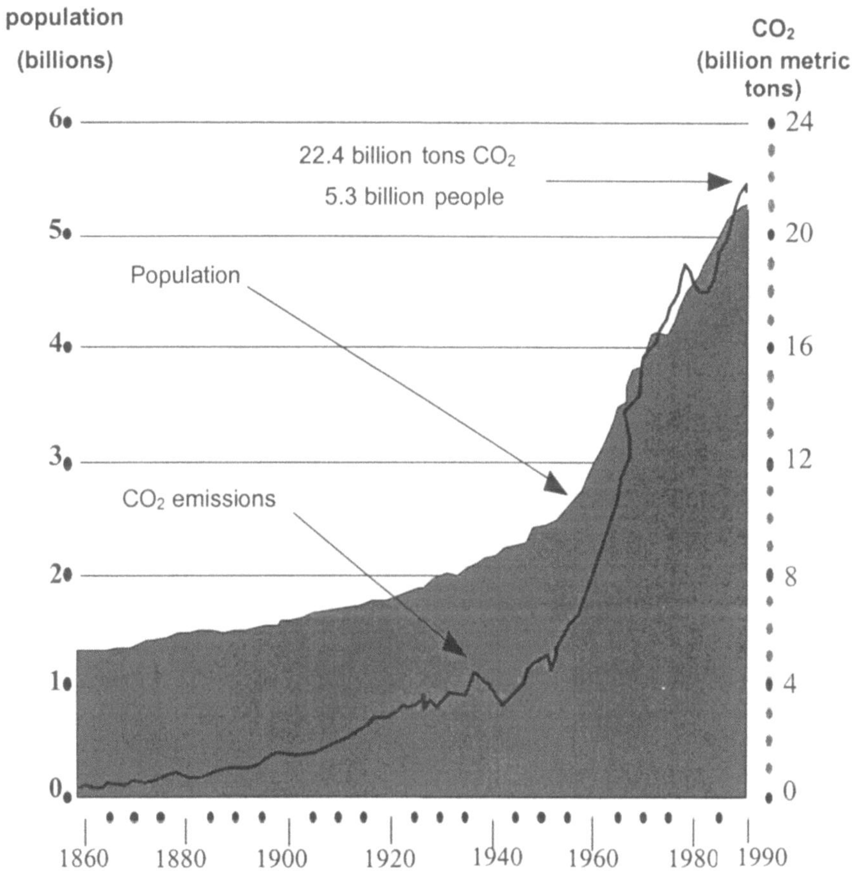

Figure 5.4 World population and carbon dioxide emissions: 1860-1990

5.3.1 Fossil resources

The major source of carbon dioxide emission in the energy sector is the power sector that uses fossil fuels for generation of electricity. The fossil fuels coal, oil and gas now provide about 60% of the world's electricity and are likely to remain important energy sources for many decades.

Hydropower and nuclear power provide the balance 40% (23% from hydropower and 17% from nuclear power). Of the fossil fuels, coal is globally used more for the generation of electricity than any other purposes.

The environmental impact of electricity generation from all fossil fuels is due largely to their relatively low energy density compared with nuclear power. In order to generate one gigawatt year of electricity (1 gigawatt = 10^9 watts) one needs to mine and transport 3.8 million tonnes of coal or extract and transport 2.2 million tonnes of oil (Roberts *et.al.*,1990).

Fossil fuel reserves and resources

Proven reserves are those quantities that, under existing conditions, can be extracted with reasonable certainty. Table 5.1 shows proven recoverable fossil (1990) resources distributed among the different regions of the world. The former USSR and the United States lead the world in proven recoverable reserves of coal. The countries of south west Asia surrounding the Persian Gulf control 57% of proven recoverable petroleum reserves, of which Saudi Arabia alone controls about 26%. The former USSR is estimated to control about 42% of the world's proven recoverable reserves of natural gas, with the Persian Gulf states controlling another 25%.

Coal

The enormous quantities of coal that are extracted and transported have considerable environmental effects. The largest effect is probably that of open cast mining. Open cast mining prevalent in many countries, directly affects human population in and around the mining areas by displacing them from their homes, affecting their livelihood, causing adverse effects to agriculture and creating serious air pollution problems mainly from particulate matter. Every country has to formulate strict standards at all stages of open cast mining operations and subsequent restoration and after care work including resettlement and redistribution even after mining operations cease to have minimal environmental impact.

Deep coal mining does not have the obvious and direct effects on the environment that are inevitable with open cast mining. However, two most important environmental problems associated with deep mining are soil disposal and the possibility of subsidence. Chances of accidents and risk of life to mines are also greater in deep coal mining.

The major environmental problem associated with use of coal is the disposal of ash generated. A typical coal fired power station (2000 megawatt) produces over 2000 tonnes of fly-ash a day, most of which is very fine pulverised fuel ash. Some of this is used commercially, mainly for road fill and building materials; the rest has to be disposed off - but how and where are the issues that need serious examination.

Table 5.1. Proven commercial energy reserves, 1990 (fossil fuel proven, petajoules (pj))

	Coal Reserves	R/P[a]	Oil Reserves	R/P[a]	Natural gas Reserves	R/P[a]	Total Reserves
World	20,299,986	209	5,639,794	45	5,004,802	52	34,578,702
Developing countries	7,445,859	163	5,030,292	68	2,358,035	96	14,344,306
Oil-exporting	632,901	911	4,780,005	78	2,058,015	106	6,982,402
- OPEC countries	582,951	1,51	4,397,271	91	1,904,014	116	6,394,357
- Non-OPEC countries	50,950	164	382,734	30	154,001	50	587,684
Oil-importing	6,811,907	152	250,287	19	300,020	58	7,362,264
- African countries	1,724,559	341	14,864	24	27,017	827	1,766,440
- Asian and Oceanian countries	263,260	333	43,017	12	44,153	28	350,432
- Latin-American countries	4,824,137	123	192,405	22	228,849	65	5,245,391
Industria-lised countries	16,978,127	239	609,502	12	2,646,768	37	20,234,397
OECD	10,112,867	231	264,443	10	507,729	14	10,858,039
- North America	6,469,377	242	179,204	10	304,933	11	6,953,513
- Western Europe	1,663,400	144	77,531	10	178,443	21	1,919,375
- Pacific	1,980,089	358	7,708	7	24,353	21	2,012,150
Non-OECD	6,856,260	252	345,059	14	2,139,039	62	9,349,358
- Central Europe	1,533,950	161	10,099	15	21,094	13	1,565,143
- Former Soviet Union	5,331,310	300	334,960	14	2,117,945	65	7,784,215

[a] R/P (in years) is the ratio of proven reserves in 1990 production rate.
Source: World Resources, 1994.

Oil reserves

The world's present proven reserves and proven recoverable reserves are estimated at approximately 128 and 134.8 billion metric tonnes respectively, of which Oil and Petroleum Exporting Countries (OPEC) accounted for more than 70%. The main direct environmental impacts of oil resources comes from the fractions that get lost by spillage or accidents. The world currently consumes nearly 3000 million tonnes of oil a year. Most of the oil is transported by ocean going tankers and about 2 million tonnes a year enter the marine environment as a result of routine operations and accidents. Besides the unpleasantness of oil on beaches and economic impacts on tourism, the main direct effects are on aquatic life and coastal flora and fauna. The oil spillage during the Gulf War was one of the worst examples of environmental impact. In addition, the operation of oil refineries results in the emission of several thousand tonnes of airborne and liquid effluents each year. The environmental impacts of burning oil to produce electricity do not differ greatly from those of burning coal except that particulate emissions are considerably smaller and there is no ash to dispose (Roberts *et al.*, 1990).

Natural gas

The world's present proven reserves and proven recoverable reserves of natural gas are estimated at 111,900 and 128,852 billion m^3 respectively. Natural gas is relatively abundant globally. The largest deposits are in the former USSR, followed by OPEC countries in Asia.

Environmental impacts associated with gas production and distribution are small. The risk of accidents associated with the transport and storage of liquefied gas, however, is significant.

Impact of fossil fuels on environment

Fossil fuels used in industry, power plants, domestic homes and road transport are the largest sources of atmospheric pollution. Oxides of sulphur and nitrogen emitted from coal-fired, and to a lesser extent oil-fired power stations, besides having local effects, can be transported by atmospheric circulation under favourable meteorological conditions over long distances across country boundaries leading to acid precipitation. Such problems were very acute particularly in Europe. The problem has been considerably solved through international agreements and various pollution control measures, by the individual countries.

5.3.2 Renewable energy resources

The Intergovernmental Panel on Climate Change (IPCC) has identified three response strategies to reduce emissions of carbon dioxide and other greenhouse gases from human activity. They are:

- efficiency improvements in energy supply, conversion, and end uses;
- fuel substitution by energy sources lower in greenhouse gas emissions; and

- reduction in greenhouse gases by removal, recirculation or fixation.

Renewable energy sources are capable of contributing significantly in each of these areas. The renewable energy sources are: biomass, solar energy, wind energy, hydropower and geothermal energy.

The World Commission on Environment and Development in its report 'Our Common Future', has stressed the need to develop the renewable energy potential, which should form the foundation of the global energy structure during the 21st century. The necessity to incorporate more renewable energy resources in the global energy structure will stimulate the development of a range of new energy technologies and may have important social effects (World Commission on Environment and Development, 1981).

Nearly all-renewable energy sources are directly or indirectly derived from the radiation of the sun. Direct: Solar radiation converted into heat (actively or passively), electricity etc. Indirect: biomass, hydropower, wind power, wave power, tidal power, geothermal power, radioactive decay of isotopes). Solar energy, wind energy and hydropower (renewable energy sources) do not have any direct affect on the emission of greenhouse gases. Biomass energy is also considered renewable because resources that are produced by forests and shrubs can be grown/cultured through a planned programme of tree plantation. However the mismanagement of biomass resources have resulted in complete destruction of these resources in certain places like the Amazon forests.

During operation, most of the renewable energy sources do not emit carbon dioxide, except for Ocean Thermal Energy Conversion (OTEC) technology and geothermal plants. As long as fossil fuels are used in construction and the production of building materials, a certain quantity of carbon dioxide will be emitted. A comparative analysis of carbon dioxide emissions of different electricity generating technologies, including renewable ones, is given in the Table 5.2. For biomass, a wood plantation feeding a power plant is assumed.

Table 5.2 Carbon dioxide emission of electric technologies (tonnes of carbon dioxide per GWh)

Technologies	**Fuel extraction**	**Construction**	**Operation**	**Total**
Conventional coal plant	1.0	1.0	962.0	964.0
Oil fired plant	-	-	726.2	726.2
Gas fired plant	-	-	484.0	484.0

Technologies	Fuel extraction	Construction	Operation	Total
Ocean thermal energy conversion	NA	3.7	300.3	304.0
Geothermal steam	0.3	1.0	55.5	56.8
Small hydropower	NA	10.0	NA	10.0
Boiling water reactor	1.5	1.0	5.3	7.8
Wind energy	NA	7.4	NA	7.4
Photovoltaics	NA	5.4	NA	5.4
Solar thermal	NA	3.6	NA	3.6
Large hydropower	NA	3.1	NA	3.1
Wood (sustainable forest)	-1509.1	2.9	1346.3	-159.9

NA : Not applicable; (-) missing or inadequate data
Source: U.S. Department of Energy.

Biomass

Biomass has been the prime cooking and heating fuel, serving millions of households in developing countries. Directly or indirectly it provides our food and has given fossil fuel reserves to the earth.

The annual production of biomass is of the order of 100 billion tonnes of dry matter per year on land, and at least 50 billion tonnes per year in the ocean. To make a comparison: the world grain production equals 1.8 billion tonnes/year and the wood harvest is roughly 1.5 billion tonnes/year. The present consumption of biomass for energy purposes is estimated at 40 EJ/year ($1.4*10^{12}$ watt), or 12% of the world consumption of commercial energy. China consumes about 500 million tonnes of biomass annually and accounts for about 80% of the energy used in rural China. In India the current annual demand of fuelwood is in the order of 240 million m^3 (World Resources, 1994).

However, biomass can not be viewed only as an energy source. It provides the most readily convertible natural store of wealth on earth, such as food, shelter and clothing for people; in developing countries it also makes available the main means for cooking food and heating dwellings. The incorporation of biomass material into the soil is essential for maintaining soil productivity. In addition, biomass in the form of vegetative cover prevents soil erosion, and other forms of environmental degradation.

Energy from biomass

Energy from biomass can be discussed under two broad categories-subsistence energy services and developmental energy services. Using biomass for fuel for household purposes accounts for a very large percentage of total biomass consumption by the developing countries. Such fuels include fuelwood, twigs and branches of trees, cow dung cakes, biogas, rice husk, nut husks and other agricultural wastes. Their energy value varies considerably depending on the type of biomass (Table 5.3).

Table 5.3 Energy value of certain non-commercial fuels and wastes

Fuel	*Energy content in MJ/kg*
Wood (air dry)	15.0
Wood (oven dry)	20.0-21.0
Charcoal	31.0
Bagasse (30% moisture)	14.6
Barks	11.9
Coffee grounds	13.4
Corn Cobs	19.2
Dung Cakes	8.8-14.0
Garbage	19.6
Nut hulls	18.0
Rice husks	17.6
Paper	17.6
Saw dust & Shaving	11.3
Cotton Stalks	15.8
Coconut husks	18.1
Coconut shells	20.1

Source: Mwandosya and Luhanga, 1984.

Fuelwood need in selected countries

Various surveys for energy use in domestic activities show a demand between 1 m^3 and 12 m^3 per person each year, which works out to be equivalent to between 70 and 140 million tonnes of coal equivalent. Such a large demand for fuelwood has put a tremendous pressure on forest resources.

In India, it has been calculated that a minimum of 1.4% growth rate in energy availability will be needed for achieving a 1% growth rate of GNP. However, the energy mix, made available, is itself often the cause of both short term and long term environmental damage. The fuelwood deficit is growing day by day. Taking into account the present level of consumption of forest products and the current productivity of forests, India needs a minimum of 0.47 ha of forestland for every individual. The existing forest area on this basis would be adequate only for a population of 150 million (in contrast to the present population of above 900 million). The task of improving the productivity of forests and mobilising alternate sources of energy is thus urgent.

Other major categories of biomass are agricultural residues and agro-industrial residues. The estimated total production of agricultural residues (based on a study conducted by the National Productivity Council during 1984-86 covering fifteen major states of India) is nearly 320 million tonnes annually. Agricultural residues account for approximately 88% of total biomass residues. Besides use as fuel energy source, this can be used for other purposes as well.

Agro-industrial residues include bagasse and rice husk, which comprise about 80% of the total. They are applied in low-grade thermal plants, but the efficiency can be increased through gasification.

Biomass production and utilisation prove to be labour intensive in terms of man-days involved per cycle per hectare of land. The most significant contribution of the biomass programme is in terms of recycling of carbon dioxide without additional emissions of greenhouse gases.

Solar energy source

Solar energy being free from serious environmental problems offers a great potential for renewable energy. At the earth's outer atmosphere, the supply of energy from the sun is 1368 watts of energy per square metre and on the earth's surface the maximum amount of solar energy available on a clear day is 1,000 per square metre.

The solar energy flow figures clearly indicate that the solar energy source is very large and can supply all energy needs of mankind. For example, the situation in the Netherlands, with a land area of 33,900 km^2 is as follows:

Average solar energy on Dutch land area	3.6 TW
Energy use in the Netherlands (1987:2.7 EJ/yr)	0.85 TW

(1 terawatt (10^{12} watt) corresponds to $31.5 * 10^{18}$ joules/yr or 31.5 EJ/yr).

If solar energy could be converted with an overall efficiency of 20%, the supply of half of the present Dutch energy consumption would require 6% of the land area for utilising Solar Technologies.

Some of the renewable energy technologies and applications

Some of the presently used solar technologies are listed below (La Porta, 1990):

Technology and typical size system	Application
1.	
Photovoltaic conversion with solar technologies	Water-heating
Active solar (1 to 500 kW and more)	Pool-heating
Flat plate collectors	Space-heating and -cooling
Evacuated tube collectors	Industrial process heat
Seasoned storage systems	Crop drying
2.	
Passive solar (1 to 500 kW and more)	Day lighting
Solar gain and control	Space heating
Thermal storage and distribution	Space cooling
Natural vibrations	
3.	
Solar thermal (4 kWE to 10 MWE)	Industrial process heat
Parabolic troughs	Building energy needs
Parabolic dishes	Electricity
Central receivers	Irrigation pumping
Solar ponds	Chemical and fuel production
	Desalination
	Destruction of toxic wastes
4.	
Photovoltaic (watts to megawatts)	Consumer electronic
Flat plate collectors	Remote power stations
Concentrating collectors	Utility grid power

Passive solar building design, using sunlight and natural ventilation has extensive potential in countries with temperate climates. For example in the United States in 1985, buildings utilised 37% of primary energy supplies. Heating, cooling, ventilation and lighting consumed 54.4% of the primary energy needed for residential buildings. Considering the national fuel mix for energy supply to residential buildings in the United States, and the geographical distribution of building inventory, a conservative estimate calculates 6.7 million solar-design buildings would eliminate annually the following levels of atmospheric emissions (La Porta, 1990):

- 67.5 million tonnes of carbon dioxide
- 284,000 tonnes of sulphur dioxide
- 165,000 tonnes of nitrous oxide
- 10,000 tonnes of particulate matter

Wind energy

Wind has the longest history in meeting societal needs among renewable energy sources. Harnessing the wind for ships and boats to sail across the seas and oceans, and to produce energy for pumping water, grinding grain and providing mechanical energy for other tasks is a centuries old practice. During the 18th century Denmark reportedly had 10,000 operating windmills, five times the number it has today. Wind energy is clean, causes no air pollution, and produces no radioactive wastes or greenhouse gases. As with solar energy, wind is an intermittent source, limited by the lack of storage technologies.

The magnitude of wind energy resource is highly site specific and wind power varies as a function of the wind speed influenced by surface turbulence and local terrain. In order to get an accurate picture of wind resources, a detailed 'Wind Atlas' needs to be compiled. California (United States) and Denmark have prepared such atlases. Such a map is useful in identifying wind rich regions. Many excellent sites are as yet unexploited.

Wind technology development has come a long way. Wind electricity generated with new turbine technology in California already produces enough electricity for San Francisco and other neighbouring areas. Even without subsidies, the cost of wind electricity from better sites is now competitive with that of fossil fired plants in some regions. The potential for wind power far exceeds that of hydropower, which currently supplies the world with one-fifth of its electricity. Indeed it now seems likely that the growth in wind generation capacity will exceed nuclear generating capacity.

Three countries, Denmark, the Netherlands and Germany, each have plans to develop a minimum of a thousand megawatts of wind energy capacity by 2005. China aims to reach the same goal by 2000.

Hydropower

Hydro electric power generation results in no direct emission of greenhouse gases or air pollutants. However, the construction of dams results in displacement of people in and around the site selected. As a result, large hydropower projects face protests from environmental groups in different countries. Moreover, clearing of forests for construction of dams results in indirect greenhouse gas emission.

Small hydel power on the other hand does not cause widespread displacement of local population and has the least impact on biodiversity and forests. It has an added advantage of being environmentally benign. Small hydel power can be used both for mechanical applications as well as for generation of electricity. Power from small hydel projects can provide energy for remote rural and hilly areas and is most suitable for developing countries. Many countries are now looking for opening/establishing mini hydel projects in remote areas for generation of electricity, and for agricultural purposes. In India so far 127 such small hydel power stations have been commissioned each with a capacity up to 3 MW, and 115 more similar projects (with capacity up to 3 MW) are under construction.

5.3.3 Nuclear energy resources

The fundamental process on which nuclear power depends is the fission of uranium. The fission process involves the release of very large amounts of energy. The energy released when one atom of carbon (coal) combines with one molecule of oxygen is $7 * 10^{-19}$ joules. The energy released from the fission of one uranium nucleus, $3.2 * 10^{-11}$ J, is about fifty million times greater. In terms of weight of fuel, the fission of one tonne of uranium would be equivalent to burning of 2.7 million tonnes of coal.

Large uranium reserves are found in South Africa, Nigeria, Canada, the United States, Brazil, Australia and the USSR. Uranium is widely distributed in the earth's crust with an average concentration of 203 parts per million. It is concentrated well above normal levels in some rocks such as granite (0.001 to 0.002% uranium) and phosphate rocks (0.01 to 0.02%) but economic recovery is seldom possible below 0.05%. The annual production is now about 40,000 metric tonnes a year with a projected rise to 100,000 tonnes by the year 2000. Since, a typical uranium content is 0.3%, the production of 40,000 tonnes of uranium involves mining 20 million tonnes of ore (Roberts *et al.*, 1990).

Nuclear power plants operate in an essentially closed cycle. They do not contribute any greenhouse gases to the atmosphere and are also free from emission of particulates, volatile hydrocarbons, stratospheric ozone depleters and acid rain precursors. Nuclear power plants however, pose a serious problem of operational accidental hazards. Lung cancer is an occupational health hazard due to uranium exposure with miners being the most affected people. There have been two accidents involving nuclear reactors, which resulted in extensive damage to the reactor core and the consequent release of radiation to the environment. One took place at the Three Mile Island (Pennsylvania, USA) and the other at Chernobyl (Ukraine, former USSR). These two accidents led to a lowering of public confidence in the nuclear energy sources. Moreover nuclear waste and its disposals has drawn considerable public attention and concern due to the inherent danger to environment and health.

5.3.4 Minerals

The production and consumption of minerals are central to many modern industrial processes. The United States, the countries under the former USSR, Japan, and countries of the European community consume the majority of the world's metals. Only six developing nations- Algeria, Brazil, China, India, the Republic of Korea, and Mexico, constitute the top ten consumers of selected metals. Five countries - the constituents of the former USSR, South Africa, The United States, China and Australia-hold over half of the world's reserves of the 15 metals viz., scarcity metals: cadmium, mercury and uranium; medium scarcity metals: nickel, tin, zinc, copper and lead. Abundant metals: aluminium (bauxite), iron and steel.

The consumption of minerals has increased appreciably during recent decades. Great changes have taken place in the use of metals. For example, the use of aluminium has increased by eightfold since 1970. Copper and zinc have increased by about one quarter, while lead and tin are being used less compared to other metals. The use of ceramics in industry and of glass fibre in communication technology is very promising. Rare earth metals such as lanthanum and cerium are being used increasingly in the optical and electrical engineering industry.

A comparison of the world reserve to production ratio for nine major metals vary about 20 years (zinc, lead, mercury) to well over 100 years (for iron, aluminium). Extraction, refining, dispersive use and the disposal of industrial minerals and metals may cause considerable local environmental problems. Mining also degrades lands, creating quarries, vast open pits and a huge amount of solid waste. During 1991, for example more than 1000 million metric tonnes of copper ore were extracted to obtain 9 million metric tonnes of metal.

A careful accounting of environmental impacts resulting from extraction and processing of minerals needs to be done world-wide. Preliminary studies reported from various countries indicate considerable toxicity in soils, ecosystem and food chain, particularly in many industrial countries. Dust from mining, acidic gases from smelting and refining contributes to air pollution; leaching from abandoned mines and disposal of chemicals used in refining is a source of water pollution in the mining regions.

Again improper use of minerals, for example asbestos insulation, lead plumbing, gasoline additives, lead and chromium based paints can threaten human health.

5.4 Population

Rapid population growth contributes significantly to environmental damage posing a serious threat to climate change. An increase in population results in an increase in production and consumption and generates a demand for fuelwood, fodder and commercial logging for timber resulting in deforestation. Deforestation in turn leads to a decrease of biodiversity and results in increased global loading of carbon as forests mitigate global warming by playing an important role in sequestering carbon. An increasing rate of population growth results in a significant increase in industrial and transport sector, which in turn increases the greenhouse gas emissions leading to climate change.

5.4.1 Historical growth

World population touched its first billion mark early in the nineteenth century. It took another century (1830-1930) to add the second billion. Then the population increased very rapidly in a very short span of time (30 years to reach 3 billion, 15 years to reach 4 billion and 12 years to reach the 5 billion mark).

The ever shortening time needed to add successive billions is likely to continue in the near future, and the total human population is expected to reach 6 billion by 1998 (10 years). A stable world population is not expected until 2200, when the total could be over 11 billion, twice the current level (UNFPA, 1992). Currently, the four most highly populated countries in the world (China, India, United States and the former Soviet Union) together make up almost half of the worlds population. Figure 5.5 shows thc world population by region. India's population, that at present is 900 million, may be the most populous country in another few decades.

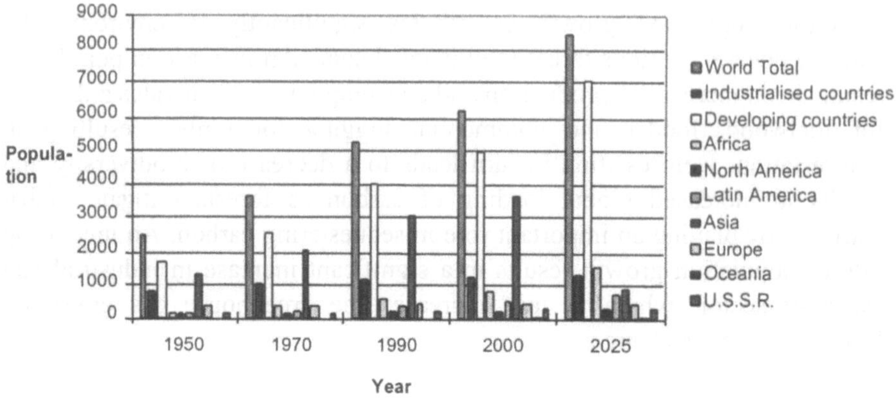

Figure 5.5 World population (in millions) by region (1985-1990)

5.4.2 Birth rate

Birth rate and death rate are the two factors governing the rate of population growth. The average number of births per women in the South fell from 6.1 to 3.9 between 1965 and 1988 - this is still more than twice as high as in the developed nations where the average number of birth per women fell from 2.8 to 1.8 in the same period. Within the developing countries there are also widely differing patterns. In East Asia, for example, the average number of births per women fell from 6.2 to 2.7 as a result of successful family planning policies (primarily in China). In sub-Saharan Africa, on the other hand, fertility remains around 6.7 births per women.

Socio-economic condition and cultural factors are two key factors influencing birth rate. People particularly in developing countries living at economic margin have many children as part of a survival strategy and social security. Children, for them means more hands to work. Cultural forces severely limit a woman to opt for fewer births. Community attitudes on the other hand, pressure a woman to keep having children until she gives birth to a son.

Rate of fertility depends on the age structure and level of fertility. On the surface, the world population age pyramid, or the distribution according to sex and age of the population appears balanced enough, but it hides very diverse situations in different areas of the world. Young populations predominate in Southern Asia, Latin America, the population is much older, with a significant middle aged bulge. By 2000, more than half of the world's population in the developing world will be under the age of 25. The number of women in the reproductive ages would therefore increase tremendously in developing countries, which in turn would lead to continued population growth.

5.4.3 Death rate

The indicator of the number of deaths in a population is given by Crude Death Rate. The number of deaths per annum and the mid-year population (P) is used for calculating CDR.

Life expectancy at birth is the average number of years a new born baby would live if patterns of mortality prevailing for all people at that time of its birth were to stay the same throughout its life. Considerable improvements have been made in the past 30 years in reducing the mortality rate. As a result, life expectancy at birth in developing countries has increased from about 50-52 years to 61-62 years. At the same time, in developed countries, life expectancy has increased from 67 to 70 years for men and from 75 to 79 years for women (Table 5.4). Infant mortality rate has gone down from 28 in 1950 to 10 by 1990. This decline in death rate and infant mortality rate can be attributed to special public health efforts such as eradication of small pox, malaria, childhood immunisation programmes and improvement in public health services. Improvements in social and economic conditions such as increase in income, nutrition and education has also played a significant role in declination of death rate.

Table 5.4 Life expectancy at birth (1985-1990)

Country	**Lowest**	**Country**	**Highest**
Afghanistan	41	Japan	78
Sierra Leone, Guinea	42	Canada, Hong Kong, Spain, Netherlands, Norway, Sweden	77
Ethiopia, Mali	44	Australia, France, Greece, Italy	76
Angola, Niger	45	Belgium	57

Source: "Exploding the Population Myth: Consumption versus Population, Which is the Climate Bomb?", p.29, July 1993

In spite of these, enormous health problems remain, and absolute levels of mortality in developing countries remain unacceptably high. Besides premature mortality, disability compromising of polio related paralysis, blindness and psychosis also play a significant role in maintaining high level of mortality in developing countries.

Improved health contributes to economic growth. It reduces production losses and helps in channelling resources for useful purposes. Freedom from illness ensures more monetary gains. Farmers in disease prone areas in developing countries at times forego higher output in return for less income. For poor people, income is directly proportional to health. Healthier workers earn more because they are more productive in their jobs-similar gains are likely among farmers, particularly marginal farmers in the developing countries. In Paraguay, for example, farmers in malarious areas choose to grow crops that are of lower economic value but that can be produced outside the malaria season.

5.4.4 Future population projections

Rapid population growth is the dominant feature of global demographics, and will continue to be so at least for the next 30 years. The 1993 global population of 5.57 billion is projected to increase to 6.25 billion in 2000; 8.5 billion in 2025; and 10 billion in 2050; significant growth will probably continue until about 2150, and level off at about 11.6 billion (UNPFA, 1992).

Developed countries

In the developed countries, population growth has slowed or stopped altogether and fertility is at or below replacement level. Growth is only about 0.4 per year, as is the case in the United States. Some countries in Europe are already experiencing a population decline due to a low birth rate rather than emigration

Developing countries

Growth rate (1990-1995) is estimated at 3% for Africa, with Asia at 1.9% and Latin America 2.1%. Fastest growth rates are in the poorest countries. The 47 countries officially designated by the UN as "least developed" accounted for 7% of the global increase in 1950 but 13% by 1990. Fertility has fallen by 60% in East Asia in same period, but only 25% in South Asia and least in Africa. By the year 2000, the 125 cities in the developing countries with more than one million people will increase to some 300 million. By 1990s no less than 83% of the world's population increase is expected to take place in town and cities. Rural areas have not received adequate developmental benefits and this is leading to 20-30 million of the world's poorest people migrating annually to the urban centres (towns and cities).

The impact of the population of a nation upon its environment is given by Ehrlich and Holdren's I = P *A*T identity; in which I is the impact of any population or nation upon environmental sources and sinks; P is the product of its population, A is the affluence (measured by per capita consumption of resources) and T is the damage done by the particular technologies that support that affluence. However due to lack of statistics that allow the consumption and technology factors to be readily measured scientists substitute per capita energy consumption to give a measure of the effect each person has on the environment. The average rich-nation citizen used 7.4 kilowatts (kW) of energy in 1990. The average citizen of a poor nation, by contrast, used 1 kW. There were 12 billion people in the rich nations, so their total environmental impact, as measured by energy use, was 1.2 billion * 7.4 kW = 8.9 terawatts (tW). Some 4.1 billion people lived in poor nations in 1990, hence their total impact was 4.1 tW. The relative small population of rich people therefore accounts for roughly two third of global environmental destruction. Thus the northern population must be stabilised as fast as possible. But since the total population and energy use per capita in the developing countries is about to increase tremendously, it is also important to stabilise the population in developing countries.

The root cause of increasing population particularly in developing countries being poverty and lack of education, the effort should be devoted to social and economic development. The focus should be on literacy, woman literacy in particular which leads to multitude of changes in society allowing women to have greater power and awareness through education and employment.

5.5 Land use

Forests help to regulate the climate by playing a significant role in the atmospheric cycles of substances like carbon, nitrogen and oxygen. They protect soils from erosion and are involved in the water balance and in the hydrological cycle. The increase in population has led to the transfer of forestlands to agricultural use. Moreover the increasing demand for fuelwood, fodder, timber etc. has resulted in large-scale depletion of forest resources consequently degrading the land.

5.5.1 Deforestation

Deforestation not only leads to land and environmental degradation but also contributes significantly to greenhouse warming. The vegetation and soils of unmanaged forests hold 20 to 100 times more carbon per unit area than agricultural systems. It has been estimated that in the period between 1850 and 1985, the total release of carbon to the atmosphere from changes in land use (primarily deforestation), is about 115 GtC, with an error limit of about ±35GtC (1 GtC = One billion metric tonnes of carbon).

Although the greatest release of carbon in the nineteenth and early twentieth centuries were from land use in the temperate zone (maximum 0.5 GtC per year), the major source of carbon during past several decades has been from deforestation in the tropics, with a significant increase occurring since 1950. Estimates of the flux in 1980 range from 0.6 to 2.5 GtC per year (IPCC, 1990).

Land demand for food production and pasture

In spite of a growing awareness of the need to stabilise human population, a global annual population growth (see Section 5.3) of 1.6 to 1.7% is still taking place. This would lead to a net increase of more than 90 million people annually and will add to the existing pressure on land to increase the arable land for food production and pasture. It is also estimated that 30 to 50% of the earth's land area is degraded due to improper management. In particular, both the conversion of forestland for agricultural practices has led to increase in soil erosion over the past 25 years. The rate of soil erosion is almost imperceptible (1 mm of soil loss in a storm amounts to 15 tonnes per hectare of land), and significantly exceeds its floor renewal rate (2.5 cm/500 years and at best 1 tonne per hectare per year). The rate of soil erosion in temperate countries is 10 to 20 times the soil renewal rate, while in the tropics it is almost 20 to 40 times. Both the above factors add to the land demand for food production and pasture particularly in the tropics.

The total cultivated land area expanded from about 1.3 billion in 1965 to nearly 1.5 billion hectares in 1990. Over the same period the population increased to more than double, indicating that, a unit of land now supports more people (3.4 persons per ha), compared to 25 years ago (2.4 persons per ha).

The technological revolution concerning food production has helped the densely populated countries of the South and Southeast Asia to keep pace with the increase in population in contrast to many African countries. (See Land Use in Asia (Box 5.3)).

Box 5.3

Land Use in Asia

Arable land resources in Asia are facing intense pressure from farmers seeking to maintain self-sufficiency in food. Forests and marginal lands are suffering from serious degradation for a variety of reasons, including excessive conversion to agriculture, commercial logging and over exploitation of firewood and fodder.

The total land area in Asia is about 2 billion hectares. About 390 million hectares (20%) has been brought under cultivation and classified as cropland; about 500 million hectares (25%) is classified as forest or woodland; permanent pasture land accounts for another 500 million hectares (25%); and the rest, (30% unused grassland and built up areas. Table 1 and 2 show grasslands and forest areas in Asia.

Table 1 Countries in Asia with extensive grassland

Country	***Extent of grassland (million hectares)***	***Percentage of total land area***
China	*319*	*34*
Mongolia	*124*	*80*
Afghanistan	*30*	*46*
India	*12*	*4*
Indonesia	*12*	*7*
Pakistan	*5*	*6*
Nepal	*2*	*15*

Table 2 Forest areas in Asian countries

Country	***Forests ('000 hectares)***	***Percentage of total land area***
With more than 100 million hectares		
- China	*123,600*	*13*
- Indonesia	*113,433*	*63*
With high percentage of forested land		
- Papua New Guinea	*38,230*	*84*
- Malaysia	*19,100*	*58*
- Cambodia	*13,372*	*76*

Country	*Forests ('000 hectares)*	*Percentage of total land area*
- *Lao P.D.R.*	*12,800*	*55*
- *Korea*	*6,485*	*66*
- *Bhutan*	*2,605*	*55*
- *Solomon Islands*	*2,560*	*91*
- *Fiji*	*1,185*	*65*
- *Vanuatu*	*914*	*75*
With low percentage of forested land		
- *Mongolia*	*13,915*	*9*
- *Afghanistan*	*1,900*	*3*
- *Maldives*	*1*	*3*
- *Pakistan*	*3,500*	*5*
Others		
- *India*	*66,736*	*22*
- *Myanmar*	*32,418*	*49*
-*Thailand*	*14,240*	*28*
- *Philippines*	*10,550*	*35*
- *Vietnam*	*9,800*	*30*
- *Nepal*	*2,480*	*18*
- *Bangladesh*	*1,950*	*15*
- *Sri Lanka*	*1,747*	*27*
- *Western Samoa*	*134*	*47*

Land Degradation

Estimates are that nearly 20% of the vegetated area in Asia has been affected by human induced land degradation. Table 3 shows estimates of land affected by soil degradation in selected Asian countries.

Table 3 Estimates of land effected by soil degradation in selected Asian countries

Country	*Estimated degraded land ('000 hectares)*	*Percentage of total land area*
South Asia		
- *India*	*148,100*	*50*
- *Pakistan*	*15,500*	*17*
- *Bangladesh*	*989*	*7*
- *Sri Lanka*	*700*	*11*
East Asia		

Country	***Estimated degraded land ('000 hectares)***	***Percentage of total land area***
- China	*280,000*	*30*
- Indonesia	*43,000*	*24*
- Thailand	*17,200*	*34*
- Vietnam	*15,900*	*50*
- Lao P.D.R.	*8,100*	*35*
- Philippines	*5,000*	*17*
- Myanmar	*210*	*3*
- Western Samoa	*32*	*32*
- Tonga	*3*	*5*

Deforestation in Asia

Among all tropical regions, East Asia experienced the highest rate of deforestation during 1981-90 (1.4% per year) Figure 5.6 The major causes of deforestation include commercial logging, conversion to agricultural land, and demand for fuelwood and fodder.

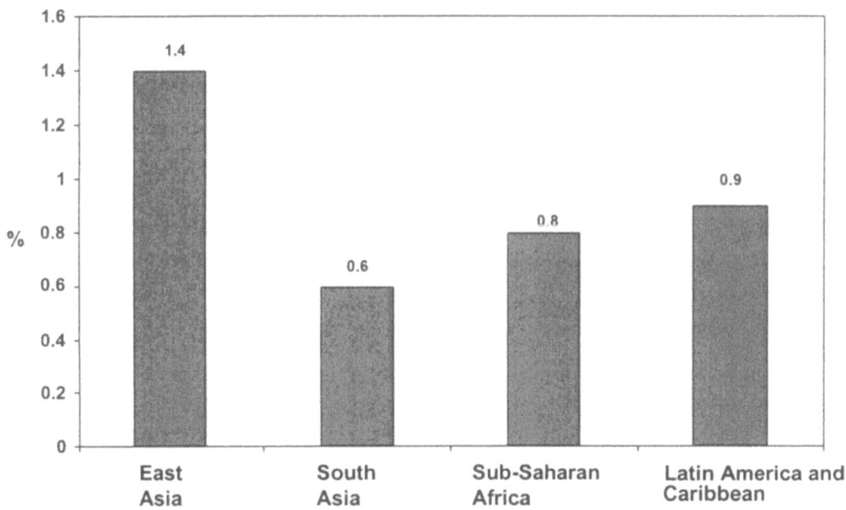

Figure 5.6 Deforestation (annual percentage), 1981-90

Source box: Brandon, et al. (1993).

Resource to sustain population growth

In the coming 30 years food production will have to increase by at least 1.4 to 2.1% per annum, depending on the population growth rate and the average per capita income. The rise in food production could be realised through increase in the land use. Potentially the land base can be doubled. Data reveals that, the rate at which new land is taken for production, has slowed down considerably to about 0.4% per annum. Moreover land is lost due to urbanisation and infrastructural developments. Proper land use policies have not been developed (Swaminathan, 1991).

Fuelwood

Fuelwood comprises the major source of energy in developing countries where in many cases it accounts for 50% or more of total national energy use. There are two main user sectors: (a) rural and urban domestic use, and (b) rural and urban processing and small industries. Households are the largest consumers, accounting for over 80% in the majority of countries.

Fuelwood takes up 77% of the 1 billion cubic meters of wood that Asia harvests per year (data of 1988). The poor people meet their energy requirements through fuelwood and other biomass like grass, crop residue, cattle dung etc. which constitutes important sources of natural fertilisers. Data for Africa and Asia demonstrate that the mean fuelwood use of 1.42 m^3/capacity/yr in rural Africa is higher than that of rural Asia where it is 0.83 m^3. Most African countries with extensive semi-arid zones (for example, the Sudan, Mali, Niger, Tunisia, Lesotho and Botswana) have lower rural per capita fuelwood consumption than the mean of 1.42 m^3 for Africa.

Commercial cutting of forest proves to be a major cause of deforestation as compared to rural firewood gathering. Another important contributor to deforestation in developing countries is cattle which grazes on forest land. This problem seems to be very acute in India, which has 15% of the world cattle population, but has only four percent of the global land area.

Industrial use of fuelwood

Urban and industrial enterprises add considerably to the demands for urban fuelwood. In Kenya, industries and commercial enterprises account for as much as 23% of the fuelwood demand. In Central America, they are estimated to account for as much as 18% of the total fuelwood demand. In Brazil, charcoal supplies 40% of the energy needs of the steel industry. The copper industry in Zambia uses some 20,000 tonnes of charcoal annually as a reducing agent in its copper refineries. Many developing countries face energy crises, as the forests recede; rural people, especially women, have to walk several kilometres a day in search of fuelwood.

Shifting cultivation

Modern agriculture with its new technological inputs is trying to maximise production on a short-term basis. These short-term gains are proving to be limiting in the long run. As a result, renewed interest is focused on traditional agricultural practices like shifting cultivation/slash and burn or jhum cultivation, which has proved to be sustainable for centuries. Shifting cultivation proves advantageous in humid tropics. This type of agricultural practice proves sustainable when there is optimum time lag between two successive croppings.

It is estimated that 30-80 million people are involved in shifting cultivation in Asia alone affecting 75-120 million hectares of land. The practice is widely prevalent in the north-eastern and drier central states of India, Kalimantan island in Indonesia, the central highlands in Philippines, and parts of Myanmar, Thailand and Bangladesh.

Any kind of developmental planning needs to have a strong traditional foundation. Depending on ecology, socio-economic situation and other criteria developmental strategies have to be location specific as this approach would assure people's participation, acceptance and appreciation.

Development of plantation

Plantation is primarily for industrial use, although in recent years, they have been established for non-industrial purposes also, such as household energy needs and agro-forestry. Plantation usually grows faster than natural forests and is established to produce desired single species, mainly exotic, selected for wood quality, growth and stem form, disease resistance and manageability. There are now 100 million hectares of temperate plantations and 35 million hectares of tropical and subtropical plantations. Together they account for less than 3% of the total forest area.

As measured by wood increment per unit area, well-managed plantations are many times productive than most natural forest systems and the plantation forests of many countries (for example, Australia, Chile, Kenya, New Zealand and South Asia) are now an important component of their productive forest resources. Managed natural forests in the tropics can be expected to yield between 2 m^3 and 8 m^3 per hectare annually, whereas man-made plantations in the tropics can yield between 10 m^3 and 60 m^3 per hectare annually depending on site, species and intensity of management.

Forests also have been exploited to develop tea, coffee, natural rubber, palm oil, and cacao plantations. Export sales of crops that trace their origins to forests in the tropics and sub-tropics approached $24 billion in 1991.

Tropical hardwood plantation

Plantations of softwood or eucalyptus plantation usually produce products viz., industrial sawnwood and pulpwood but are not found suitable to replace high quality hardwood found in natural forests. Such plantation of eucalyptus can reduce pressure on natural woodland particularly in the arid and semi-arid region: plantation of tropical hardwood can reduce pressure on tropical moist forest. Unfortunately with a few notable exceptions (such as teak) valuable tropical hardwood species are difficult to grow in plantation. Present plantations use large-scale monoculture. For example, coniferous plantations mostly under colonial rule before the 1950s and 60s, were exploited much faster than these forests could re-generate, resulting in widespread deforestation. In the post independent era of these countries, the national government had very little forest left to declare them as reserve forests, and save them for regeneration. The case of Nigeria is one such example (Box 5.4).

Wood production by plantation seldom takes less than four years, traditionally at least 25 years in the tropics, and at least twice that period in the temperate regions. Initial investment is high in the plantation forestry compared to natural forest management. For example, in Côte d'Ivoire, plantation-grown timber reportedly costs \$7.40 per cubic meter as against \$5.60 per cubic meter for timber derived from managed natural forests.

Development of plantation technology

Country specific/site specific research needs to be undertaken to improve the present plantation technology. The principal technical requirements for successful plantation forestry are:

- Availability of genetic resources of potentially useful species, and research in genetic improvement.
- Raising of planting stock under favourable nursery condition.
- Informed scientific observation
- Man power development

Box 5.4
Plantation Forestry in Nigeria

The tropical forests of Nigeria have been largely over-exploited and are in great danger of being wiped out. In the wake of the almost complete disappearance of Nigeria's natural forests and growing demand on timber the Forest Department has embarked on a policy of establishing tree plantation mainly of fast growing exotics such as teak. Since the early 1960s, the total area under forest plantation through out the country has increased from nearly 3000 ha in 1961 to 150,000 ha in 1978. Most of these plantations came up in the humid climatic zones where the expected rate of growth/yield is much higher than in the other areas (dry areas). The trend of tree plantation continued in the 1980s (Table 1).

Table 1 Plantations established in different regions of Nigeria in the 1980s

State	Period	Plantation area (in hectares)
Cross River and Akwa Ibom	1982-1983	237.71
Lagos	1980-1985	2,727.00
Lmo	1983	80.70
Oqum	1982-1983	1,729.00
Oyo	1980-1983	1,593.20

In the developing countries integration of other fast growing varieties with plantation forestry is likely to be the best means of achieving successful plantation production.

Also, large areas of the native rain forest ecosystem in south-western Nigeria have been cleared and replanted with exotic species, especially teak, for economic reasons. Though tree plantation may be economically desirable to provide woodpulp for paper industry and other industrial use, from the ecological point of view, should plantation forestry be pursued at the expense of conserving the native rain forest ecosystem is a question that needs careful consideration. The issues of tropical forest conservation are of global concern and the rapid destruction of tropical forests has been a major concern as such deforestation has already added considerable carbon dioxide to the atmosphere (at a rate of about 1-2 billion tonnes of C per year).

Source: Aweto, 1990.

5.5.2 Urbanisation

Population growth (see Section 5.4) has induced a spatial redistribution of the people globally. An ever-increasing proportion of the world population lives in urban areas in particular, in metropolitan centres. Rural-urban migration in demographic terms is the dominant cause of urbanisation in low-income countries, this percentage doubled between 1950 and 1990 (Table 5.5). Nearly a third of Asia's population currently lives in cities and towns. In urban areas, environmental problems are largely caused by high concentrations of domestic and industrial wastes that, in the absence of adequate collection treatment and disposal, overwhelm the assimilative capacity of the environment. In addition ecosystems are destroyed as urban and industrial development demand land in environmentally sensitive areas. Given high population densities, high incomes and large industrial concentrations, cities are now faced with serious environmental problems of water and air pollution.

Table 5.5 Degree of urbanisation (%)

Region	1950	1970	1990
World	29	37	43
Low-income countries	17	25	34
High-income countries	54	67	73
Bangladesh	4	8	14
Kenya	6	10	24
Colombia	37	57	68
U.S.A.	64	74	74
France	56	71	73
Great Britain	84	89	93

Source: UN, 1989.

5.5.3 Burning

Forest fires not only destroy standing crop, productive capacity of soil, but also endanger wildlife. During the period 1978-1981 Mediterranean countries experienced 82% of the total number of fires in Europe, accounting for almost 99% of the burnt area (excluding the former Soviet Union). Although large forest areas have been lost in developing countries, statistics on fires in Asia, Africa and Latin America are not systematically collected. In 1982-1983, fires burned for several months on the island of Borneo damaging 3.5 million hectares. Regulated or controlled burning is a very useful tool in forestry practices for the accomplishment of certain specific purposes.

Forest fires can be categorised under three heads: ground, surface and crown. Ground fires are rare and occur in the peaty layers beneath the litter with generation of intense heat but no flame. Surface fires are the most common type and occur near the ground in the litter. Crown fires occur on the treetops, they consume foliage and kill the trees.

There are three main causes of forest fires. They are: natural, unintentional/accidental by the human activity, and deliberate/intentional. Natural fires occur mostly due to lightning, rubbing of bamboo's. Unintentional/accidental fires occur due to careless throwing of matchsticks and burning cigarettes from picnic spots. Deliberate/intentional fires would include annual burning of fire lines in the forest or burning slash for raising new plantation.

Prescribed burning

There are three main forms of controlled burning: for natural regeneration, for protection (burning of the fire lines) and for raising of new plantation.

- Natural regeneration: Controlled burning for purposes of natural regeneration is associated mainly with three species in India: teak, sal and chir. As prescribed burning for purposes of regeneration is controlled and consists mostly of ground fires carried out during winter months, the material burnt is usually litter lying on the forest floor, grasses, shrubs, and other undergrowth.
- Protection Fires: The Forest Department/Authorities carry out annual burning of fire lines, as a measure of controlling against any unregulated fires that may cause damage to the forests. Burning of fire lines being in the nature of controlled fires carried out during the cold season, the materials burnt usually consist of litter on the forest floor, grasses, shrubs and other undergrowth.
- Raising new Plantations: It is customary to burn the forest, during the winter season, in all plantation areas prior to new sowing/planting in the next season. After all the utilisable materials have been extracted from the area, the residual material and debris is allowed to dry, and later these are collected in heaps and burnt. This cleans up the area to facilitate various planting operations and keeps the plantation area relatively free from weeds for sometime.

5.6 Agriculture

5.6.1 Agricultural activities

Agriculture and climate are inextricably related. The dependence of agriculture to climate change has only been established over the last few decades. Agricultural activities including clearing of forests, burning plant matter, cultivating rice, raising livestock and using fertilisers, leads to increased concentration of greenhouse gases in the atmosphere. Present agricultural practices contribute towards 17% of the global warming.

Biomass burning

Land clearing, an agriculture-related activity, has made significant contribution to the greenhouse gas build-ups. When land is cleared, carbon dioxide is released through the burning or decomposition of plant matter and the oxidation of organic matter in the soil. Cleaning forestland for agricultural purposes results in increased carbon emissions, because forest ecosystems store 20-100 times more carbon per unit area than cropland. Moreover burning of biomass for agriculture and pasture farming result in significant carbon dioxide emissions. In 1990, the carbon dioxide emissions caused by forest fires amounted to 987 million tonnes, while Savannah fires (for agricultural purposes) produced as much as 2,047 million tonnes of carbon dioxide. During biomass burning carbon monoxide is also released and accounts for a total of 350 million tonnes of carbon monoxide emission per year. Other direct and indirect greenhouse gases such as methane, nitrous oxide, nitrogen dioxide and hydrocarbons are also released during biomass burning. In addition to contributing to atmospheric carbon dioxide biomass burning contribute approximately 10-20% of total annual methane emission 5-15% of nitrous oxide emissions 10-35% of nitrogen oxide emissions and 20-40% of carbon monoxide emissions (Burke et al., 1989).

Rice cultivation

Flooded rice fields are responsible for the production of methane by anaerobic decomposition. The level of production of methane is affected by the particular growth phase of the rice plant, temperature, irrigation and water management practice, fertiliser usage, presence of organic matter, rice species under cultivation and number and duration of rice harvests. Of the global methane budget of about 500 Tg per year (Tg CH_4/yr), an estimated 60-70 Tg/yr is from rice, 65-70 Tg/yr from domestic animals and 50-100 Tg/yr from biomass burning. Methane's global warming potential is twenty times more than that of carbon dioxide.

As the population continues to grow the demand for rice will also continue to increase. In order to meet this demand more intensive cultivation methods will be required which will eventually lead to higher emissions.

Cattle farming

Methane emissions from cattle farming amounts to between 60-100 Tg of CH_4 /yr annually. The volume of methane produced depends on the type and the quantity of animal feed used. Intensive animal husbandry, involving the use of correspondingly high-energy animal feed leads to maximum methane production rates. Energy balance studies have shown that up to 12% of the animal feed energy is lost in the form of methane emissions (Economica Verlag, Verlag C.F. Muller, Climate change: A threat to global development).

Fertilisers

The increase in world-wide fertiliser use has had enormous benefits in terms of food production. Although at the macro-level it is difficult to separate the relative impact on production of fertilisers and other inputs, studies of Asian rice production indicate that fertilisers as a whole have contributed about 20% to new varieties, irrigation and capital investment. On the whole global food production per capita has risen by 14% since 1964.

On the other hand over-use and abuse of agro-chemicals impose heavy costs on a nation's economy and can quickly erode the ecological foundations of a thriving agro-ecosystem. Excessive use of fertilisers results in runoff of nitrogen and phosphates thereby contaminating soil and water resources. Many rivers in the United Kingdom, despite considerable year to year variation, show progressive increase in nitrate concentration over the past 30 to 40 years. Several important rivers have mean values greater than 50 mg/l (European community limit). In the United States, although many rivers are also showing increasing trends in nitrate contamination, the majority are still substantially below 45 mg/l nitrate in ground water. Nitrate pollution of water is a health hazard, particularly in infants and the percolation of nitrogen to the under ground water sources can impose a heavy cost on a community or region.

Nitrogen fertilisers are responsible for increase in nitrous oxide emissions from 0.2 to 0.3% in the past decade (Economica Verlag, Verlag C.F. Muller, Climate change: A threat to global development). Nitrous oxide plays an important role in atmospheric and ozone chemistry. At high altitudes it photo-decomposes to form nitrogen oxides which contribute to ozone depletion. Because atmospheric life times of nitrous oxide are long (between 130-150 years), it has a high ozone depletion potential next to the chlorofluorocarbons' ozone depletion potential. Apart from depleting ozone, nitrous oxide, being a radiatively active gas, is responsible for global warming leading to climate change.

Nitrous oxide is only present in small amounts in the atmosphere. The pre-industrial atmospheric concentration of nitrous oxide was 288 parts per billion by volume (ppbv) while the present day concentration is 310 ppbv. The increase in nitrous oxide emissions is mainly due to the increased consumption of nitrogen fertilisers. The global nitrogen fertiliser consumption of 70.5 Tg N in 1984/85 is increasing at 1.3%/year in industrialised countries and at 4.1% per year in developing countries.

In 1978/1979 developed countries used about three-quarters of world consumption of fertilisers and the US alone consumed over one-third of all pesticide used. The balance is shifting rapidly, however, as it must in the interest of global food security. In the developing countries average fertiliser rates are a great deal lower. The annual application to arable and permanent crops in Asia is currently about 30 kg N/ha, in Latin America 15 kg N/ha and in Africa only 4 kg N/ha, and much of the land receives no inorganic fertiliser at all. The use of fertilisers has been growing more than twice as fast in developing as in developed countries (averaging 10% a year between 1960/1970 and 1978/1979). The FAO projects future growth by the year 2000 to be between 7.5 and 8% per year, twice the projected increase in crop production.

With this rising consumption fertiliser is likely to become an increasingly significant source of nitrous oxide in future. It is therefore necessary to manage the use of fertiliser so as to steadily reduce available damage to soil, water, plants, animal and human health.

Pesticides

Pesticides are chemical agents used for controlling insects, pests, weeds, fungus etc. Use of pesticides in controlling pests has resulted in the disruption of ecosystems because of the death of non-target species, accumulation of pesticide residues in the environment and in food, and the built up of pesticide resistance in target species.

Pesticide that is not lost by volatilisation or in runoff, enters the soil where it is eventually degraded or it percolates to the ground water table. The time required for degradation varies greatly according to the type of pesticide. Some products break down readily in soil, others, particularly the organochlorines, can resist degradation for decades and are hence non-biodegradable.

Finally the proportion of pesticide that is not degraded will leach below the root zone and eventually end up in the ground water. Most pesticide was assumed to be volatilised or degraded, leaving ground water unaffected. But this assumption was shattered when evidence of widespread ground water contamination came to light in the USA in the 1970s and in the UK in the 1980s. Insecticides are also responsible for impacting human health. Estimates suggest that in the developing countries some three million suffer from single short-term exposure with 220,000 deaths; death rates vary from 1% to 9% in cases undergoing treatment. Over 7000,000 people a year suffer from chronic long-term exposures.

The use of chemicals in pest control has a poor reputation due to the deleterious environmental effects of a number of pesticides noted during the last two decades. So far the most common side effects have been: mortality in non-target species, sometimes regulation resulting in temporary and occasionally long-term changes in the abundance of species and the diversity of ecosystems; a reduced reproductive potential in birds (known for DDT), fish and other organisms; and the development of resistance in non-target species.

Scientists of the developing countries should look for alternatives to maximise the use of safe, cheap and simple pest management techniques including those traditionally used by the Third World farmers. Furthermore they should look for a way to integrate these techniques with the use of the safest possible chemical pesticides, where necessary, into a strategy which could deal with pest problems as a part of a system - including people, crops, beneficial insects, fish and livestock, as well as pests and chemicals. Such an ecological approach constitutes the best use of pesticides. It is also the basis of the idea of Integrated Pest Management (IPM). IPM can be drawn upon a number of different pest management methods. These include the careful and selective use of pesticides as well as biological and cultural controls, physical controls, the use of pest resistant plant varieties and a number of other techniques (Bull, 1982).

5.6.2 International trade

A global economy totally and irreversibly interlocked with the earth's ecology is the new reality of the late 20th century. This reality is nowhere more evident than in the relationship between trade and the natural environment. According to the OECD, trade policies can contribute to environmentally adverse patterns of production, unsustainable exploitation of natural resources. In fact the impact of trade and trade related policies on the environment is significant and growing rapidly. The three major reasons for this are: tariff and non-tariff barriers often distort global patterns of production in ways that accelerate the degradation of environment and natural resources; and has most severe impact on developing countries.

Since the end of Second World War, seven rounds of multilateral trade negotiations have been initiated and successfully completed under the auspices of the General Agreement on Tariffs and Trade (GATT). The eighth, known as the Uruguay Round, that formally began in September 1986 at Punta Del Este in Uruguay, was concluded as Final Act Embodying the Results of Uruguay Round of Multilateral Trade Negotiations 1994 ushering liberalisation of international trade, particularly in agriculture.

The far reaching changes sought by the Uruguay Round of GATT and the North American Free Trade Agreement (NAFTA) in matters of world agricultural trade have profound environmental and social implications. At present agricultural food commodities account for 10% of the value of world trade. The industrial countries account for much of the commerce in primary products (such as plywood, logs, coffee, fish, food grain, bauxite and alumina, iron ore, crude oil, natural gas, coal etc.), but it is the developing countries that are most dependent on such trade and most vulnerable to its environmental costs.

According to the UNDP (the United Nations Development Programme is one of the specialised agencies of the UN dealing with developmental needs of developing countries and countries in economic transition), developing countries now loose some $ 100 billion in agricultural trade per year as a result of quotas, tariffs, and other trade barriers. With the new GATT rules this situation may be considerably improved. But the farmers of the developing countries are still very apprehensive of the new GATT agreement.

One concern very often sounded is the possibility of adverse environmental consequences as a result of actions needed for a considerable increase in food production to feed an expanding population particularly in the developing countries. Much will however depend on macro and sectoral policies which may either help or hinder an environmentally sound pattern of resource utilisation, the availability of environmental friendly technology, and above all, a liberal international trade regime that allows a relatively free flow of food and agricultural commodities across countries, states and regions.

One of the important provisions in the Final Text of the Uruguay Round (1994) GATT agreement is the application of sanitary and phyto-sanitary measures, to protect human, animal or plant life or health. However, it is required that these measures are not applied in a manner which would constitute a means of arbitrary or unjustifiable discrimination between members where the same conditions prevail or a disguised restriction on international trade.

In the Uruguay Round 1994, it is broadly agreed that standards for sanitary and phyto-sanitary provisions should be harmonised. The harmonisation here means the establishment, recognition and application of common sanitary and phyto-sanitary measures conforming to international standards necessary to protect human, animal or plant life or health and consistent with the relevant provision of the GATT 1994.

5.7 Transport

Ever-increasing urbanisation and industrialisation, accompanied by economic development and improved standard of living, have increased the demand on the transport sector. As a result, pollution from the transport sector is on a rise both in absolute terms as well as in relative terms compared to other energy consumption sectors. Growing traffic volume has diluted the emission reductions achieved in the past 15 years.

The transport sector, which includes motor vehicles, aircraft and ships, is a significant source of air and noise pollution. The main contributors to air pollution in the transport sector are carbon monoxide, nitrogen oxides, sulphur dioxide, chlorofluorocarbons and carbon dioxide. Transport is the source of about 90% of total emissions of carbon monoxide in OECD countries.

Carbon monoxide induces severe health problems and indirectly contributes to the increase of the greenhouse effect. Fossil fuels in the transport sector emit about 20% of the global carbon emissions. Figure 5.7 shows the transport related carbon dioxide emissions from different regions/countries.

About two third of the global nitrogen oxide is emitted by motor vehicles. The most damaging effect of nitrogen oxides is in the formation of photochemical smog. 40-50% of total hydrocarbon emissions originates from the transport sector and plays an important role in the formation of photochemical oxidants. These air pollutants, when exposed to intensive solar radiation, undergo chemical reactions to form tropospheric ozone, which accounts for 7% of the manmade greenhouse effect. Sulphur dioxide, released by combustion of fossil fuels in aircraft, leads to an increased concentration of sulphate aerosols in the lower stratosphere, thereby contributing to ozone depletion. Chlorofluorocarbons from car air-conditioners also contribute to global warming and are one of the main contributors to the depletion of the ozone layer.

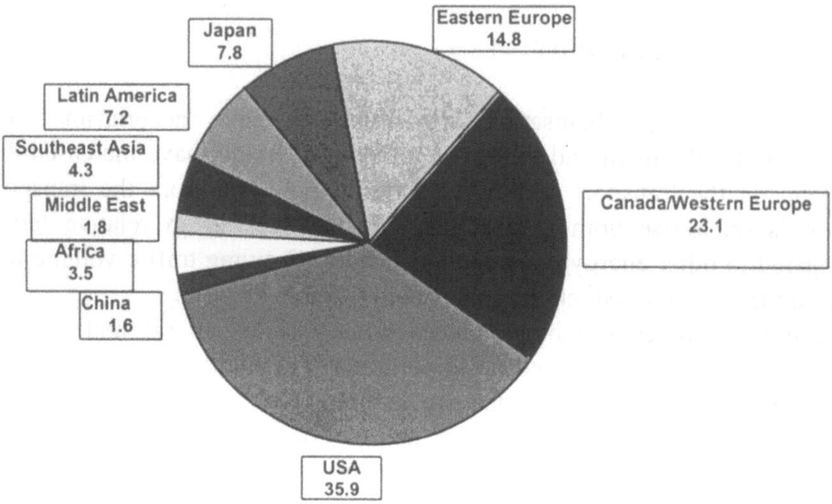

Figure 5.7 Breakdown of global transport-related CO_2 emissions (percentage shares)

5.7.1 Road transport

Motor vehicles are one of the major causes of environmental degradation in the form of air and noise pollution. Emissions from road transport including heavy trucks are about 14% of the global total. Much of the increase in global emissions has been caused by the rapid growth in ownership of two and four wheeled motor vehicles especially cars.

Between 1970 and 1987, the number of cars multiplied by a factor of two in Western Europe, a factor of 1.6 in North America and a factor of 3.4 in Japan. There are about 150 million cars (plus 44.5 million goods vehicles) in North America and 1303 million cars (plus 14.5 million goods vehicles) in Western Europe (Barde and Button, 1990). The total automobile fleet emits 0.725 billion tonnes of carbon annually, about one tenth of the global annual carbon dioxide emissions from human activities. While the most abundant pollutant from cars is carbon dioxide, other greenhouse gases such as methane and nitrous oxide are also produced.

Emissions of carbon dioxide from cars are directly related to the amount of fuel burnt and the carbon content of the fuel. Unlike for other pollutants, such as sulphur dioxide and nitrous oxide, there is no add-on abatement technology to reduce carbon dioxide emissions. The only options are to increase the fuel efficiency or to change to fuel with lower carbon content.

Total emissions from automobiles should consider all stages of the life cycle, including raw material extraction and processing, and transport of material. Considering all these processes, a car produces 59.1 tonnes of carbon dioxide and 26.5 tonnes of wastes over its life span. If the emissions from refining, transportation and distribution of fossil fuels are included, carbon dioxide emissions from this sector of economic activity amount to around 20 % of the total global anthropogenic emissions of carbon dioxide. The transport sector is projected to double over the next 20 years; it is therefore apparent that road transport will continue to be one of the world's major contributors of carbon dioxide (Elsworth, 1991).

Apart from emitting carbon dioxide, the road transport sector is also responsible for emission of nitrogen oxides, carbon monoxide and organic compounds, which in turn are involved in the formation of tropospheric ozone (an indirect greenhouse gas). Diesel vehicles also emit particulate matter, which plays an important role in global warming.

Vehicles equipped with air-conditioners contribute another important greenhouse gas, chlorofluorocarbon. About 40% of the US production of chlorofluorocarbons and 30% of European production is used in the air conditioning and refrigeration sector. Mobile air-conditioning accounted for 56,500 metric tonnes of chlorofluorocarbons. These are finally released into the atmosphere, and being inert, they have a long atmospheric lifetime of about a 100 years.

Effects on human health

Emissions from vehicles are very close to the ground and thus have a direct effect on human health. It is difficult, however, to assess the extent of these health impacts from air pollution in urban areas around the world. Information on air quality levels is scarce, and so is the direct evidence of adverse health effects. Noise is one of the most resented problems caused by transport and is also a major public health issue. In OECD countries about 130 million people (16% of the population) are exposed to 'unacceptable' noise levels i.e. above 65 dB(A), caused by various transport systems including aircraft. Until stringent measures are enforced the noise pollution from the transport sector, is likely to deteriorate in the future (Barde and Button, 1990).

Energy efficiency in road transport

Transport accounts for 15% of the total energy consumption in the European community, 26% in the US and between 30 % (Kenya) to 50% in developing countries. On an average, road transport in developing countries consumes 82% of the total transport energy. In most developing countries the road transport sector accounts for over one-half of the total oil consumption and one third of commercial energy use. Within the road sub-sector, cars consume about two thirds of the energy in the developed countries. In developing countries, conversely, buses and trucks account for more than 70 % of energy consumed on the road, as opposed to 20 to 25% in developed countries.

Large energy savings are possible in the road transport sector, particularly in automobiles. Annual oil consumption in developing countries is estimated to be 833 million tonnes of oil equivalent, of which almost half is used in automobiles. This dependence on liquid fuels in the transport sector has now become a significant burden on the balance of payments in most oil-importing countries. Fuel efficiency of motor vehicles, particularly in the developing countries, can be considerably improved. This will require considerable investment, which most of the developing countries cannot afford.

Scenario in developed and developing countries

The automobile production continues to grow rapidly. In 1994, motor vehicles fleets grew at an average rate of 8% in Eastern Europe and at an average rate of 18% in Asia Pacific region. Developed countries, like North America, Europe, Japan and Oceania, own 81% and produce 88% of all automobiles. The United States alone account for 35% of car ownership and 25% of production. Thus the major share of emissions still comes from developed countries; the USA alone account for 48% of all emissions and all OECD countries together account for 77%.

In developing countries automotive pollution is more serious than in industrial countries as many vehicles are in poor condition and the fuels used are of low quality. Moreover, a high percentage of the population moves and lives in open air and is thus exposed to automotive pollutants. In developing countries leaded fuels are still widely used. As a result lead accounts for up to 9% of airborne toxic pollutants. Despite these constraints the demand for automobiles is increasing in these countries. Automobile-sales in Brazil have doubled in the last five years, hitting 1.4 million in 1995. In Malaysia sales of new vehicles increased by 25% in the last year alone. In South Korea car-sales have jumped 100% to 2 million a year. China's new car sales are expected to jump nearly 30% in the next year to 750,000 and to pass 1 million by the end of the decade (World Watch, Jan-Feb 1996).

With this escalating demand for automobiles across the developing countries their auto industries are growing rapidly. Developed countries are now looking to developing countries as potential new markets for automobile sales. Without aggressive policy measures the rapid growth in automobiles on global scale will increase the emissions and result in further degradation of human health and environment.

Potential for reducing emissions

Emissions of carbon dioxide and other pollutants are influenced by a number of factors, like fuel used, engine and vehicle design, maintenance, driver behaviour, traffic conditions, vehicle speeds and the total kilometres driven. The greatest potential for reducing emissions is through a combination of improved vehicle technology and reduced traffic demand.

A typical diesel car proves to be 20% more efficient in terms of global warming if particulate matter is not taken into consideration. Diesel cars are unlikely to reduce greenhouse gas emissions very significantly. To bring down the emissions of greenhouse gases from vehicles, many alternative fuels, such as compressed natural gas (CNG) mainly methane, biofuels like ethanol or vegetable oil easters and alternative technologies like electric vehicles are being proposed.

5.7.2 Air transport

Air transport impacts the environment in three possible ways: (i) gaseous emissions from aircraft engines (ii) noise pollution and (iii) air craft maintenance operations which contaminate soil and underground water.

Impact of air transport on the global environment

Aircraft presently releases 2-3% of global emissions of carbon dioxide and nitrous oxide from fossil fuels and this fraction is predicted to grow rapidly. Aircraft also emits a mixture of other pollutants including soot, carbon monoxide, hydrocarbon and water vapour. About half of these emissions are released into the atmosphere at an altitude of 8-12 km (particularly into the stratosphere over the mid- and high- latitudes). These are of special concern because they produce ozone by photochemical reactions and hence contribute to global warming. Stratospheric water vapour emissions on the other hand affect the earth's climate by adding to the man made green house effect.

The combustion of aircraft fuel releases sulphur in the form of sulphur dioxide. The impact of sulphur dioxide emissions is much greater at higher altitude than on the earth's surface because of longer lifetimes, lower background concentration and higher radiative activity. The sulphur dioxide, released in lower stratosphere, leads to a substantial increase in sulphate aerosol concentration, thereby contributing to ozone depletion. It is estimated that about 150 megatonnes of carbon dioxide emitted by aircraft globally contributes about 2% to global warming. In addition, nitrogen oxide emissions from aircraft near the ground and in the lower atmosphere produce ozone near the tropopause and the earth's surface. This ozone may considerably increase the human induced global warming (Barrott, 1993).

Potential for reducing emissions

The demand for civil air transport, which uses over 80% of aviation fuel, is predicted to grow at 5% per year, doubling in less than 15 years. The long-term growth potential is vast because of the current low per capita demand in the developing countries. As the demand for air travel increases, the emissions will also rise. To stabilise or reduce emissions, demand management may be required.

Technological breakthrough, like introduction of fuel efficient aircraft and operational changes, like increasing load factors, can reduce pollution by 30 %. Most air freight goods and other urgent commodities could be carried by less polluting surface or water modes. Business travel accounting for 20-30% of passenger demand could be reduced by increased use of telecommunications. Leisure travel constituting 70-80% of passenger demand, could be encouraged to use less polluting modes of transport wherever possible (Barde and Button, 1990). Thus reducing the growth of demand, introducing less energy intensive technologies in aviation and operational measures can reduce green house gas emissions.

Air quality in the vicinity of airports can be improved by replacing diesel transit vehicles with electric ones and modifying surface access to airports. Problems associated with disposal and recycling of wastes and hazardous substances produced by catering, maintenance and other airline and airport operations can be solved by correct product designs using biodegradable products.

5.7.3 Rail transport

As already discussed, the road transport contributes considerably to greenhouse gases and ozone depleting agents, thereby affecting the global climate. Such problems associated with the transport sector can be lessened by emphasising diverse transport modes with railways playing a significant role. Railways have numerous environmental advantages compared to other modes of transport. Table 5.6 lists some advantages of railways compared to road and air transport.

Table 5.6 Advantages of rail over highway and air transport

Category	Examples
Greater Energy Efficiency	An intercity passenger train is three times as energy efficient as commercial air and six times as efficient as a car with one occupant.
Less Dependence on Oil	Switching 5% of U.S. highway driving to electrified rail would save more than one-sixth the amount of oil imported annually from the Middle East.
Less Air Pollution	For every ton of goods moved one kilometre, freight rail emits one-third the nitrogen oxide and carbon monoxide, and one-tenth the volatile organic compounds and diesel particulates emitted by heavy trucks.
Lower Greenhouse Emissions	For every ton of goods switched from roads to rail in the United Kingdom, the amount of carbon emitted per kilometre would drop by 88%.
Less Road & Air Traffic Congestion	Conservative estimates suggest that without Amtrak (the U.S. intercity passenger railroad), air passengers on the New York City/Washington, D.C. route would increase 36%.
Fewer Injuries and Deaths	Between 1964 and 1992, more than 3 billion trips were made on Japan's bullet trains without a single fatality; the equivalent volume of road travel over that period killed nearly 2,000 people.
Less Land Paved Over	Two railroad tracks can carry as many people an hour as sixteen lanes of highway. Some 500 kilometres of the French TGV high-speed rail system could fit into the area occupied by a single large airport.
Local Economic Development	In the Washington, D.C. area during the eighties, 40% of new building space - worth $3 billion - was built within walking distance of a Metro (subway) stop.
Sustainable land use patterns	Rail corridors help encourage compact, efficient land use. Rail-based cities, such as Paris, Stockholm and Toronto, have accommodated new growth, while remaining liveable and avoiding sprawl and excessive car dependence.
Greater social equity	The majority of the world's people can afford neither an automobile nor an airline ticket; rail is a vital option for people who are disabled or too young or old to drive.

Source: World Watch Paper, 118, 1994.

Electric trains

Compared to highway- or air-travel, diesel powered trains produce far less harmful emissions such as carbon monoxide and volatile organic compounds (VOCs), as well as far less carbon dioxide. Electrified rails offer even greater reductions. They do not only operate without creating pollution and noise, but are also energy efficient. Minimising energy use minimises air pollution, so expanding the role of electrifying railways would help improve air quality.

However, if electricity for power trains is produced from coal combustion, the net reduction in emissions will be small. In the UK carbon emission, attributable to electric trains, while lower than those from road or air transport, are still roughly the same as those from trains. Western Europe now has the highest concentration of electrified railways, ranging from 99% of intercity rail in Switzerland to 55% in Italy and 26% in the UK. 35% of Japan's intercity network is electrified. With the exception of China and India, most of the developing countries have marginal rail services, which are not maintained well. China and India however have large efficient national rail networks. China transports 80% of its intercity passengers travel by train. In India, rail transport carries 46% of freight and is six times more efficient than shipping or truck transport (World Watch paper, 1994).

Use of diesel and electric trains will bring down the emission of greenhouse gases from burning coal. Economists generally agree that in case of railways, infrastructural improvements turn out to be sound investments, if environmental and social criteria are taken into consideration. Pollution related health hazards and traffic congestion are minimised. Oil dependence and road accidents are reduced.

5.7.4 Marine transport

Marine transport also contributes to global carbon dioxide emissions through the use of fossil fuels, but its contribution is relatively small compared to surface transport. It is however responsible for the spillage of oils and petroleum products, thereby degrading the marine ecosystem. Most of the world's oil is transported by oil tankers and at least two million tonnes a year enter the marine environment as a result of routine operations and accidents. Besides these discharges of oils and petroleum products, dumping of radioactive wastes, industrial effluents, etc. into sea, also causes marine pollution and affects the marine and coastal ecosystems.

5.8 Discussion

While earth's climate has changed over the millennia, what is causing concern globally at present, is the predicted faster rate of warming (0.2°C per decade) of the earth's surface and the atmosphere, due to enhanced emissions of greenhouse gases caused by human activities. This will have effects on human life, earth's ecosystems, water resources, human habitat, wildlife biodiversity, forests and agriculture, and may also cause changes in the temporal and spatial distribution of rainfall and sea level rise, due to thermal expansion of sea water and melting of land glaciers.

The atmospheric concentrations of two important greenhouse gasses, water vapour and ozone, are not included in Box 5.1. Water vapour has the largest greenhouse effect, but its concentration in the troposphere is determined internally within the climate system, and, on a global scale, is not effected by human sources and sinks. The concentration of water vapour will increase in response to global warming and further enhance it; this process is included in the climate models. The concentration of ozone is changing, both in the stratosphere and troposphere, due to human activities, but it is difficult to quantify the changes from present observations. The rest of the greenhouse effect is due to gases at very low concentrations also known as atmospheric trace gases (CO_2, CH_4, N_2O and CFCs).

Discussion on the causes of human induced greenhouse gases so far has conclusively brought out that the two main human activities viz., (a) burning of fossil fuels in various economic activities such as generation of electricity, district heating, cooking, lighting, industrial activities, including chemical industry, transportation, deforestation, agriculture and (b) land use changes, have emitted significant amounts of carbon dioxide and have also added considerable methane, nitrous oxide and chlorofluorocarbons into the atmosphere.

Carbon dioxide is the most significant greenhouse gas man contributes to the atmosphere, accounting for slightly more than half of the human enhancement of the greenhouse effect (IPCC, 1990). Current levels of CO_2 emissions from human activities are in the order of 7 GtC/yr, including emissions from fossil fuel combustion, cement production and land use changes. The industrialised nations were responsible for ¾ of these emissions, with the OECD countries accounting for 25% of the total CO_2 emission. The greenhouse gas emission levels vary widely among nations. The global average annual per capita emissions of CO_2 from fossil fuels is at present 1.1 tonnes of carbon. In addition, a net amount of 0.2 tonnes per capita is emitted annually from deforestation and land use changes etc. Per capita emission of CO_2 from fossil fuels ranges from 5.3 tonnes of carbon in the United States of America to 2.4 tonnes of carbon in Japan and 0.2 tonnes of carbon in India. CO_2 is removed from the atmosphere by a number of processes by biota and oceans that operate on different time-scales. It has a relatively long residence time in the atmosphere in the order of 50-200 years or more.

The total estimated emission of methane to the atmosphere is 525Tg CH_4 per year, out of which about 350 Tg CH_4 are from human activities and 175 Tg CH_4 from natural sources (e.g. natural wetlands, termites, oceans and freshwater, CH_4 hydrate destabilisation). The major sink for atmospheric CH_4 is the reaction with OH (hydroxyl group) in the troposphere. Global concentrations of OH are dependent upon the intensity of UV-B radiation and the concentrations of gases such as water vapour (H_2O), CO, CH_4, NO_x, NMHC and O_3. The OH-concentration in the atmosphere may have declined during the twentieth century due to changes in atmospheric concentration of these gases. The reaction with OH in the troposphere, the major sink for CH_4, results in a relatively short atmospheric lifetime of about 10 years.

Human activities responsible for nitrous oxide in the atmosphere are deforestation, land use changes, combustion of fossil fuels, biomass burning and the use of nitrogen fertiliser for agriculture. Natural sources of N_2O are the oceans and the soils. Total emissions are in the order of 4.4-10.5 Tg N per year, out of which about 4.3 Tg N per year is from natural sources. There is no sink mechanism in the troposphere for N_2O. The major sink for N_2 is photolysis (atmospheric chemical reaction in the presence of UV-B) in the stratosphere, resulting in a relatively long atmospheric life time of about 150 years.

5.9 Conclusions

The use of energy by burning fossil fuels and other industrial activities emits about 70 % of the greenhouse gases and the other 30 % is emitted by deforestation, agriculture and land use changes. Generation of electricity by using fossil fuels and transportation are the two key problem areas in global warming. Greenhouse gas emissions from transportation alone now account for 27-30 % of the overall total energy consumption, and seem to rise still further. In OECD countries energy for transportation nearly tripled during the period 1980-1990, and this sector now has the largest final energy use. The great concern now has to be that such unsustainable travel systems will spread world-wide. Central to these worries is the level of automobile ownership; there are currently around 500 million cars and present trends indicate 1 billion cars by 2010 and 2 billion by 2025.

The other important factor, which must be touched upon in the conclusion, is the unprecedented growth in human population and the consequential effects on the greenhouse emissions. The world's population is growing faster than ever before, with 93-95 million people added each year. Our current population of 5.6 billion will grow to 6.2 billion by the end of the century. Nearly all of this growth will be in South Asia, Africa and Latin America; more than 50% of this growth will be in South Asia and Africa, the poorest regions of the world. In addition to the greenhouse gas emissions from the human activities, the unprecedented growth in human population will have profound effects on our physical environment.

We are to make progress towards development for alleviating poverty. Some ways must be found to give hope to the 'bottom billion' poorest people. Some means must also be found to meet the aspirations of three billion others that are neither poor nor very affluent. The challenge, therefore, is to raise the living standard of 80% of the world's population. We need development, but development without sustainability defeats its own purpose. Climate change is a case in point. More people mean more consumption of electricity, more industrial production, more land claimed for agriculture and a higher demand for fuel wood. Escalating numbers of people contribute, both directly and indirectly, to increasing emissions of greenhouse gases into the atmosphere. The new approach to development must look for ways to slow down population growth by recognising how closely population, consumption patterns, technology, and the environment are interlinked.

There is an urgent need to shift our economic activities from fossil to non-fossil fuel energy sources, to improve energy efficiency and to take up large scale afforestation activities, including sustainable agriculture to bring down GHG emissions from burning of fossil fuels and to increase CO_2 sink capacity and its reservoir globally.

References

Aweto, A.O. Plantation, Forestry and Forest Conservation in Nigeria. The Environmentalist, 10(2), 1990.

Barde, J.P. and Button, K. Transport Policy and the Environment. Earthscan Publications Ltd., London, 1990.

Barrott, M. Aircraft Pollution Control. In Procedings of the Climate Action Network UK: Transport & Climate Change Conference, p17, 1993.

Brandon, Carter and Ramankutty. Towards an Environmental Strategy in Asia. World Bank Discussion Paper 224, 1993.

Bull, D. A growing problem: Pesticides and the third world poor, OXFAM, 1982.

Burke, L., Daniel, M.K. and Lashof, A. Greenhouse Gas Emission Related to Agriculture. Prepared for the Annual Meeting Proceedings of the Agronomy Society of America, California; February 1989.

Economica Verlag, Verlag C.F. Muller, Climate Change - A threat to Global Development

Elsworth S. (Ed). The Environment Impact of the Car, Greenpeace International, July, 1991.

Food 2000: Global Policies for Sustainable Agriculture. A Report of the Advisory Panel on Food Security, Agriculture, Forests and Environment, to the World Commission for Environment and Development; Zed Books Limited, 1987.

India 1993, Published by the Director, Publication Division, Ministry of Information and Broadcasting, Government of India, Jan. 1994.

IPCC - Climate Change - The IPCC Scientific Assessment, Cambridge University Press, 1990.

IPCC - Climate Change 1995: Impacts, Adaptations and Mitigation of Climate Change: Scientific-Technical Analysis. Report of Working Group II of the Intergovernmental Panel on Climate Change, Cambridge University Press, London and New York, 1996.

Report on Energy Audit of Integrated Steel Plants, Ministry of Industry, Government of India, New Delhi, 1989.

Holman, C. No Easy Fixes for Transport, HOT NEWS - Transport Special Report, The newsletter of the Climate Action Network, UK issue 8, Winter 1993.

J.T. Hougton, Jenkins, G.J. and Ephraums, J.J. (Eds).: Climate Change: The IPCC Scientific Assessment. Cambridge University Press, 1990.

Laporta, C. Renewable Energy: Recent Commercial Performance in the USA as an Index of Future Prospects, in 'Global Warming, the Greenpeace Report' (Ed) Leggett, Jeremy, p 229, OUP, 1990.

Lowe, M. D. Back on Track: The Global Rail Revival, Ed. Carole Douglas: World Watch Paper, 118, April 1994.

Mackenzie, J.J. The Keys of the Car: Electric and Hydrogen Vehicles for the 21st Century.

Maryland, G., T.A. Boden, R.C., Smith, S.F. Huang, P. Kanciruk and T.R. Nelson, 1989. Estimates of carbon dioxide emissions from fossil fuel burning and cement manufacturing, CDIAC. Oak Ridge National Laboratory, U.S.A. May 1989 Report No. ORNL/CDIAC-25 and fact sheet 30 of the Information unit of climate change May 1993, UNEP, Switzerland.

Mwandosya and Luhanga, "Energy Planning, Energy Technologies and Development: In: Bioenergy 84, Vol V, p.30 (Eds) Eqnuas, Ellegard, Keefe and Kristofferson. Elsevier Applied Science Publishers, 1984.

O' Neil, William: Change of Gear in the World's Shipping lanes: Environment Strategy Europe, 1993.

Our Common Future. The World Commission on Environment and Development. Oxford. Oxford University Press, 1981.

Ottow, Johannes, C.G. Pesticides-contamination, self purification and fertility of soils. Journal of Plant Rescarch & Development. 21.p 161, 1984.

Pimental, D. (1978). 'Socioeconomic and legal aspects of pest control. E.H. Smith & D. Pimental (eds.) Pest Control Strategies, Academic Press.

Quinet, E., The Social Cost of Land Transport, Paris: OECD, 1990.

Robers, L.E.J. P.S. Liss and P.A.H. Saunders. Power Generation and the Environment. Oxford University Press, 1990.

Schipper Lee and Stephen Meyers. Energy Efficiency and human activity past trends, future prospects. Cambridge University Press, P III, 1990.

Swaminathan M.S. 'Agriculture and Food System'. Proceedings of the Second World Climate Conference, J. Jager & H.L. Ferguson (Eds). Cambridge University Press, 1991.

US Environmental Protection Agency. The Potential Effects of Global Climate Change on United States, Report to Congress, 1989.

UN, World Population Prospectus 1988. New York, United nations, 1989.

UNCTAD. "Combating Global Warming: Study on a global system of tradeable carbon emission entitlements", UNCTAD/RDP/DFP; p.4, UN, New York, 1992.

UNPFA. The State of World Population 1992, New York, 1992.

WHO Resources Institute. World Resources 1990-91. Oxford University Press, p 107, 1990.

World Resources Institute. World Resources 1994. Oxford University Press, p 36, 1994.

World Watch. Vol 9, Number 1, January-February 1996.

Chapter 6

IMPACT ASSESSMENT OF CLIMATE CHANGE

M.L. Parry and P. Martens

6.1 Introduction

Climate changes are associated with a multitude of effects: climate change will shift the composition and geographic distribution of many ecosystems (e.g. forests, deserts, coastal systems) as individual species respond to changed climatic conditions, with likely reductions in biological diversity. Climate changes will lead to an intensification of the global hydrological cycle and may have impacts on regional water resources. Additionally, climate change and the resulting sea-level rise can have a number of negative effects on energy, industry, and transportation infrastructure, agricultural yields, human settlements, human health, and tourism. Although there is a wide range of systems and sectors upon which changes of climate may have an effect (e.g. from mountain glaciers to manufacturing) the approaches applied in climate impact assessment are broadly similar. The purpose of this chapter is to give an overview of the current state of knowledge of potential impacts of climate change and to outline the array of methods available for assessing these impacts.

6.2 Methodology of impact assessment

This section, which deals with general approaches and methods to climate impact assessment is based upon the Technical Guidelines on Climate Impacts and Adaptations Assessment of the Intergovernmental Panel on Climate Change (Carter *et al.*, 1994; Parry and Carter, 1998). More specific methods, especially the types of models, will vary according to their suitability to the particular task.

6.2.1 Approaches to the assessment of impacts

Climate impact assessments are generally conducted according to one of three general methodological approaches: impact, interaction, and integration (Kates, 1985).

a) Impact approach

The simplest approach follows a straightforward 'cause and effect' pathway whereby a climatic event acting on an exposure unit has an impact (Figure 6.1). This impact approach is usually adopted for studies of individual activities or organisms in order to establish 'dose- response' functions, for example in the direct biophysical effects of climate on organisms or activities (e.g., on plants, animals, heating demand, water supply, etc). It can be thought of as an 'If-Then-What' approach: if the climate were to alter like this then what would be its impacts? In adopting the approach it is assumed that the effect of other non-climatic factors on the exposure unit can be held constant. Where this assumption is justified, the approach can be informative. However, the narrow focus on the effects of climate alone is also a major weakness of this approach.

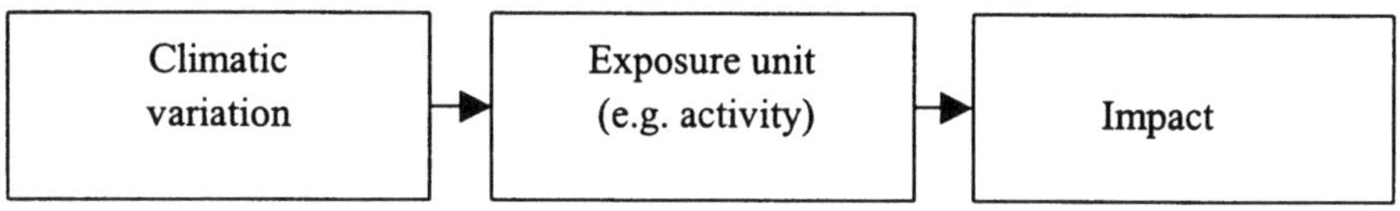

Figure 6.1 Impact approach in climate impact assessment (after Kates, 1985)

b) Interaction approach

The interaction approach is appropriate where climate is only one of a set of factors that is influenced by the exposure unit (Figure 6.2). For instance, a change in climate may lead to a shift in natural vegetation zones. However, this shift in zones may itself influence the climate through changes in fluxes of gases to and from the atmosphere, and through changes in surface reflectivity. The interaction approach may also allow for feedbacks that may enhance or damp an effect.

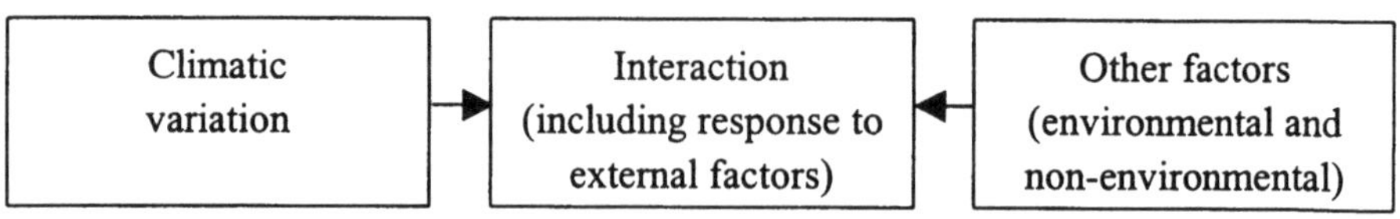

Figure 6.2 Interaction approach in climate impact assessment (after Parry and Carter, 1988)

A study method that fits closely into the structure of the interaction approach is one which works from the bottom upwards and has been termed the adjoint method (Parry and Carter, 1988; Parry, 1990). In simple terms this can be thought of as a 'What-Then-If' approach: What points of a system are sensitive to what types of climatic change and then what might the impacts be if those changes in climate were to occur? It differs from the impact approach, in that the significance of the climatic event is evaluated according to the climate-sensitivity of the exposure unit.

c) Integrated approach

An integrated approach is the most comprehensive treatment of the interactions of climate and society. It seeks to encompass the hierarchies of interactions that occur within sectors, interactions between sectors, and feedbacks, including adaptation, which serves to modify impacts and scenarios alike. A scheme of this approach is given in Figure 6.3. Much more will be said about integrated assessments, which are a relatively recent development, later in this book (Chapter 7).

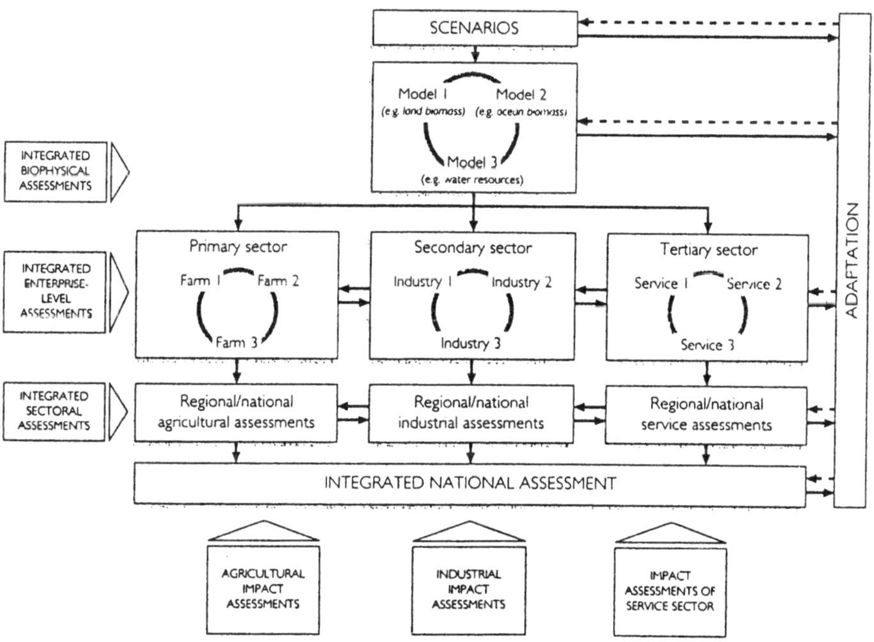

Figure 6 .3 Integrated climate impact assessment (after Parry and Carter, 1988)

6.2.2 The selection of methods for impact assessment

The selection of an appropriate method is an important step in a climate impact analysis. A variety of analytical methods can be adopted in climate impact assessment. These range from qualitative descriptive studies, through more diagnostic and semi-quantitative assessments to quantitative and prognostic analysis. Any single impact assessment may contain elements of one or more of these types, but whatever methods are selected, these should be clearly set out and explained. Four general methods can be identified: model projections, empirical studies, expert judgement and experimentation.

a) Model projections

A main focus of recent work has been on simulation modelling, using an array of mathematical models to scan the future. In order to distinguish them from 'climate models', which are used to project future climate, the term 'impact model' has now received wide currency. Below, the major classes of impact models are described.

Biophysical impact models

Biophysical impact models are used to evaluate the physical interactions between climate and an exposure unit. There are two main types: (i) Process-based models and (ii) Empirical-statistical models. The use of these in evaluating future impacts is probably best documented for the agricultural sector (e.g. see WMO, 1985), on the hydrological aspects of water resources (e.g. WMO, 1988), ecosystems (e.g. Bonan, 1993), and health (Martens, 1998), but the principles can readily be extended to other sectors.

Process-based models make use of established physical laws and theories to express the interactions between climate and the exposure unit. In this sense, they attempt to represent processes that can be applied universally to similar systems in different circumstances. For example, there are well-established methods of modelling leaf photosynthesis which are applicable to a range of plants and environments. Usually some kind of model calibration is required to account for features of the local environment that are not modelled explicitly, and this is generally based on empirical data. Nevertheless, there are often firmer grounds for conducting explorative studies with these process-based models than with empirical-statistical models. The major problem with most process-based models is that they generally have demanding requirement for input data, both for model testing and for simulating future impacts. Furthermore, theoretically based models are seldom able to estimate system responses successfully without considerable effort to calibrate them for actual conditions. For example, crop yield may be overestimated by process-based yield models because the models fail to account for all of the limitations on crops in the field at farm level.

Empirical-statistical models are based on statistical relationships between climate and the exposure unit. They range from simple indices of suitability or potential (e.g. identifying the temperature thresholds defining the ice-free period on important shipping routes), through univariate regression models (e.g. using air temperature to estimate energy demand) to complex multivariate models, which attempt to provide a statistical explanation of observed phenomena by accounting for the most important factors (e.g. estimating crop yields on the basis of temperature, rainfall, sowing date and fertiliser application). Empirical-statistical models are usually developed on the basis of present-day climate variations. Thus, one of their major weaknesses in considering future climate change is their limited ability to estimate effects of climatic events that lay outside the range of present-day variability. Furthermore, a statistical relationship between factors does not imply causation and lacks an understanding of the important causal mechanisms. However, where models are founded on a good knowledge of the determining processes and where there are good grounds for extrapolation, they can be useful tools in climate impact assessment. Empirical-statistical models are often simple to apply, and less demanding of input data than process-based models.

Economic models

Economic models of many kinds can be employed to evaluate the implications of climate change for local and regional economies. To simplify their classification, it is useful to distinguish between three types of economic models according to the approach used to construct them, and three scale of economic activity that different model types can represent.

Thrcc broad classes of economic models can be identified: (i) Programming models, (ii) Econometric models, and (iii) Input-output (IO) models.

Programming models have an objective function and constraints. The objective function represents the behaviour of the producer (e.g. profit maximising or cost minimising). These models can be static or dynamic. For example, programming models have been designed to estimate changes in land use under altered conditions of supply or demand. The land use not only depends on the yield, but also on the price. The model therefore needs to account for changes in food prices due to climate induced changes in food production as well as yield changes due to climate change.

Econometric models consist of supply and/or demand functions which use as independent variables prices and a number of 'technical' variables. Conventionally, econometric models do not state any decision rules.

However, in the last decade a new set of econometrically specified models has emerged: the so-called dual models. These assume decision rules such as profit maximazing or cost minimizing of producers and utility maximizing or expenditure minimizing of the consumer. The bulk of the econometric models are static, although among the few examples of dynamic models are the so-called adaptive models.

Input-output (IO) models are developed to study the interdependence of production activities. The outputs of some activities become the inputs of others and vice versa. These input-output relationships are generally assumed to be constant, which is a weakness of the approach, since reorganisation of production or feedback effects (such as between demand and prices) may change the relationship between activities. More recently, dynamics versions of IO models have been developed, but these still lack many of the dynamic aspect of economic behaviour. Nonetheless, the approach is relatively simple to apply and the data inputs are not demanding. Moreover, these models are already in common usage as planning tools.

Three scales of economic activity are commonly represented by economic models: (i) Firm-level, (ii) Sector-level, and (iii) Economy wide.

Firm-level models, which are often programming models, depict a single firm or enterprise. Typical examples include farm level simulation models, which attempt to mirror the decision processes facing farmers who must choose between different methods of production and allocate resources of cash, machines, buildings and labour to maximise returns. These models are sometimes referred to as microsimulation models.

Sector-level models encompass an entire sector or industry. It is quit common for such models to consider a firm as representative of the average of the entire section under study. They can be programming models or the econometric type, that depict production.

Economy-wide models, sometimes referred to as macro-economic models, link changes in one sector to changes in the broader economy, dealing with all economic activities of a spatial entity such as a country, a region within a country or a group of countries.

Economic models are useful tools to estimate the likely effects of climate change on (measurable) economic quantities such as income, Gross Domestic Product (GDP), employment and savings. However, most of the economic models fail to account for non-market effects of climate change. For example, many inputs to production are directly affected by climate change (e.g. land, water, labour (human health)) but are not contained in most (macro)economic models. Economic models are also widely used to consider the relative cost-effectiveness of mitigation and adaptation options, along with associated economic, social and environmental impacts of these options.

Integrated assessment models

In general, integrated assessment models (see also Chapter 7) try to describe quantitatively as much as possible of the cause-effect relationships of climatic changes (vertical integration) and interactions between different issues (horizontal integration), including feedbacks and adaptations. In practice, since the knowledge base is insufficient to permit us to conduct a full-integrated assessment, only partial integration is possible. Two main types of integrated models can be identified: (i) macroeconomic-oriented models and (ii) biosphere-oriented models.

Macroeconomic-oriented models seek to estimate the likely monetary costs and benefits of GHG-induced climate change in order to evaluate the possible policy options for mitigation or adapting to climate change. This macro-economic modelling approach (see above) has been applied to certain aspects of the greenhouse problem for many years. It is, however, more suited to studies of global rather than regional impacts of climate change. In particular, the methods have been used to compute the development paths for emissions of carbon dioxide and other greenhouse gases in the atmosphere (the driving force for climate change). An example of these global models of emissions is DICE (Nordhaus, 1992) (see Chapter 7).

Biosphere-oriented models attempt to model the sequence of cause and effect processes originating from scenarios of future GHG emissions, through atmospheric GHG concentrations, radiative forcing, global temperature change, regional climate change, possible regional impacts of climate change and the feedbacks from impact to each of the other components. Regional impacts can be aggregated, where appropriate, to give global impacts which can then be used in evaluating the likely effectiveness of global or regional policies. The approach is derived from the applied natural sciences, especially ecology, agriculture, forestry and hydrology, where climate impact assessment has evolved from site or local models in combination with geographical information system (GIS) technology. Examples include two related models: ESCAPE (European focus) and MAGICC (global) (Rotmans *et al.*, 1994; Hulme *et al.*, 1995), and versions of a global model: IMAGE 1.0 (Rotmans, 1990) and IMAGE 2.0/2.1 (Alcamo, 1994; Alcamo *et al.*, 1998).

b) Empirical analogue studies

Observations of the interactions of climate and society in a region can be of value in anticipating future impacts. The most common method employed involves transfer of information from a different time or place to an area of interest to serve as an analogy. Four types of analogy can be identified: (i)

Historical event analogies, (ii) Historical trend analogies, (iii) Regional analogies of present climate, and (iv) Regional analogies of future climate.

Historical event analogies use information from the past as an analogue of possible future conditions. Data collection may be guided by anomalous climatic events in the past record (e.g., drought or hot spells such as the thirties drought in the U.S. Great Plains or the 1970s drought in Sahel) or by the impacts themselves (e.g., periods of severe soil erosion by wind). However, the success of this method depends on the analyst's ability to separate climatic and non-climatic explanations for given effects.

There are several examples of historical trends that may be unrelated to greenhouse gases but which offer an analogy of GHG-induced climate change. Long-term temperature increases due to urbanisation are one potential source for a warming analogue (as yet seldom considered by impact analysts). Another example is past land subsidence, the impacts of which have been used as an analogue of future sea level rise associated with global warming. Finally, the El Niño/Southern Oscillation (ENSO) phenomenon (one of the strongest natural climatic fluctuation, occurring at irregular time intervals of several years and influencing weather patterns of large areas of the world) could be used to explore the effects a change in the frequency of weather anomalies likely to occur as climate changes.

Regional analogies of present climate refer to regions having a similar present-day climate to the study region, where the impacts of climate on society are also judged likely to be similar. To justify these premises, the regions generally have to exhibit similarities in other environmental factors (e.g., soils and topography), in their level of development and in their respective economic systems.

Regional analogies of future climate work on the same principle as analogies for present-day climate, except that here the analyst attempts to identify regions having a climate today which is similar to that projected for the study region in the future. For example, a model of grass growth in Iceland has been tested for species currently found in northern Britain, which is an analogue region for Iceland under a climate some 4°C warmer than present (Bergthorsson *et al.*, 1988).

c) Expert judgement

A useful method of obtaining a rapid assessment of the state of knowledge concerning the effects of climate on given exposure units is to solicit the judgement and opinions of experts in the field (Stewart and Glantz, 1985). Of course, expert judgement plays an important role in each of the analytic methods described in this section. On it own, this method is widely adopted by government departments for producing position papers on issues requiring policy responses. In circumstances where there may be insufficient time to undertake a full research study, literature is reviewed, comparable studies identified, and experience and judgement are used in applying all available information to the current problem. Examples are the U.K. Department of Environment Climate Change Impacts Review Group (CCIRG, 1996) and the Dutch Programming Study on human health impacts (Martens, 1996).

The use of expert judgement can also be formulated into a quantitative assessment method, by classifying and then aggregating the responses of different experts to a range of questions requiring evaluation. This method was employed by the National Defence University's study of Climate Change to the Year 2000, which solicited probability judgements from experts about climatic change and its possible impacts (NDU, 1978,1980). The pitfalls of this type of analysis include problems of questionnaire design and delivery, selection of representative samples of experts, and the analysis of experts' responses.

d) Experimentation

In the context of climate impact assessment experimentation has only a limited application. It is not possible physically to simulate large-scale systems such as the global climate, nor is it feasible to conduct controlled experiments to observe interactions involving climate and human-related activities. Only where the scale of impact is manageable, the exposure unit measurable, and the environment controllable, can experiments be usefully conducted. Up to now most attention in this area has been observing the behaviour of plant species under controlled conditions of climate and atmospheric composition (e.g. Strain and Cure, 1985; van de Geijn *et al.*, 1993). In the field such experiments have mainly comprised gas enrichment studies, employing gas releases in the open air, or in open and closed chambers including greenhouses.

There are other sectors in which experimentation may yield useful information for assessing impacts of climate change. For instance, field research on mosquito habitats and the relationship between transmission dynamics and climate factors may provide clues as to the spread of vector-borne diseases like malaria under altered climatic conditions. The information obtained from the experiments, while useful in its own right, is also invaluable for calibrating models which are to be used in projecting impacts of climate change (see above).

6.3 Assessments of impacts in different systems and sectors

In the following section a summary is given of the nature of the assessed impacts for a range of exposure units - coastal zones, agriculture, water resources, natural ecosystems and human health. In each case it should be emphasised that the impacts are generally extremely complex, varying greatly from one geographic area to another, and that summarising such varied effects involves much generalisation. Reference to the Second Assessment Report (SAR) by the Intergovernmental Panel on Climate Change (IPCC) (1996) will, however, provide a pointer to regionally specific studies that may be consulted for further debate. At the moment of publication of this book, the Third Assessment Report is being prepared by the IPCC.

6.3.1 Sea-level rise, coastal zones and small islands

Coastal zones and small islands are characterised by highly diverse ecosystems, such as mangroves, coral reefs, and seagrasses, that are important as a source of food and a habitat for many species. They also support a variety of economic activities, including fisheries and aquaculture, tourism, recreation, and transportation.

For many low-lying coastal areas of the world, the effects of accelerated sea level rise (ASLR) associated with global climate change may result in catastrophic impacts in the absence of adaptive response strategies to cope with an increased vulnerability of populations and economies to storms, storm surges and erosion. Assuming the IPCC 1990 scenarios of a sea-level rise of 25 to 80 cm for the year 2100, with a best estimate of 50cm, recent studies summarised by the second IPCC assessment (IPCC, 1996) indicate that:

- While some marshes and mangroves may be under threat from sea-level rise over the next century, human development (e.g. a high rate of population growth and economic development) is the major threat at present.
- Projected increases in seawater temperature are a major threat to coral reefs, and thus to coral atolls and reef islands. An example of effects in the Marshall Islands is given in Box 6.1.
- Sea-level rise is likely to cause extensive submergence of most deltaic coastlines especially where sediment availability is reduced by upstream dams and other human activities.

- On average, while some 46 million people are estimated to experience flooding due to storm surge every year under present conditions, a 50 cm and 1 metre rise in sea level would increase this number to almost 92 million and more than 118 million people, respectively.
- Preliminary costing of protective measures includes the following estimate for OECD (Organisation for Economic Cooperation and Development) countries facing a sea-level rise of 50 cm by 2100: US$520 million per year, based on the assumption that sea walls and protective infrastructure costs are discounted by 2% per year. The overall conclusion is that it is cost-effective to protect most coastlines in developed countries, but this could be at the expense of valued natural habitats such as wetlands and sandy beaches.
- In less developed regions, particularly the threatened low-lying island states (e.g., Marshall Islands, Maldives) and delta nations (e.g., Bangladesh, Nigeria, Egypt, and China) the options are limited and likely to be very costly due to increasing human settlement and agriculture in coastal areas.

Box 6.1

Effects of climate change on coastal environments of the Marshall Islands

The Republic of the Marshall Islands consists of 34 atolls and islands in the Pacific Ocean with a majority elevations below 2-3 metres above mean sea level. A vulnerability analysis case study for Majuaro Atoll was conducted to provide a first order assessment of the potential consequences of sea-level rise during the next century. Assuming a 1 m sea-level rise effects include (1) an approximate 10-30 percent shoreline retreat with a dry land loss of 160 acres out of 500 acres of the most densely populated part of the atoll; (2) a significant increase in severe flooding by wave runup and overtopping of half of the atoll from even normal yearly runup events; (3) flood frequency increases dramatically; (4) a reduction in the freshwater lens area which is important during drought periods; (5) a potential cost of protecting a relatively small portion of the Marshall Islands or more than four times the current GDP (Gross Domestic Product); (6) a loss of arable land resulting in increased reliance on imported foods.

Source: Hotthus et al., 1992, summarised by Carter et al., 1994

The IPCC has identified a range of options to respond to possible sea-level rise. These options fall broadly into three categories: retreat, accommodate and protect. The three strategies each may incorporate a number of specific activities based either on management or on engineering. These are summarised in Figure 6.4. The appropriateness of the different options will vary among and within countries and the IPCC recognise that different socio-economic sectors may prefer competing adaptation options for the same areas.

Present Situation

(Planned) Retreat

Emphasis on abandonment of land and structures in highly vulnerable areas and resettlement of inhabitants

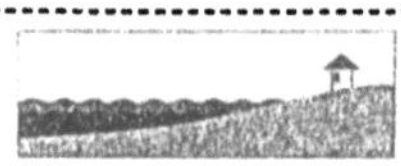

- Preventing development in areas near the coast
- Conditional phased-out development
- Withdrawal of government subsidies

Accommodate

Emphasis on conservation of ecosystems harmonized with the continued occupancy and use of vulnerable areas and adaptive management responses

- Advanced planning to avoid worst impacts
- Modification of land use, building codes
- Protection of threatened ecosystems
- Strict regulation of hazard zones
- Hazard insurance

Protect

Emphasis on defense of vulnerable areas, population centers, economic activities, and natural resources

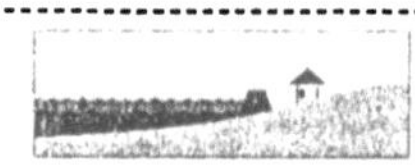

- Hard structural options
  - Dikes, levees, and floodwalls
  - Sea walls, revetments, and bulkheads
  - Groins
  - Detached breakwaters
  - Floodgates and tidal barriers
  - Saltwater intrusion barriers
- Soft structural options
  - Periodic beach nourishment (beach fill)
  - Dune restoration
  - Wetland creation
  - Littoral drift replenishment
  - Afforestation

Figure 6.4 Response strategies to sea level rise (IPCC, 1990)

6.3.2 Impacts on food and fibre production

Three broad classes of effects due to increased greenhouse gases may influence the quality of agricultural and forestry yields:

- Direct effects on plant growth through changes in photosynthetic rate and in water use efficiency resulting from elevated concentrations of CO_2;
- Direct effects from changes in temperature, water availability, radiation and extreme weather events; and
- Indirect effects through changes in distribution, frequency and severity of pest and disease outbreaks, incidences of fire, weed infestations, or through changes in soil properties.

Fisheries respond to direct climate change effects such as increases in water temperature and sea level and changes in precipitation, freshwater flows, climate variability and currents. They are also indirectly effected through changes in food supply and the expansion in ranges of red tides and other biotoxins, which could lead to increased contamination of fisheries.

Agriculture

Many studies suggest that high atmospheric CO_2 concentrations enhance plant growth. CO_2 increases photosynthesis, dry matter production and yield, it decreases stomatal conductance and transpiration and improves water efficiency. Different plant types respond differently to CO_2. C_3 plants make up the majority of the species globally, especially in cooler or wetter habitats. They included all important trees and most crop species, such as wheat, rice, barley, cassava, and potato. It is called C_3 because the first component into which CO_2 is incorporated in photosynthesis is a compound with three carbon atoms. C_4 plants tend to grow in warmer, more water-limited regions, and include many tropical grasses and the agriculturally important species maize, sugarcane, and sorghum. Here, a four carbon compound incorporates CO_2.

C_3 species are more sensitive to atmospheric CO_2 concentrations than C_4 plants Increased concentrations of CO_2 are estimated to lead to increased productivity in C_3 crops (e.g. rice and wheat). The mean value response is a thirty- percent productivity increase under doubled CO_2 conditions (~ 550 ppm), although the range is from -10 to +80%. However, these estimates represent responses under optimal conditions of water and temperature and are likely to vary considerably under field conditions according to the availability of plant nutrients, insects, weeds and diseases, in addition to weather conditions.

Effects of altered weather on crop plants are likely to take two forms: firstly through the altered range of suitability of crops to areas which are characterised by given climate, soils and terrain; secondly through changes in the productivity of given crops at particular places. The former type of effect (a spatial shift in suitability) is expected to vary considerably between different regions of the world. In summary, the evidence is that projected changes in temperature and rainfall may contribute to a poleward extension of suitability of crop plants that are currently limited by resources in high-middle and high latitude (e.g. on the Canadian prairies, in northern Europe, in northern Russia, northern Japan, southern New Zealand, southern Argentina and southern Chile etc). At high mid latitudes (circa 50°) a 1°C increase across the growing season is estimated to lead to a poleward spatial shift of about 200 km. The geographic shifts of agricultural zones that may result from changes in thermal resources may be substantial. To illustrate, in western Europe it has been estimated that increasing thermal resources would enable the commercial production of major crops such as grain maize, sunflower and soybean in new regions northward of their current climate limit (see Figure 6.5 and Table 6.1).

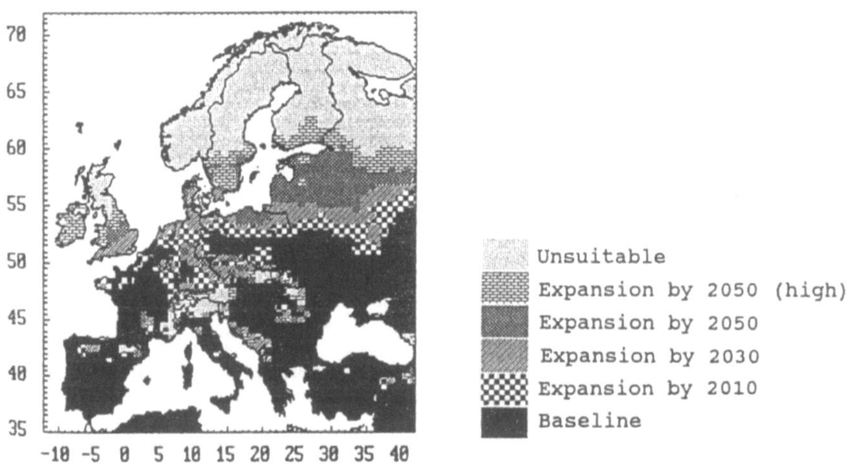

Figure 6.5 Changes in the thermal limit of grain maize relative to the baseline (1951-1980) climate for the year 2010, 2030, and 2050 (including a high estimate) (Kenny *et al.*, 1993)

Table 6.1 Likelihood of regional temperatures at average altitude being suitable for crop production in the 2010s and the 2030s based on GISS (Goddard Institute for Space Studies) scenarios A and C (Carter *et al.*, 1991)

		Likelihood of suitability	
	High	**Moderate**	**Low**
2010s			
Grain maize	S Netherlands	S England	E Denmark
	NE Germany	SE Belgium	N Poland
	C Czechoslovakia	Luxembourg	Lithuania
	N Austria	NW Germany	
	NE Poland	N Netherlands	
Early sunflower/soy	C/E Germany	N France	S Netherlands
	S Poland	S Germany	Belgium
	W Ukraine		C/W Germany
	W Romania		SW Yugoslavia
Later sunflower/soy	S France	SW France	N Hungary
	C Yugoslavia		
	S Ukraine		
2030s			
Grain maize	S England	S Ireland	N Ireland
	Benelux	C England	N England
	E Denmark	W Denmark	S/C Sweden
	N Germany	S Sweden	C Austria
	N Poland	Latvia	S Finland
	Lithuania		Estonia
	Moscow region		
Early sunflower/soy	N France	S Netherlands	S England
	Belgium	C Germany	E Denmark
	W/E Germany	SE England	N Germany
	NE Switzerland		W Czechoslovakia
	NE Austria		N Poland
	S Poland		Lithuania
	W Romania		Moscow region
	W Ukraine		S Sweden
Later sunflower/soy	NW Spain	S/C France	N Portugal
	SW France	N Hungary	N/C France
	C Yugoslavia		E/C Germany
	C Ukraine		E Austria
	W/E Hungary		S Poland
			C Romania
			W Bulgaria
			N Ukraine

Box 6.2
Impacts on agriculture in the Midwest USA

A temporal analogue was employed as the climate scenario, specifically the decade 1931-1940 in the MINK region (Missouri, Iowa, Nebraska and Kansas). Overall, this period was one of severe drought - both drier and warmer than average in the region, consistent in sign with GCM (General Circulation Models; see Chapter 3) projections. These conditions were assumed to occur in the present and also in the year 2030, along with an increase in CO_2 concentration of 100 ppm (to 450 ppm). Four sets of conditions were assumed to occur in the present and also in the year 2030, along with an increase in CO_2 concentrations of 100ppm (to 450 ppm). Four sets of conditions were investigated: (1) the current baseline, which referred to the economic situation in the early 1980s, with 1951-1980 as the climatological baseline (2) climate change imposed on the current baseline (3) a baseline description of the economic structure of the region in the future without climate change (including population, economic activity and personal income) and (4) imposition of climate change on the future baseline (including feedbacks between sectors, such as projected extent of irrigated agriculture given scenarios of future water supply).

In the MINK region of 2030 with a climate like that of the 1930s the main results of the study are: Crop production would decrease in all crops except wheat and alfalfa, even accounting for CO_2 effects. However, impacts on agriculture overall would be small given adaptation, though at the margins losses could be considerable. Unless the climate-induced decline in feedgrain production falls entirely on animal producers in MINK (which would lead to an overall loss to the regional economy of 10 percent), the regional economic impacts of the climate change would be small. This is because agriculture, while important relative to other regions of the US, is still only a small (and diminishing) part of the MINK economy. In economic terms, in the absence of on-farm adjustments and CO_2 enrichment, the analogue climate would reduce the value of 1984-87 crop production in MINK by 17 percent. The CO_2 effect would reduce the loss to 8 percent, and on-farm adjustments would reduce it further to 3 percent.

Source: Rosenberg, 1992.

At middle and lower latitudes, agricultural productivity is likely to be more sensitive to changes in water availability. Projections of changes in available moisture, based upon experiments with general circulation models, are at present very uncertain. As a consequence, it is not yet possible to identify those middle and lower latitude regions which are likely to see significant increases or decreases in productive potential in agriculture. In the absence of model-based projections some studies have adopted the temporal analogue approach using weather data for anomalous warm and dry periods, such as during the 1930s in the U.S. Great Plains. A recent study has estimated that a recurrence of these weather conditions under technological conditions that might characterise mid-western agriculture in the future (2030) might lead to a 10% overall loss of regional economic output. Details of this study are given in Box 6.2.

At the global level the indications are the overall food production might be maintained relative to base-line production in the face of climate change but that access to food particularly in developing countries might be effected. Recent studies have indicated that under doubled-equivalent CO_2 equilibrium conditions (which are currently projected to occur in about the middle of the next century), and assuming a modest degree of farm-level adaptation (e.g. minor shifts in planting dates and changes in crop variety), the net effect of climate change is to reduce global cereal production by up to five percent (Rosenzweig and Parry, 1994). This global reduction could largely be overcome by more major forms of adaptation such as the widespread installation of irrigation.

More importantly, however, climate change will tend to increase the disparity in cereal production between developed and developing countries. While production in the developed world may possibly benefit from climate change, productivity may decrease in low latitude, particularly semi-arid, areas which are largely characterised by developing nations. Adaptations at farm level seem to reduce this disparity only a little. As a consequence, cereal prices and thus the population at risk of hunger in developing countries could increase. Even a high level of farmer adaptation in the agricultural sector would not entirely prevent such adverse effects (Figure 6.6).

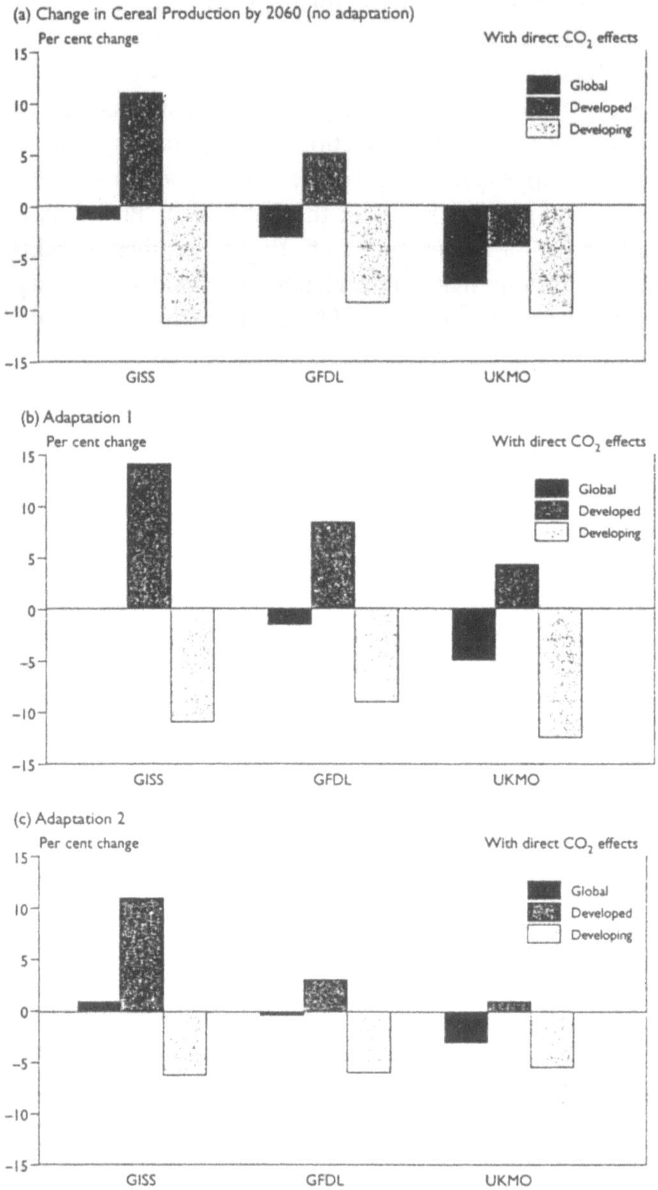

Figure 6.6 Change in cereal production by 2060 under altered climate with no adaptation to climate change (a), with minor farm-level adjustments, such as changes in planting date, in amounts of irrigation and in choice in crop varieties that are currently available (b), and with major adjustments, such as expansion of irrigation systems and the development of new cultivars. (Rosenzweig and Parry, 1994).

Fisheries

Marine fishing generates about 1% of the global economy, but coastal and island regions are far more dependent on fishing. About 200 million people world-wide depend on fishing and related industries for livelihood. Any effects of climate change on fisheries will occur in a sector that is already characterised on a global scale by massive over-capacities of usage. Climate change impacts are likely to exacerbate existing stress on fish stocks, notably over-fishing, diminishing wetlands and nursery areas, pollution and UVB radiation.

Should climate develop according to IPCC scenarios, marine fisheries production is projected to remain the same at the global level, while higher latitude freshwater and aquaculture production are estimated to increase (IPCC, 1996). Probably more significant would be effects on the geographical distribution of major fish stocks: longer growing seasons and faster growth rates may increase stocks in high latitudes, although these may be offset by negative factors such as changes in established reproductive patterns, migration routes and ecosystem relationships. The most sensitive aspects of fisheries include freshwater fisheries in small rivers and lakes, particularly at high latitudes, and fisheries in estuaries which may be effected both by altered river flow and by sea level rise.

Forestry

The current rate of increase in demand for forest products (about two percent per year) is estimated to have a much larger effect on forest resources than future climate change (IPCC, 1996). For example, projections indicate that the growing stock of tropical forests will have declined by about half due to non-climatic reasons related to human activities by the middle of the next century. Slight increases in temperature and enhanced atmospheric CO_2 concentrations might increase forest productivity and also extend the area where tropical forests can potentially grow, but this is unlikely to affect the rate of projected decline to any significant degree. The same estimations hold for temperate-zone forests, the conclusion being that non-climate related driving forces may reduce forest cover to half as much land in 2050 as today, but that climate change may increase harvests slightly. Boreal forests are likely to undergo irregular and large-scale losses of living trees due to climate change. These could initially generate additional wood supply from salvage harvests, but could severely reduce standing stocks and wood-product availability over the long term (IPCC, 1996).

6.3.3 Impacts on water supply and use

A variety of non-climate and climate-related factors affect water use today and are likely to alter water use in the future. For example, projected changes in demand due to factors such as population growth and increased need for crop irrigation are major trends that need to be factored into estimates of possible climate change-induced effects on water resources.

Table 6.2 Per capita water availability (m^3 / per cap) in 2050 for the present and for three transient climate conditions

Climate Country	**GFDL 1990**	**UKMO Change-2050**	**MPI 2050**	**No 2050**	**2050**
Cyprus	1,280	770	470	180	1,100
El Salvador	3,670	1,570	210	1,710	1,250
Haiti	1,700	650	840	280	820
Japan	4,430	4,260	4,800	4,480	-
Kenya	640	170	210	250	210
M'gascar	3,330	710	610	480	730
Mexico	4,270	2,100	1,740	1,980	2,010
Peru	1,860	880	830	690	1,020
Poland	1,470	1,200	1,160	1,150	1,140
Saudi Arabia	310	80	60	30	140
South Africa	1,320	540	500	150	330
Spain	2,850	2,680	970	1,370	1,660

Notes

1) Assumptions about population growth are from the IPCC IS92a scenario based on the World Bank (1991) projections; the climate data are from the IPCC WGII TSU climate scenarios (based on transient model runs of Geophysics Fluid Dynamics Laboratory (GFDL), Max-Planck Institute (MPI) and UK Meteorological Office (UKMO)).

2) The results show that in all developing countries with a high rate of population growth, future "per capita" water availability will decrease independently of the assumed climate scenario.

Source: IPCC (1996).

Current understanding suggests that climate change could have major impacts on regional water supplies, but general circulation models at present only provide outputs on a large geographic scale and different models do not agree on the likely range of changes in average annual precipitation for any given basin or watershed. In addition, changes of climate may affect water availability in several different ways, for example, through changes in precipitation, changes in sea level and changes in vegetation (which may affect run-off). An example of this complexity is given in Box 6.3. This shows, for Egypt, the wide range of different estimated impacts for different scenarios of future climate change.

Box 6.3
An integrated assessment of impact of climate change on the water resources of Egypt

A digital elevation model of the Nile Delta was developed for determining land loss due to sea level rise. A physically based water balance of the Nile Basin was used to evaluate river runoff. This was linked to a simulation model of the high Aswan dam complex to determine impacts on Lake Nasser yields. Process-based agronomic models (incorporating direct effects of elevated CO_2) were used to estimate crop yields and crop water requirements.

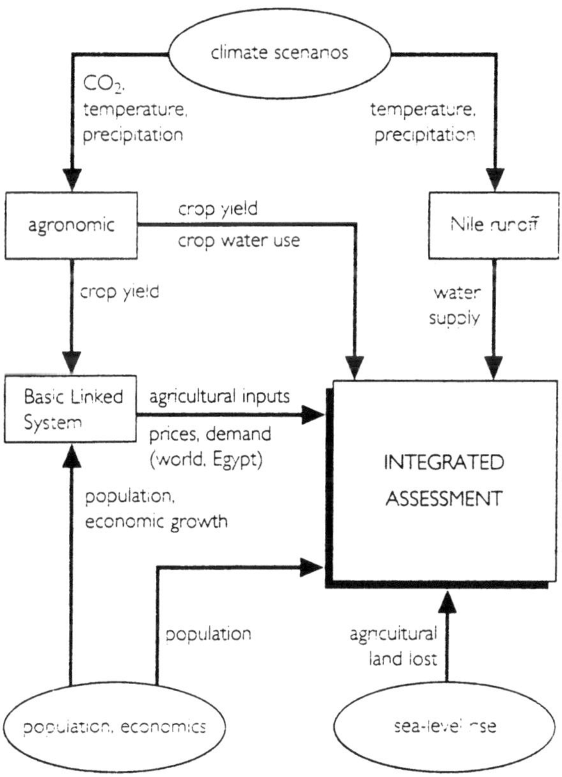

Figure 6.7 An integrated assessment of impact of climate change on the water resources of Egypt.

Box 6.3 Continued

Cropping patterns, under different climatic scenarios, were determined using the Egyptian food supply and trade model, one component of an international food trade model, the Basic Linked System (BLS), which was run at a global level (Figure 6.7). Impacts were estimated as the difference between simulations for 2060 without climate change, based on projections of population, economic growth, agricultural production, commodity demand, land and water resources and water use (Base 2060), and simulations with changed climate according to several climatic scenarios.

Table 6.3 provides a summary of the impacts of three scenario climates on each sector together with the integrated impacts on economic welfare (the consumer-producer-surplus). The agricultural water productivity index is an aggregate measure of impacts on agriculture: total agricultural production (tonnes) divided by total agricultural use (m^3). The results illustrate how impacts on individual sectors are affected by impacts on other sectors. For example, under the GISS scenario, despite an 18% increase in water resources, the 5% loss of land and 13% reduction in agricultural water productivity leads to a 6% reduction in economic welfare.

Table 6.3 A comparison of sectoral with integrated impacts for three climatic scenarios (percentage of change from 2060 Base results)

		Sectoral impacts			***Integrated impact***
Climatic scenario	***Land area***	***Food demand***	***Agricultural water productivity index***	***Water resources***	***Consumer -producer surplus***
UKMO[1]	-5	-3	-45	-13	-23
GISS[2]	-5	-1	-13	+18	-6
GFDL[3]	-5	-1	-36	-78	-52

Notes
1) United Kingdom Meteorological Office model.
2) Goddard Institute for Space Studies model.
3) Geophysical Fluid Dynamics Laboratory model.

The results also demonstrate how individual sectoral assessments may give a misleading view of the overall impact, which is better reflected in the integrated analysis. For instance, under the GISS scenario, while the impact on water resources is positive, the integrated impact is actually a 6% decline in economic welfare.

Source: Strzepek and Smith, 1995.

From an array of such studies (though most of them not as detailed as that for Egypt) the following broad conclusions have been drawn (IPCC, 1996). Reductions in water availability could result in chronic shortages in regions and nations that are already water-limited, and for which there is a considerable competition among users (see Table 6.2 for details). Water resources in arid and semi-arid zones are particularly sensitive to climate variations because the volume of total runoff and infiltration is low, and because relatively small changes in temperature and precipitation can have large effects on runoff.

High latitude regions may experience increased runoff due to an increased precipitation, but lower latitudes may experience decreased runoff due to the combined effects of increased evapotranspiration and decreased precipitation. Even in areas where precipitation is projected to increase, higher evaporation rates may lead to reduced runoff. Areas where annual runoff increases or where there are fewer but more intense precipitation events may experience more flooding episodes, and higher lake and river levels. Effects on run-off could be exacerbated in regions where there is a reduction in the amount of vegetation (e.g., through deforestation, overgrazing, logging).

In some continental areas, mountain regions are the most important sources of water. Changes in the spatial pattern of precipitation and temperature could lead to altered river flows regimes in mountain watersheds. For example, changes in the timing of snow-pack melting and/or a slight shift in precipitation patterns in the mountains will considerably change the cycles of hydrological discharge in many river basins. These changes would have consequences for water storage and delivery systems and hydro-electricity production.

6.3.4 Impacts on terrestrial and aquatic ecosystems

The Second Assessment Report of the IPCC (1996) concludes that increases in mean temperature of only 1°C are sufficient to cause changes in the growth capacity of many plant species, and hence in the composition of plant ecosystems. We do not know, however, whether these changes in ecosystems will be a linear or non-linear function of mean temperature change. Global models project that a substantial fraction (14-65% with a global mean of 34%) of the existing forested area of the world will undergo major changes in vegetation types (i.e. transform from one equilibrium vegetation class to another) in response to climate change.

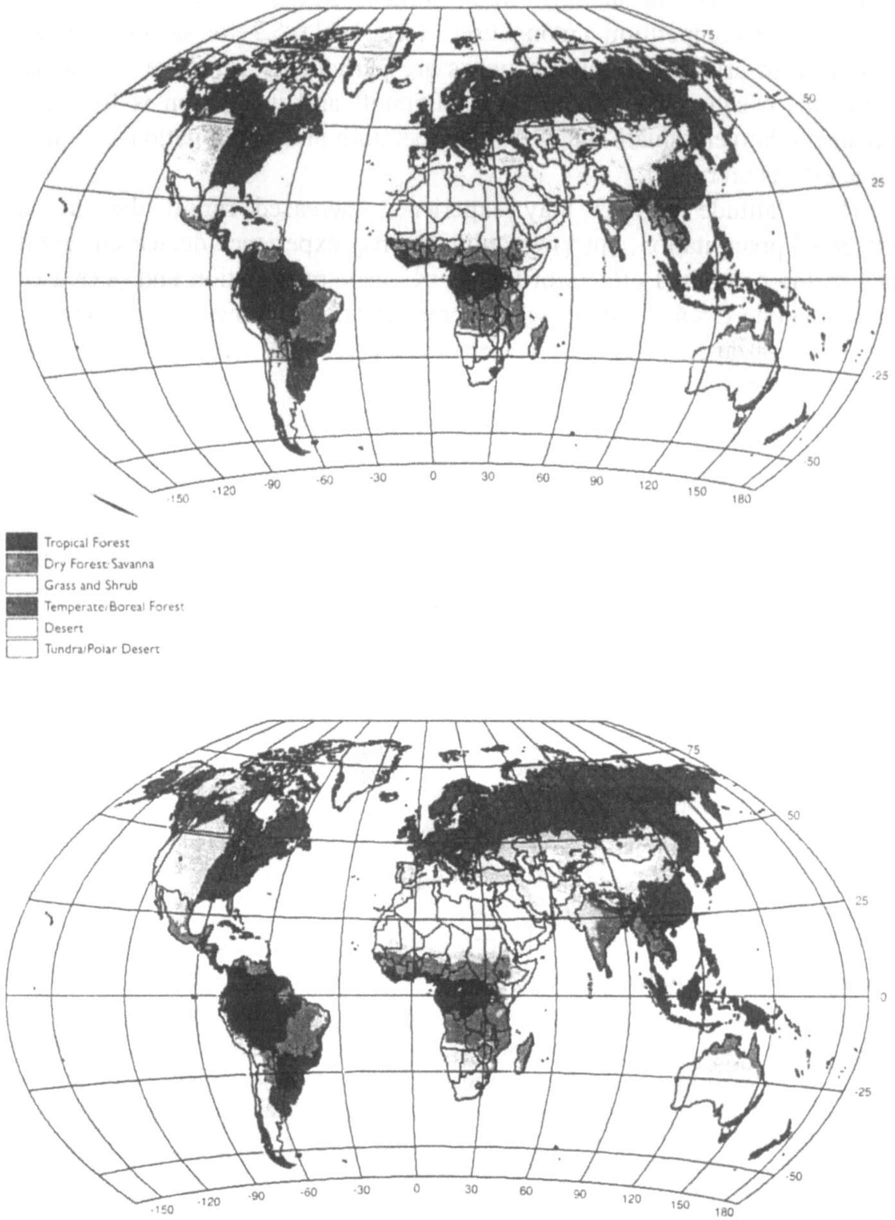

Figure 6.8 Potential distribution of natural vegetation according to the BIOME model (Prentice *et al.*, 1992) for present climate (top) and climate as projected by the GFDL GCM (bottom) (Source: IPCC, 1996).

A global average warming of 1-4°C over the next 100 years would be equivalent to a poleward shift of isotherms of approximately 160-640 km. This compares to vegetation migration rates that are generally believed to be on the order of 4-200 km per century, based on the paleo-environmental record. Migration is dependent on a combination of environmental factors, and on vectors responsible for seed dispersion and successful establishment. Species composition of vegetation communities is therefore likely to change, and entire vegetation types may even disappear, while new assemblages of species and hence ecosystems maybe established. At the global level this may lead to an altered geographical distribution of vegetation zones (or biomes). Figure 6.8 shows the distribution for one particular general circulation model experiment (GFDL), though several others and (some with quite different distributions) could be shown.

The distribution of rangelands (i.e. the unimproved grasslands, shrublands, savannahs, hot and cold deserts and tundra), which occupy 60% of the earth's land surface, may be particularly sensitive to change both as a result of direct climate impacts on species composition as well as indirect factors such as changes in wildfire frequency and land-use. These effects will be greatest in temperate rangelands and tundra). The degradation of fragile ecosystems in arid, semi-arid, and dry sub-humid zones ('desertification') is not primarily a function of observed fluctuation in rainfall or temperature, but rather of land-use mismanagement. However, because droughts may trigger or accelerate desertification and land degradation, and since the frequency and magnitude of such droughts may be altered by climate change, the distribution of arid and semi-arid biomes may also show substantial alteration as a result of climate change.

At the more regional and local level the changes detailed above are the result of altered species distribution, resulting from geographical shifts of climatic suitability and altered competition. Box 6.4 describes a typical local-scale bio-geographical study of effects of climate change on particular plant distributions, in this case in Norway.

Box 6.4
Effect of climate change on natural terrestrial ecosystems in Norway

The objectives of this study were to examine the probable patterns of ecological change in Norway under a changed climate, with a particular emphasis on identifying plant species and communities sensitive to or at risk of climate change.

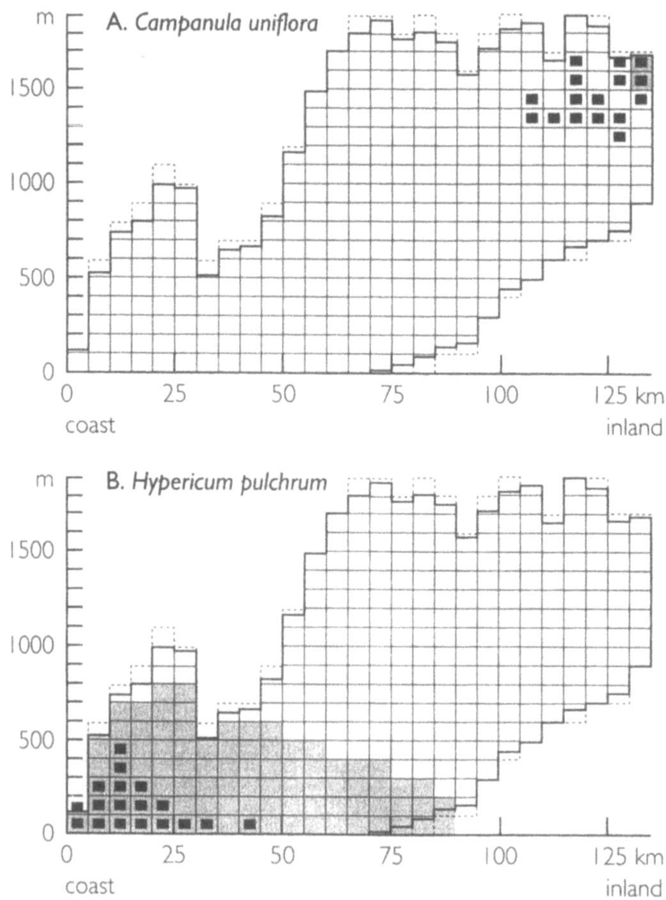

Figure 6.9 Altered distribution of two plant species in central Norway (see text)

Box 6.4 Continued

Correlative models, which are based on spatial coincidence of vegetation species and climate variables under present-day climate, were used to project distributions for an equivalent doubling of CO_2, based on a subjective composite of results from several GCMs from the Norwegian region.

The effect of climate change on species distribution were estimated using a narrow west-east transect through central Norway, giving altitude on the vertical axis and distance from the Atlantic on the horizontal axis. Dashes boxes indicate variations in altitude at site locations within the transect. Figures A and B illustrate the sensitivity of two species: Campanula uniflora (a rare alpine and continental species) and Hypericum pulchrum (a frost sensitive coastal species) to the climate changes described by the scenario. Solid squares indicate the current and shaded squares the projected distribution of the species.

The analysis suggests that rare northern or Alpine species may be threatened by extinction (Figure A), both due to shifts in climate and changes in snow cover and runoff. Temperate and oceanic zone species would be favoured under the change climatic regime (Figure B), but anthropogenic or natural barriers could delay their colonisation.

Source: Holten and Carey, 1992.

6.3.5 Human health

The various potential health effects of global climate change upon human health can be divided into direct and indirect effects, according to whether they occur predominantly via the impacts of climate variables upon human biology, or are mediated by climate-induced changes in other biological and biogeochemical systems. Figure 6.10 summarises some of the important potential effects of climate change upon human population health.

In healthy individuals, an efficient regulatory heat system enables the body to cope effectively with thermal stress. Temperatures exceeding comfortable limits, both in the cold and warm range, substantially increase the risk of (predominantly cardio-pulmonary) illness and deaths. Directly, an increase in mean summer and winter temperatures would mean a shift of these thermal related diseases and deaths. Studies in large urban populations in North America, North Africa and China indicate that the annual number of heat-related (summer) deaths would represent several thousand extra deaths annually. This may be partially offset by fewer cold-related deaths, although the balance would vary by location (and would also depend on adaptive responses). An increased frequency or severity of heat waves will also have a strong impact on these diseases. If extreme weather events (droughts, floods, storms, etc.) were to occur more frequently, increases in rates of deaths, injury, infectious disease and psychological disorder would result.

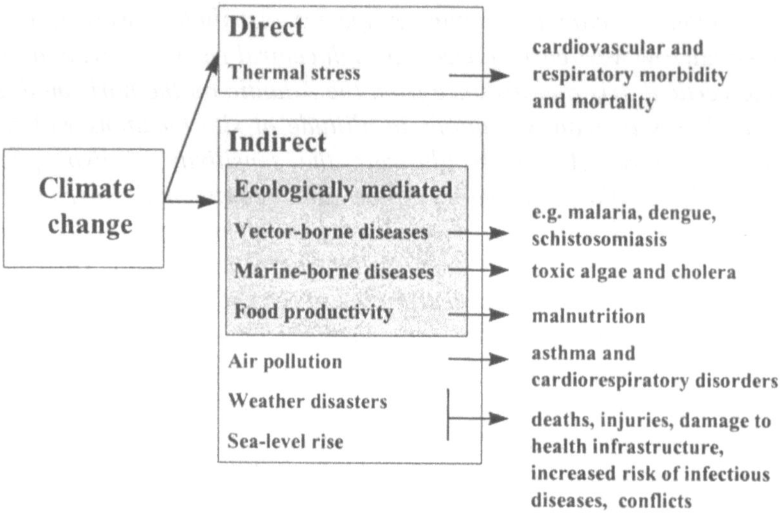

Figure 6.10 Ways in which climate change can effect human health (Martens, 1998)

One of the major indirect impacts of global climate change upon human health could occur via effects upon cereal crop production. Cereal grains account for around two-thirds of all foodstuffs consumed by humans. These impacts would occur via the effects of variations in temperature and moisture upon germination, growth, and photosynthesis, as well as via indirect effects upon plant diseases, predator-pest relationships, and supplies of irrigation water (see also Section 6.3.2.). Although matters are still uncertain, it is likely that tropical regions will be adversely affected (Rosenzweig and Parry, 1994), and, in such increasingly populous and often poor countries, any apparent decline in agricultural productivity during the next century could have significant public health consequences.

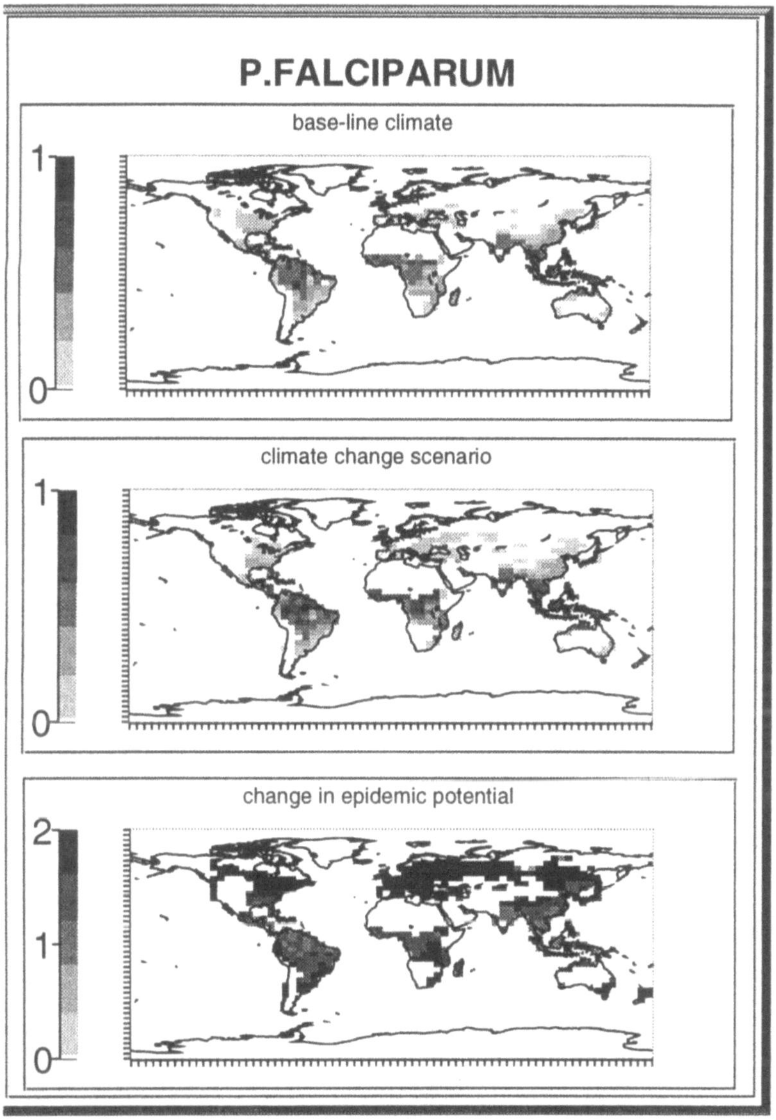

Figure 6.11 Potential malaria risk areas for baseline climate conditions (1931-1980) and for a global mean temperature increase of ~1.2°C (based on the climate patterns generated by the ECHAM1-A GCM) and changes in malaria risk (epidemic potential) for *P. falciparum*, (see also Box 6.6), relative to baseline climate (Martens, 1998; Martens *et al.*, 1995 a, b, 1997)

Box 6.5
Modelling the effects of climate change on vector-borne diseases

(Partial) integrated mathematical models, to estimate climate change impacts on malaria, dengue and schistosomiasis, were developed by Martens and colleagues. These models link GCM-based climate change scenarios with a module that models the relationship between climate variables and the Epidemic or Transmission Potential index. Although these models do not taken into account all the possible impacts of climate on vector-borne disease transmission, they do consider how temperature directly affects the transmission cycle of these diseases. Also the indirect effect of changes in moisture (and the consequent changes in vegetation patterns) was estimated.

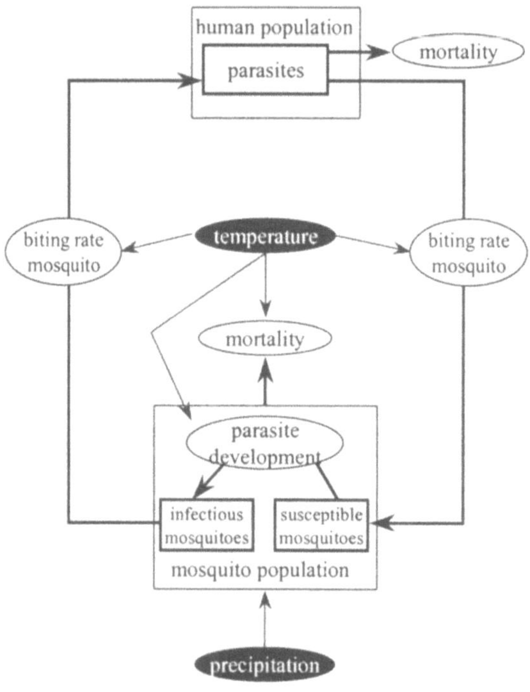

Figure 6.12 Diagram of temperature and precipitation effects on main population and rate processes involved in the life cycle of malaria and dengue, two vector-borne diseases transmitted by mosquitoes.

Box 6.5 Continued.

Simulations with the vector-borne diseases models show an increase of the populations at risk of malaria, dengue and schistosomiasis. There would be an increase of the risk of local transmission of the three vector-borne diseases in developed countries, associated with imported cases of the disease. However, given the fact that effective control measures are economically feasible in these countries, it is not to be expected that human-induced climate changes would lead to a return of a state of endemicity in these areas.

In the current highly endemic areas, the prevalence of infection is persistently high, and will probably only be marginally affected by these climate-induced changes. In view of their high potential receptivity and the immunological naivety of the population, the highest risks for the intensifying of transmission of malaria, dengue and schistosomiasis reside in the hitherto non- or low-endemic regions on the altitude and latitude fringes of disease transmission. An aspect of particular importance is the increase in epidemic potential at higher altitudes within endemic areas such as the Eastern Highlands of Africa, the Andes region in South America, and the Western mountainous region of China.

Source: Martens, 1998; Martens et al., 1995a,b, 1997

A further important indirect effect on human health may well prove to be a change in the transmission of vector-borne diseases. Temperature and precipitation changes might influence the behaviour and geographical distribution of vectors (Figure 6.11), and thus change the incidence of vector-borne diseases, which are major causes of morbidity and mortality in most tropical countries (see Box 6.5). Increases in non-vector borne infectious diseases, such as cholera, salmonellosis, and other food and water-related infectious diseases could also occur, particularly in (sub-) tropical regions, due to climatic impacts on water distribution, temperature and the proliferation of micro-organisms.

Many health impacts could also result from deterioration in physical, social and economic circumstances caused by rising sea levels, climate-related shortages of natural resources (e.g. fresh water), and impacts of climate change on population mobility and settlement. Conflicts may arise over decreasing environmental resources. The climate change process is associated with air pollution, since fossil fuel combustion produces various air pollutants. Furthermore, higher temperatures would enhance the production of various secondary air pollutants (e.g. ozone and particulates).

As a consequence, there would be an increase in the frequency of allergic and cardio-respiratory disorders and deaths caused by these air pollutants.

However, many uncertainties remain. Table 6.4 gives a survey of what is known and the uncertainties in relation to the health impacts discussed above.

Table 6.4 Summary of known effects and uncertainties regarding health impacts of climate change

Health effect	**Known effects**	**Uncertainties**
Thermal stress	* Mortality (especially cardiopulmonary) increases with cold and warm temperatures * Older age groups and people with underlying organic diseases are particularly vulnerable * Mortality increases sharply during heat waves	* The balance between cold and heat related mortality changes * The extent with which heat waves take their toll among terminal patients * The role of acclimatisation of people to warmer climates
Vector-borne diseases	* Climate conditions (particularly temperature) necessary for some vectors to thrive and for the micro-organisms to multiply within the vectors are relatively well known	* Indirect effects of climate change on vector-borne diseases, such as changes in vegetation, agriculture, sea-level rise, migration, etc. * Effects of socio-economic development, resistance development, etc.
Water/food-borne diseases	* Survival of disease organisms (and insects which may spread them) is related to temperature * Water-borne diseases most likely to occur in communities with poor water supply and sanitation * Climate conditions affect water availability * Increased rainfall affects transport of disease organisms	* For many organisms the exact ambient conditions in which they survive and are transmitted are not known * Interaction with malnutrition is not well understood
Food production	* Temperature, precipitation, solar radiation and CO_2 are important for crop production * Crop failure may lead to malnutrition * Under-nourishment may increase susceptibility for infectious diseases	* Variations in crop yield due to climate change are poorly understood * Effect of climate on weeds, insects and plant diseases are not well known * Interaction between nutritional status and diseases is poorly understood

6.4 Adapting to climate change

Many of the impacts outlined above could be reduced in the event of extensive adaptation. However, most impact assessments are conducted to evaluate the effects on an exposure unit in the absence of any responses which might modify these effects.

Six types of adaptation for coping with the negative effects of climate have been identified (Burton *et al.*, 1993):

Prevention of loss, involving anticipatory actions to reduce the susceptibility of an exposure unit to the impacts of climate (e.g., controlled coastal zone retreat to protect wetland ecosystems from sea level rise and its related impacts).

Tolerating loss, where adverse impacts are accepted in the short term because they can be absorbed by the exposure unit without long term damage (e.g., a crop mix designed to minimise the maximum loss, to ensure a guaranteed minimum return under the most adverse conditions).

Spreading or sharing loss, where actions distribute the burden of impact over a larger region or population beyond those directly affected by the climatic event (e.g., government disaster relief).

Changing use or activity, involving a switch of activity or resource use from one that is no longer viable following a climatic perturbation to another that is, so as to preserve a community in a region (e.g., by employment in public relief works).

Changing location, where preservation of an activity is considered more important than its location, and migration occurs to areas that are more suitable under the changed climate (e.g., the re-siting of a hydro-electric power utility due to a change in water availability).

Restoration, which aims to restore a system to its original condition following damage or modification due to climate (for example, an historical monument susceptible to flood damage). This is not strictly an adaptation to climate, as the system remains susceptible to subsequent comparable climatic events.

Different criteria can be used for organising the information about possible alternative types of adaptation. For instance, detailed tables have been used to catalogue traditional adjustment mechanisms for coping with inter-annual climatic variability in self-provisioning societies (Jodha and Mascarenhas, 1985; Akong'a *et al.*, 1988). Other methods of cross-tabulation have been employed in formal procedures of resource management. For example, alternative water resource adaptation measures in the United States are commonly analysed according to both the type of measure and its scope.

Four groupings of strategy have been identified (Stakhiv, 1994):

- Long range strategies, generally pertinent to issues involving mean changes in climate (e.g., river basin planning, institutional changes for water allocation).
- Tactical strategies, concerned with mid-term considerations of climatic variability (e.g., flood proofing, water conservation measures).
- Contingency strategies, relating to short-term extremes associated with climatic variability (e.g., emergency drought management, flood forecasting).
- Analytical strategies, embracing climatic effects at all scales (e.g., data acquisition, water management modelling).

It is worth nothing that society in general, and each resource use sector in particular, already contends with contemporary climatic variability and the wide range of natural hazards (e.g., floods, droughts, storm surges and hurricanes) and the variety of opportunities (e.g., a benign period of weather; an unbroken 'snow season'; a mild winter) this brings. As a first approximation, it is probably fair to assume that most of the current measures employed in resource management to deal with climatic variability will be equally feasible, even if not comparatively cost-effective, under a different climate regime. It follows, therefore, that the adaptation measures which ought to be selected now are those which are beneficial for reasons other than climate change and, for most part, can be justified by current evaluation criteria and decision rules. This is sometimes referred to as the 'no regrets' strategy.

6.5 Discussion

It is clear that the sensitivity to climate change varies greatly between different systems and sectors. As a result, there is a very wide array of adaptation strategies that could serve both to reduce the negative effect of climate change and to enhance the potentially positive ones. But little has yet been achieved in evaluating the efficacy and costs of these, in comparison to the more extensive evaluation of mitigation strategies to reduce greenhouse gas emissions. It is becoming clear, however, that even the most substantial, internationally agreed mitigation strategies cannot reduce likely impacts to insignificant levels. Some and probably extensive, adaptation will be necessary if the most negative effects of currently projected changes of climate are to be avoided. But how much adaptation, when, and at what cost? These are questions that must be addressed urgently by the policy community.

References

Akong'a, J., Downing, T.E., Konijn, N.T., Mungai, D.N., Muturi, H.R. and Potter, H.L. (1988). The effects of climatic variations on agriculture in central and eastern Kenya. In M.L. Parry, T.R. Carter and N.T. Konijn (eds.). *The Impact of Climatic Variations on Agriculture, Volume 2. Assessments in Semi-Arid Regions.* Kluwer, Dordrecht, 121-270.

Alcamo, J. (ed.). IMAGE 2.0: *Integrated Modelling of Global Climate Change.* Kluwer, Dordrecht.

Alcamo, J., Leemans, R. and Kreileman, E. (1998). *Global Change Scenarios of the 21st Century. Results from the IMAGE 2.1 Model.* Elsevier Science Ltd., Oxford.

Bergthorsson, P., Bjomsson, H., Dyrmundsson, 0., Gudmundsson, B., Helgadottir, A., and Jomnundsson, J.V. (1988) . The Effects of Climatic Variations on Agriculture in Iceland. In M.L. Parry, T.R. Carter and N.T. Konijn (eds.). *The Impact of Climatic Variations on Agriculture. Volume 1. Assessments in Cool Temperate and Cold Regions.* Kluwer, Dordrecht, 381-509.

Bonan, G.B., (1993). Do biophysics and physiology matter in ecosystem models? *Climatic Change,* 24, 281-285.

Burton, I., Kates, R.W. and White, G.F. (1993). *The Environment as Hazard.* Guilford Press, New York.

Carter, T.R., Parry, M.L., and Porter, J.H. (1991) Climatic warming and crop potential in Europe. *Global Environmental Change,* 1(4), 291-313.

Carter,T.R., Parry, M.L., Harawasa, H., and Nishioka S. (1994). *IPCC Technical Guidelines for Assessing Climate Change Impacts and Adaptations,* University College London, London and Centre for Global Environmental Change, Tsukuba, Japan.

CCIRG-Climate Change Impacts Review Group (1996). *Assessment of the Potential Effects of Climate Change in the United Kingdom.* Second Report of the United Kingdom Climate Change Impacts Review Group, UK Department of the Environment. HMSO, London.

Geijn, S.C. van de, Goudriaan, J. and Berendse, F. (eds) (1993). *Climate Change: Crops and Terrestrial Ecosystems.* Agrobologische Thema's 9, CABO-DLO, Centre for Agrobiological Research, Wageningen, The Netherlands.

Hotthus, P., Crawford, M., Makroro, C. and Sullivan, S. (1992). *Vulnerability Assessment for Accelerated Sea Level Rise Case Study: Majuro Atoll, Republic of the Marshall Islands*, SPREP Reports and Studies Series No. 60, Apia, Western Samoa.

Hulme, M., Raper, S. and Wigley, T.M.L. (1995). An integrated framework to address climate change (ESCAPE) and further developments of the global and regional climate modules (MAGICC). *Energy Policy*, 23, 347-356.

IPCC (1990). *Climate Change: The IPCC Scientific Assessment*. J.T. Houghton, GJ. Jenkins and J.Ephraums (eds.). Report of Working Group I of the Intergovernmental Panel on Climate Change, Cambridge University Press.

IPCC (1996). *Climate Change 1995: Impacts, Adaptations and Mitigation of Climate Change: Scientific-Technical Analysis*. Report of Working Group II of the Intergovernmental Panel on Climate Change, Cambridge University Press, London and New York.

Jodha, N.S. and Mascarenhas, A.C. (1985) Adjustment in self-provisioning societies. In R.W. Kates, J.H. Ausubel and M. Berberian (eds). *Climate Impact Assessment: Studies of the Interaction of Climate and Society*. SCOPE 27, Wiley, Chichester, 437-464.

Kates, R.W. (1985). The interaction of climate and society. In R.W. Kates, J.H. Ausubel and M. Berberian (eds.). *Climate Impact Assessment: Studies of the Interaction of Climate and Society*. SCOPE 27, Wiley, Chichester, 3-36.

Kenny, G.J., Harrison, P.A. and Parry, M.L. (1993). *The Effect of Climate Change on Agricultural and Horticultural Potential in Europe*. Environmental Change Unit, University of Oxford, Research Report No. 2, Oxford.

Martens, P. (ed.) (1996). *Vulnerability of Human Population Health to Climate Change: State-of-Knowledge and Future Research Directions*. Dutch National Research Programme on Global Air Pollution and Climate Change, Report No. 410200004 , Bilthoven.

Martens, P. (1998). *Health and Climate Change: Modelling the Impacts of Global Warming and Ozone Depletion*. Earthscan, London.

Martens, P., Niessen, L.W., Rotmans, J., Jetten, T.H., and McMichael, A.J. (1995a). Potential impact of global climate change on malaria risk. *Environmental Health Perspectives*, 103(5), 458-464.

Martens, P., Jetten, T.H., Rotmans, J., and Niessen, L.W. (1995b). Climate change and vector-borne diseases: a global modelling perspective. *Global Environmental Change*, 5(3), 195-209.

Martens, P., Jetten, T.H., and Focks, D.A. (1997). Sensitivity of malaria, schistosomiasis and dengue to global warming. *Climatic Change*, 35(2), 145-156.

NDU - National Defense University (1978). *Climate Change to the Year 2000*. Washington, D.C., Fort Lesley J. McNair.

NDU -National Defense University (1980). *Crop Yields and Climate Change to the Year 2000. Vol. 1.* Washington, D.C., Fort Lesley J. McNair.

Nordhaus, W.D. (1992). The DICE Model: Background and Structure of a Dynamic Integrated Climate Economy Model of the Economics of Global Warming. *Cowles Foundation Discussion Paper, No. 1009*, New Haven, Connecticut.

Parry, M.L. and Carter, T.R. (1988). The assessment of effects of climatic variations on agriculture: aims, methods and summary of results. In M.L.
Parry, T.R. Carter and N.T. Konijn (eds.). *The Impact of Climatic Variations on Agriculture. Volume 1. Assessments in Cool Temperate and Cold Regions*. Kluwer, Dordrecht, 11-95.

Parry, M.L. and Carter, T.R. (1998). *Climate Impact and Adaptation Assessment*. Earthscan, London.

Parry, M.L. (1990). *Climate Change and World Agriculture*, Earthscan, London.

Prentice, I.C., Cramer, W.P., Harrison, S.P., Leemans, R., Monserud, R.A. and Solomon, A.M. (1992). A global biome model based upon plant physiology and dominance, soil properties and climate. *Journal of Biogeography*, 19, 117-134.

Rosenberg, N.J. (1992). Adaptation of agriculture to climate change. *Climatic change*, 23, 293-335.

Rosenzweig, C. and M.L. Parry. (1994). Potential impact of climate change on food supply. *Nature*, 367, 1933-138.

Rotmans, J. (1990). *IMAGE: an Integrated Model to Assess the Greenhouse Effect*. Kluwer, Dordrecht.

Rotmans, J., Hulme, M. and Downing, T.E. (1994). Climate change implications for Europe. An application of the ESCAPE model. *Global Environmental Change*, 4(2) 97-124.

Stakhiv, E.Z. (1994). Water resources planning of evaluation principles applied to ICZM. In: *Preparatory Workshop on Integrated Coastal Zone Management and Responses to Climate Change*. World Coast Conference 1993, New Orleans, Louisiana.

Stewart, T.R. and Glantz, M.H. (1985). Expert judgement and climate forecasting: A methodological critique of Climate Change to the Year 2000. *Climatic Change*, 7, 159-183.

Strain, B.R. and Cure, J.D. (eds) (1985). *Direct Effects of Increasing Carbon Dioxide on Vegetation*. DOE/ER-0238, United States Department of Energy, Office of Energy Research, Washington D.C.

Strzepek, K.M. and Smith, J. (1995). *As Climate Changes: International Impacts and Implications*. Cambridge University Press, United Kingdom.

WMO (1985). *Report of the WMO/UNEP/ICSU- SCOPE Expert Meeting on the Reliability of Crop-Climate Models for Assessing the Impacts of Climatic Change and Variability*. WCP-90, World Meteorological Organization, Geneva.

WMO (1988). *Water Resources and Climatic Change: Sensitivity of Water Resource Systems to Climate Change and Variability*. WCAP-4, World Meteorological Organization, Geneva.

Worldbank (1991). *World Development Report 1991*. Oxford University Press, New York.

Chapter 7

INTEGRATED ASSESSMENT MODELLING

J. Rotmans and M. van Asselt

7.1 Introduction

The increasing mutual interplay between social, economic, and environmental issues demands integrated policies. Despite an early history of isolated regulatory initiatives related to for instance air, water, and soil over the past two decades, environmental policies have become increasingly integrated. Nevertheless, this is only a first step towards further integration of environmental, social and economic policies. The complexity of the pressing issues, like global climate change, demands an integrated approach to ensure that key interactions, feedbacks and effects are not inadvertently omitted from the analysis. The various pieces of the complex puzzle can no longer be examined in isolation. Integrated Assessment (IA) aims to fit the pieces of the puzzle together, thereby indicating priorities for policy.

There is increasing recognition and credibility for the rapidly evolving field of Integrated Assessment. Within the setting of the political arena, it is accepted that IA can be supportive in the long-term policy planning process, while in the scientific community more and more scientists do realise the complementary value of IA research.

One of the problems of IA is still the many definitions and interpretations that circulate (Weyant *et al.,* 1996; Rotmans and Dowlatabadi, 1998; Parson, 1996; Ravetz, 1997; Jaeger *et al.,* 1997). Notwithstanding this diversity, these definitions have three elements in common, i.e. interdisciplinarity, decision-support and participation of stakeholders. Thus irrespective of whatever definition is taken, IA can be described as (Rotmans, 1998):

a structured process of dealing with complex issues, using knowledge from various scientific disciplines and/or stakeholders, such that integrated insights are made available to decision-makers.

Integrated Assessment is an iterative, continuing process, where integrated insights from the scientific and stakeholder community are communicated to the decision-making community, and experiences and learning effects from decision-makers form one input for scientific and social assessment. Although participation of stakeholders is not a necessary prerequisite, the conviction in the IA community grows that participation of stakeholders is a vital element in IA. The engagement of non-scientific knowledge, values and preferences into the IA process through social discourse will improve the quality of IA by giving access to practical knowledge and experience, and to a wider range of perspectives and options.

Integrated Assessment attempts to shed light on complex issues by illuminating different aspects of the issue under concern: from causes to impacts, and from options to strategies. IA has been widely applied in the global change research area. IA emerged as a new field in global change research because the traditional disciplinary approach to global change research has been unable to meet two significant challenges central to understanding global phenomena. The first challenge is the development of an adequate characterisation of the complex interactions and feedback mechanisms among the various facets of global change. Such feedbacks and interactions are defined away or treated parametrically in traditional disciplinary research. The second challenge is that of providing support for public decision-making. IA offers an opportunity to develop a coherent framework for testing the effectiveness of various policy strategies, and estimating trade-offs among different policy options.

Performing IAs has a number of advantages. In general terms, IA can help to:

- put a complex problem in the broader context of other problems, by exploring the interrelations of the specific problem with other issues;
- assess potential response options to complex problems. This may be, but not necessarily, in the form of cost-benefit and cost-effectiveness analysis;
- identify, illuminate and clarify the different types and sources of uncertainties in the cause-effect chain(s) of a complex problem;
- translate the concept of uncertainty into the concept of risk analysis, to assist in decision-making under uncertainty;

- set priorities for research topics, also by identifying and prioritising decision-relevant gaps in knowledge.

7.2 Methods for integrated assessment

A diversity of approaches is needed in Integrated Assessment, varying from analytical methods (such as models) to participatory methods (such as policy exercises). The divergence of methods employed arises from the uneven state of scientific knowledge across different problem domains, from the differences in problem perception, and from the different perspectives of the scientists and stakeholders involved in the assessment process (see Figure 7.1).

Figure 7.1 Tool-kit for Integrated Assessment

In general, two types of Integrated Assessment methods can be distinguished: analytical methods and participatory methods. While analytical methods are often rooted in natural sciences, participatory methods, also labelled as interactive or communicative methods, stem from social sciences. The group of analytical methods is reasonably well defined and basically includes model analysis, scenario analysis and risk analysis. Their commonality is that they provide analytical frameworks for representing and structuring scientific knowledge in an integrated manner.

The group of participatory methods, however, involves a plethora of methods, varying from expert panels, Delphi methods, to gaming, policy exercises and focus groups. They have in common that they aim to involve non-scientists as stakeholders in the process, where the assessment effort is driven by stakeholder-scientist interactions.

The aim of the various methods is to facilitate the IA process as sketched in Figure 7.2.

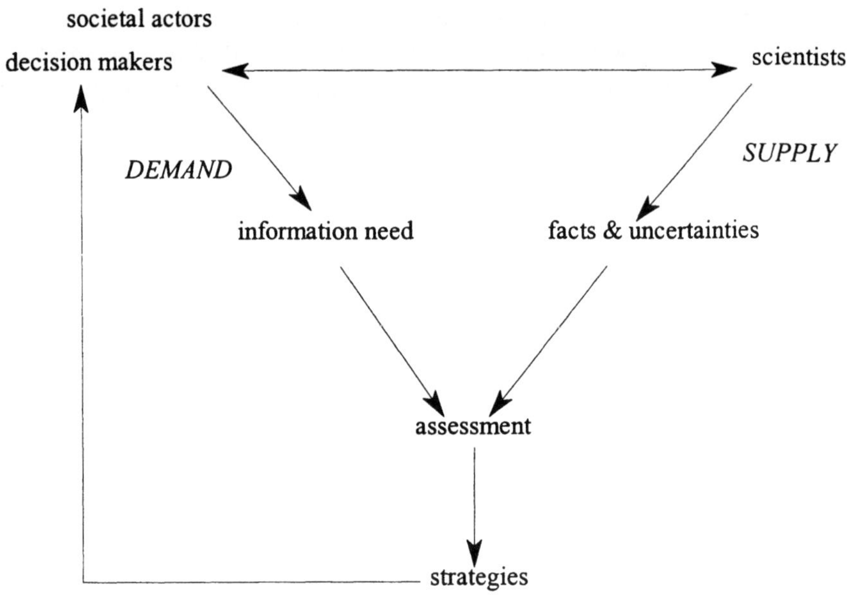

Figure 7.2 Sketch of IA process

It should be noted that this Figure is a highly simplified representation of the complex IA process. For example, it does not exclude the possibility that societal actors also supply (delivering non-scientific expertise), and scientists also demand (e.g. in framing research agendas).

In this Chapter, we will focus on the analytical method of integrated assessment modelling. Integrated Assessment models are computer simulation (including optimisation) frameworks that try to describe quantitatively as much as possible of the cause-effect relationships of a specific issue, and of the interlinkages and interactions among different issues. At present integrated assessment modelling is the dominant method of performing IA. However, many studies that are either explicitly or implicitly integrated assessments, are still qualitative in nature, without using any model. Furthermore, the complexity of the issues addressed and the value-laden character of assessment activities makes it impossible to address this process by only one unique approach. The issues under concern are generally more complex, subtle and ambiguous than can ever be expressed by formal mathematical equations in models. In particular the decision-making processes, which are sequential and involve multiple actors, and human values, preferences and choices are not, and probably never will be, captured by formal models.

Any attempt to fully represent a complex issue and its numerous interlinkages with other issues in a quantitative model is doomed to failure. Nevertheless, even a simplified but integrated model can provide a useful guide to complex issues and complement highly detailed models that cover only some parts of complex phenomena. Among the major strengths of IA models are:

- *exploration of interactions and feedbacks*: explicit inclusion of interactions and feedback mechanisms between subsystems can yield insights that disciplinary studies cannot offer;
- *flexible and rapid simulation tools*: the simplified nature and flexible structure of IA models permit rapid prototyping of new concepts and scientific insights;
- *consistent frameworks to structure scientific knowledge*: critical uncertainties, gaps in scientific knowledge and weaknesses in discipline-oriented expert models can be identified;
- *tools for communication*: IA models can be useful tools in communicating complex scientific issues to decision-makers, disciplinary scientists, stakeholders, and the general public.

Obviously, IA models also have limitations and weaknesses. Among the most important ones are:

- *high level of integration*: many processes occur at a micro level, far below the spatial and temporal aggregation of current IA models;
- *inadequate treatment of uncertainties*: IA models are prone to an accumulation of uncertainties, and to a variety of types and sources of uncertainty;
- *absence of stochastic behaviour*: most IA models describe processes in a continuous, deterministic manner, excluding extreme conditions that may significantly influence the long-term systems behaviour;

- *limited calibration and validation*: the high level of aggregation implies an inherent lack of empirical variables and parameters, and current data sets are often too small and/or unreliable to apply.

These strengths and weakness of IA models as well as the complementary value of other IA methods have to be taken into account in reading the present Chapter that describes the state-of-the-art in Integrated Assessment modelling.

7.3 IA modelling

7.3.1 History

Current projects in IA modelling build on a tradition started in the early seventies by the Club of Rome (Meadows *et al.,* 1972). In this context the first global computer simulation model, i.e. the World3 model, was developed, which described couplings between the major social and physical components of the world system. This model inspired the development of numerous global models, focusing on resource depletion, population and pollution.

The next generation of IA models explicitly addressed environmental issues. The first among these models emerged in the late 1970s from earlier energy modelling (Nordhaus, 1979; Edmonds and Reilly, 1985). Meanwhile, the RAINS model of acidification in Europe was developed in the 80s (Alcamo *et al.*, 1990).

The phenomenon of global climate change has prompted the development of a new class of IA models. The first steps to an integrated model of climate change were taken by Mintzer (1987), Lashof and Tirpack (1989), and Rotmans (1990). Since then approximately 40 IA models of climate change have been developed (van der Sluys, 1997). Recent overviews of IA modelling activities in the field of climate change can be found in Rotmans and Dowlatabadi (1998), Schneider (1997) and Parson and Fischer-Vanden (1997).

7.3.2 Model typology

Weyant *et al.* (1996) classified IA models in 2 categories:

a) *policy optimisation models,* which try to optimise key policy variables such as carbon emission control rates and carbon taxes, given certain policy goals (such as maximising welfare or minimising costs of climate policy), and

b) *policy evaluation models,* which try to evaluate the environmental, economic and social consequences of specific policy strategies.

However, some IA models characterised as policy evaluation models can also be used for optimisation experiments. Therefore, another possible classification of IA models is given by Rotmans and Dowlatabadi (1998):

a) *macroeconomic-oriented models,* which represent relatively simple, parameterised decision-analytic formulations of complex problems; and
b) *biosphere-oriented models,* which represent a more comprehensive, process-oriented description of a complex problem.

Most macroeconomic-oriented models are neo-classical models based on an equilibrium framework, using traditional economic concepts regarding optimisation and capital accumulation, largely ignoring environmental dynamics. Biosphere-oriented models, however, focus on a systems-based description of the geophysical and biogeochemical processes and feedbacks, but do not adequately represent the socio-economic system. The DICE model (Nordhaus, 1992; 1994) is a well-known exponent of the macroeconomic-oriented school, whereas the IMAGE model (Rotmans, 1990; Alcamo, 1994) is representative for the biosphere-oriented school.

Meanwhile, some attempts are underway to combine the best of both worlds, yielding a hybrid of the two categories above. These hybrid models may contain a detailed general equilibrium model for the economy that interacts with a dynamic environment, such as GCAM (Edmonds *et al.*, 1994) and the MIT model (Prinn *et al.*, 1996). However, they may also use complexity and uncertainty as guiding modelling principles for both the human and natural system, of which ICAM (Dowlatabadi and Morgan, 1993a,b) and TARGETS (Rotmans and de Vries, 1997) are examples. Therefore, the above two classes should be considered as the polar ends of a continuum populated by many IA modelling efforts.

After the first boom of global models in the early seventies, the interest in large modelling efforts nearly extinguished. In the mid 1980s we saw a revival in global modelling efforts, which we now consider as the first examples of integrated assessment models. What stimulated this phase of modelling? This is one of the key questions of Harvard's project 'Global Environmental Assessment' (Clark and Parson, 1996). This project aims to understand under what circumstances, and in what ways, IA activities have been performed, and if and how they have influenced policy processes. This first year of the project concentrated on assessment activities on in the field of climate change (e.g. Fischer-Vanden, 1997; Franz, 1997; Kandlikar and Sagar, 1997; Long and Iles, 1997). The results of this 5-year political science, historical project will be a critical reflection on the field of IA.

Box 7.1
Current integrated assessment modelling activities

Without claiming to give a complete overview of current integrated assessment models, we give a brief and incomplete description of some of the leading current integrated assessment modelling activities. Beyond the notion that integrated assessment models should involve an attempt to combine social and natural science descriptions of the climate change problem, there is little in common between the different models. Differing objectives have brought about integrated assessment model designs characteristic to each research group. In general, however, two kinds of integrated assessment models are distinguished here: smaller models which represent simple, fully parameterised decision-analytic formulations of the climate problem; and the larger models, which represent more comprehensive, process-oriented pictures of the climate problem.

Smaller integrated assessment models

Not all integrated assessment models are large. The smaller models offer the dual benefits of versatility and clarity. They often stem from macro-economic principles and are mostly used for cost-benefit or cost-effectiveness analyses. The insights gained using these smaller models have often guided the development of the more complex modelling efforts. Ideally, the cost-benefit models can be used to inform the initiation and progress of the sequence of decisions and actions related to climate change policy, although they have played a relatively minor role in the process of implementation of climate strategies.

DICE, RICE and PRICE

The DICE model (Nordhaus, 1992; Nordhaus, 1994) is a dynamic optimisation framework in which the climate change problem has been cast as how a decision-maker chooses between current consumption, investment in productive capital, and investment to reduce emissions of greenhouse gases. In Dice population growth and technological change yield productivity growth. Both of these factors of production are exogenously specified and assumed to decline asymptotically to zero.

An increasing, convex emission-control cost function is estimated from prior studies, in which reducing emissions 50% from the prevailing level at any time costs about 1% of the world economy. Current carbon emissions add to atmospheric concentrations via a fixed retention ratio, and realised temperature change is modelled by a three-box model representing the atmosphere, mixed-layer upper ocean, and deep ocean. Damage from

Box 7.1 Continued.

climate change is a quadratic function of realised temperature change with a 3°C change calibrated to cause a 1.3% world GNP loss.

In earlier analyses, Nordhaus estimated the potential impacts of an equilibrium 3°C change on various sectors of the US economy. The sum of his damage estimates indicated a damage level of 0.25% of GNP for a 3°C equilibrium climate change. Accounting for linkages or other impacts that may have been overlooked, he judged that 1 to 2% of GNP was a plausible upper bound for damages from a climate change of this magnitude. With this research he laid the foundation upon which most subsequent impact analyses have relied.

Recently two variants of the DICE model have been developed, one with multiple regions and decision-makers (RICE) another with uncertainty in key parameter values (PRICE). In the RICE model there are six or more world regions, assuming a quasi-convergence of the different economies. The optimum tax rate is selected so that the world's utility is maximised. Co-operative solutions to RICE are similar to those achieved with DICE. In PRICE, a small number of parameters (eight) are assumed uncertain and resolved with a ½ life of 30 years. The major uncertainties in this model are long-run economic growth, the emissions of non-CO_2 greenhouse gases, and the extent of co-operation among nations.

MERGE

MERGE is the integrated assessment model which Manne and Richels have developed with additional input from Robert Mendelson. At the core of this model resides a revised version of the Global 2100, now being exercised to the end of the 22nd century. It embodies a General Equilibrium model with 5 world regions, in which each region's consumer makes both savings and consumption decisions. A simple climate model represents atmospheric lifetimes of CO_2, CH_4 and N_2O, which yield global changes in radiative forcing, and equilibrium and realised global-average temperature change. Illustrative impact functions are defined separately for market and non-market components. The former is modelled as a quadratic function of realised temperature change fitted to a single judgmental point estimate. The latter is estimated as a willingness to pay in each region to avoid a specified temperature change that is a logistic function of regional income, in effect modelled as a world-wide public good (Manne, Mendelsohn and Richels, 1993).

Box 7.1 Continued.

Larger integrated assessment models

The more comprehensive integrated assessment models aim at a more thorough description of the complex, long-term dynamics of the biosphere-climate system. This dynamic description often includes a number of the many geophysical and biogeochemical feedbacks within the system. Some modelling approaches even deal with the biosphere-climate dynamics at a geographically-explicit level. On the other hand the socio-economic system in these models is usually poorly represented. The larger models do not serve the purpose of performing cost-benefit or cost-effectiveness analyses, but can provide insights into the intricate interrelationships between the various components of the human system and the biosphere-climate system. Ideally, this can lead to new priority setting in the climate change policy process.

IMAGE 2.0

The model presents a geographically-detailed, global and dynamic overview of the linked society-biosphere-climate system, and consists of three fully linked subsystems: the energy-industry system; the terrestrial environment system; and the atmosphere-ocean system (Alcamo, 1994). The energy-industry models compute the emissions of greenhouse gases in 13 world regions as a function of energy consumption and industrial production. End use energy consumption is computed from various economic driving forces. The terrestrial environment models simulate the changes in global land cover on a grid-scale based on climatic and economic factors. The role of land cover and other factors are then taken into account to compute the flux of CO_2 *and other greenhouse gases from the biosphere to the atmosphere. The atmosphere-ocean models compute the build-up of greenhouse gases in the atmosphere and the resulting zonal-average temperature and precipitation patterns. The model includes many of the important feedbacks and linkages between models in these subsystems.*

One of the main achievements of the IMAGE 2.0 model is that it is a first attempt to simulate in geographic detail the transformation of land cover as it is affected by climatic, demographic and economic factors. It links explicitly and geographically (on a grid scale of 0.5° latitude by 0.5° longitude) the changes in land cover with the flux of CO_2 *and other greenhouse gases between the biosphere and atmosphere, and conversely, takes into account the effect of climate in changing productivity of the terrestrial and oceanic biospheres.*

Box 7.1 Continued.

ICAM

The Integrated Climate Assessment Models (ICAM versions 0, 1, and 2) were developed by the Carnegie Mellon University, at the Department of Engineering & Public Policy (Dowlatabadi and Morgan, 1993a,b). The ICAM model versions have brought increasingly sophisticated and detailed descriptions of the climate change problem, at each stage quantifying the uncertainties in the model components and asking where additional research would most contribute to resolution of the climate policy dilemma. This information is then used in the next iteration of the research program defining the disciplinary research needs and the direction of refinements to ICAM.

The ICAM model versions are designed to capture the uncertainties in knowledge about the precursors, processes and consequences of climate change. The models can be used to simulate abatement activities, adaptation to a changed climate, and geo-engineering activities. The development of ICAM 2 has involved updating all previous modules, development of demographics, fuel market, aerosols, terrestrial ecology and coastal impacts modules. The spatial and temporal scales have also been refined to 5 years and 7 geo-political regions. The differentiation between high and low latitudes makes it possible to examine the gross differences in the magnitude of climate change, as well as different economic circumstances and availability of resources needed to adapt to a changed climate.

The ICAM model has been used to show: the wide range of possible future emissions, climate conditions, and impacts, and the dangers of deterministic modelling with narrow sensitivity studies; the relative importance of decision rules in policy decision making; and how key factors in determining the character of the problem and key uncertainties in making informed judgements can be identified.

MINICAM/PROCAM

The Batelle Pacific Northwest Laboratory runs a program, which involves a multi-disciplinary team with a long and sustained track record in climate related research. There are two integrated assessment models being developed within the program — MiniCAM and ProCAM. The two models are differentiated by their complexity and specificity.

MiniCAM makes extensive use of reduced form modules, characterising fewer world regions, and fewer economic activities. MiniCAM is specifically designed to be suitable for uncertainty analysis. This framework is composed of: the Edmonds Reilly Barns model for projection of economic activity and emissions of greenhouse gases and sulphate aerosols. The MAGICC model

Box 7.1 Continued.

is then used to generate a global temperature response commensurate with the emissions. The SCENGEN model is then used to arrive at regional climates based on the global climate change. The market and non-market impacts are expressed in economic terms and follow the formulation proposed in the MERGE 1.5 model of Manne Richels and Mendelson (Manne, Mendelson et al., 1994).

ProCAM is a much more complex framework making use of much more detailed models of human activity (Edmonds et al., 1994). The human activity within this framework is simulated using the Second Generation Model (SGM), which is a Calculable General Equilibrium model. Twenty regionally specific models are being developed through collaboration with regional experts. The suite of SGMs is used explicitly to allocate land-use and other resources and estimates anthropogenic emissions. Again, the MAGICC model is used to convert the emissions data to globally averaged temperature and precipitation change, and the SCENGEN model is used to map this change into regional patterns of changed climate according to the output of one or a combination of GCMs. The regional temperature and precipitation fields are employed in detailed regional agriculture, ecology, and hydrology models in order to assess regional impacts of projected climate change.

THE TARGETS MODEL

Based on the various subsystems of the whole system portraying global change, a series of highly aggregated modules are being built, interlinked and ultimately integrated. This results in an overall integrated assessment framework, TARGETS: Tool to Assess Regional and Global Environmental and Health Targets for Sustainability. The two-dimensional integration approach sketched out above is incorporated in TARGETS in the following way. The TARGETS integrated framework basically consists of a population and health model, a resources/economy model, a biophysics model, a land use model and a water model, which are all interlinked. All types of models comprise a linkage of causative, state-descriptive and impact modules, in this way representing the (vertically) integrated cause-effect chain. All causative or pressure models describe the developments in human population, resources/economy, land use and water use, respectively, where all developments are fully interlinked. In a similar way, all models describing the disturbed states of the underlying system, are coupled, just as all impact models are. By coupling the various causes, states and impacts for the various subsystems underlying the modules, the horizontal integration comes into play. The time horizon for the TARGETS model will

Box 7.1 Continued.

span about two centuries, starting at the beginning of this century, the year 1900, symbolising the end of the pre-industrial area, until the end of the next century, the year 2100, with time steps varying from one month to one year.

The TARGETS model is a composite framework of simple systems (represented by metamodels) which may show non-linear and complex, perhaps even chaotic, behaviour. This means that incremental changes in conditions of subsystems may result in considerable changes in the results of the overall system, which may not always be predictable. One of the other crucial aspects in which the TARGETS modelling approach distinguishes itself from the modelling attempts which have hitherto been published, is the treatment of uncertainty. This subject will be discussed in Chapter 8.

We would like to tentatively address the question of what inspired IA modelling based on statements of IA practitioners and our own experiences. IA-practitioners (Weyant *et al.*, 1996; Alcamo *et al.* 1996) bring the following speculations to the fore:

- scientific curiosity
- advances in computing power
- the availability of new field data for verifying models, and an increased standardisation and co-ordination of data (for example by the United Nations Environment Programme (UNEP) and the World Climate Program).
- the interest of policy-makers in the information the models can provide on environmental issues.

We would like to challenge this evaluation of our colleagues. First, scientific curiosity in general drives many, if not all, research efforts, and therefore it does not provide enough explanatory power to help us understand why the interest in IA modelling efforts revived in the mid 80s and stays alive up till now.

Secondly, in case the advances in computing power were the major cause for a boom in IA modelling, a historically unprecedented progression in computing power should have taken place in the early 80s. However, this was not the case. During the last 20 years, there was a stable increase in computing power, roughly a doubling in computing capacity each year. Only recent innovations - namely the transition of aluminium to copper and synthetic isolators in the production of chips- will cause a quadrupling of the computing power each year. Furthermore, the argument of advances in computing power would only be a strong case, if all modelling efforts made use of supercomputers. However, the majority of the models runs on a PC, not even using the full capacity of such a PC. So computing power is not a limiting factor for IA models, as it is for General Circulation Models (GCMs). This fact makes the argument that advances in computing power explain the flaw of IA modelling rather weak.

The data-argument is even weaker. One of the major problems with IA modelling is the lack of data for calibration and validation. Furthermore, due to the level of aggregation, IA models comprise a lot of variables for which there is no or hardly any data, such as the discount rate, the AEEI (Autonomous Energy Efficiency Index) and the HDI (Human Development Index). The availability of the kind of data Weyant *et al.* (1996) and Alcamo (1996) refer to, is therefore not a plausible explanation for the observed interest in IA modelling.

We, furthermore, doubt the role attributed to policy makers and their interests. If policy interest was the trigger for IA modelling, we should find traces of explicit policy interest in the mid 1980s, either in terms of support for or usage of IA models in policy circles. However, outspoken interest in integrated assessments by policy people, such as can be distracted from statements of Gibbons (1993) or, as is revealed by the involvement of EU bodies in a conference on Integrated Assessment in Toulouse (October 1996), is just from a recent date. The RAINS model is the only one which has a success story when it comes to the explicit use of IA in the policy realm (Hordijk, 1991). Parson and Fischer-Vanden (1997) argue that if interest of policy-people is the main reason for performing IA activities, then governmental agencies and formal advisory bodies would have been plausibly involved in IA activities. However, their observation is that most integrated assessments are conducted by independent researchers or research teams at universities, research institutes and non-profit organisations, rather than by governmental agencies or formal advisory bodies.

We do not doubt that *assumed* policy relevance of IA studies plays an important role as inspiration for IA activities. However, we think it is more true to reality that the researchers expect that their work will lead to policy relevant information, then that the expectations of policy people are the source of inspiration. As is concluded on the basis of interviews in a IA modelling team at RIVM, the modellers "believe that IA models have potential policy value", notwithstanding the disappointing experiences with policy-people they actually have (van Asselt, 1994). Our hypothesis is that when researchers enter upon an IA activity, they assume that at some time policy makers will use their findings. This seems a legitimate starting point for research activities. As Jacoby (1996) phrased it:

"People are more likely to undertake the difficult task of interdisciplinary research, particularly that concerned with policy issues, if they know that the results are relevant to an important social problem, and that someone is interested in receiving them."

After having challenged the suggestions of Weyant *et al.* (1996) and Alcamo *et al.* (1996), we arrive at our hypothesis concerning the question "What triggered IA modelling?", which is that the emergence of the climate issue inspired IA modelling.

The emergence of the issue of global climate change and the revived interest in large modelling efforts coincide. We argue that the climate issue par excellence lends itself for an interdisciplinary approach, natural and social scientific analysis, quantification, cost-benefit analysis and the translation of response and mitigation strategies to concrete measures for different sectors. Based on this coincidence we hypothesise that the interest in IA modelling within the scientific community recurred to a "problem-method couple". This concept is borrowed from policy sciences, where the notion "problem-policy couple" is used to evaluate policy processes (Braybrooke and Lindblom, 1963; for an example of an analysis of climate policy processes in these terms, see Dinkelman, 1995). A "problem-method-couple" indicates that because of the good match between an emerging problem and an available method to address it, both the problem and the method gained interest in the scientific community. In other words, we suppose that the climate issue and integrated assessment turned out to be well matched dancing partners.

7.3.3 IA-cycle

The present Chapter discusses the integrated assessment cycle (see Figure 7.3) in which models are used as a tool to gather insights. An IA project starts with a problem definition that implies a certain problem perception. An example of a problem definition underlying IA projects in the field of climate change is:

Will the global climate system in the future be significantly and irreversibly disturbed by human activities, and if so, what are the consequences?

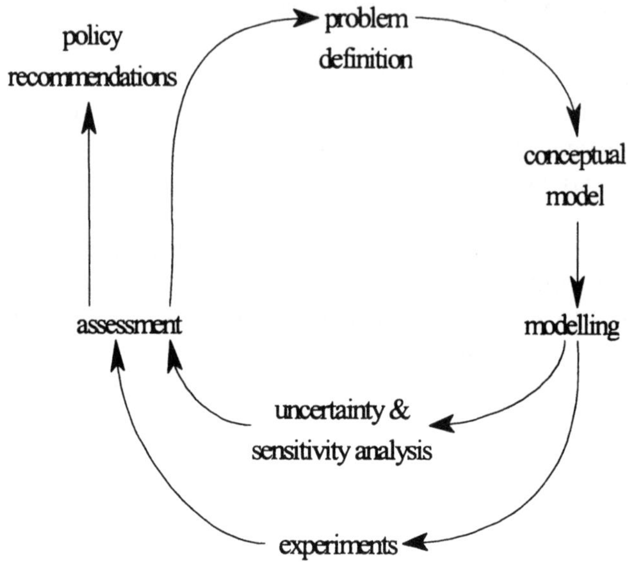

Figure 7.3 IA cycle

The first step in this cycle is to design a conceptual model. One way is to adopt a systems-oriented approach inspired by systems theory which originated in the 1960s (e.g. Forrester, 1961, 1968; Bertalanffy, 1968; Goodman, 1974). Notwithstanding the relativity of the concept 'system', the systems view always treats systems as integrated wholes of their subsidiary components and never as the mechanistic aggregate of parts in isolable causal relations (Laszlo, 1972). Complex problems like climate change and sustainable development are closely allied to the natural resilience and buffer capacity of natural reservoirs in relation to anthropogenic disturbances. Such complex issues can be represented by a set of interconnected cause-effect chains. The inextricably interconnected cause-effect chains form a complex system, the properties of which are more than just the sum of its constituent subsystems. This enables a synoptic approach that addresses the interdependencies between the various cause-effect chains.

Such a conceptual model forms the basis for a formal integrated assessment model. Data gathering, searching for valid equations in the literature, programming, calibration and validation are all part of this modelling phase. If the model is implemented and tested, experiments and sensitivity and uncertainty analysis can be performed. This enables the analyst(s) to sketch legitimate pathways into the future.

Instead of getting stuck with a whole set of future trajectories, the challenge is to synthesise these experiments and results of uncertainty and sensitivity analysis into a coherent assessment. Such an assessment involves addressing questions like 'Which trends are likely to become dominant in the course of the next century?', 'Which development will probably constitute major threats to humankind and the environment?', and 'Which societal and environmental processes seem to be of minor importance?'. The next step in the assessment involves translation of these long-term insights into short-term policy recommendations.

The insights gathered by those assessment exercises reveal a description of the issue under concern that is likely to deviate from the problem perception in the initial phase of the scientific assessment. Ideally, the new understanding, in turn, evokes an iteration of the IA cycle.

7.4 Critical methodological issues in IA modelling

We identify the following three key methodological issues that are critical in IA modelling efforts:

- Aggregation versus disaggregation
- Treatment of uncertainty
- Blending qualitative and quantitative knowledge

In the following, these methodological issues are discussed in more detail.

7.4.1 Aggregation versus disaggregation

One of the most critical issues in Integrated Assessment modelling is that of aggregation versus disaggregation. The level of aggregation of an IA model refers to the spatial and temporal resolution and the level of complexity addressed in the model. The problem of IA modelling is that it has to cope with a variety of processes that operate on different temporal and spatial levels, and differ in complexity.

First, IA modelling has to connect disciplinary processes that differ by nature: physical processes, monetary processes, information processes, and policy processes. Because of the multitude of disciplinary processes to be combined, a simple as possible representation of disciplinary knowledge is preferable. There is, however, no unifying theory how to do this. In addition, the processes to be linked are usually studied in isolation from other disciplinary fields. This isolation is needed as part and parcel of the classic model of scientific progress and discovery. However, when the constraints of isolation are removed, there is a variety of ways in which to connect the reduced pieces of disciplinary knowledge. This manifold of possible integration routes, for which there is again no unifying theory, is one of the reasons why quality control is so difficult to achieve in IA modelling.

For instance, in order to link the reduced pieces of disciplinary knowledge in a systemic way, one can use elements from classical systems analysis, or the method of system dynamics, or a sequential input-output analysis, or a correlation-based approach, or a pressure-state-impact-response approach.

Second, IA modelling has to deal with different spatial scale levels. One of the ultimate challenges in IA modelling is to connect higher scale assessments with lower scale ones. So far, there has been hardly any experience with playing around with scale levels in IA modelling. Down-scaling or up-scaling the spatial level of a model has profound consequences. This is related to the question to what extent the processes considered are generic, or distinctly spatially bound in character. In other words: does a relationship hold at larger or lower scale levels? For instance, land and water issues seem to be local/regional by nature, whereas atmospheric issues often appear to be trans-regional or even global.

In this context, an interesting approach is proposed by Root and Schneider (1995): the so-called Strategic Cyclical Scaling (SCS) method. This method involves continuous cycling between large and small-scale assessments. Such an iterative scaling procedure implies that a specific global model is disaggregated and adjusted for a specific region, country or river basin. The new insights are then used to improve the global version, after which implementation for another region, country or river basin follows. This SCS method can also be used for conceptual validation of models.

Methodologically, a local or regional modelling approach is more attractive than a global approach, for a number of reasons (Costanza and Tognetti, 1998) in terms of:

- *data quality*: more reliable data is available locally;
- *manageability*: fewer interactions, and fewer causes and impacts have to be considered;
- *communication*: greater political interest;
- *reciprocity*: to test the reciprocal relationship between local/regional processes and global processes.

Third, IA modelling is faced with a multitude of temporal scales. Short-term needs and interests of stakeholders have to be considered. However, biogeochemical processes usually operate on a long time scale, whereas economic processes operate on short- to medium time scales. Another challenging aspect of IA modelling is to interconnect long-term targets as specified as a result of analysing processes operating on longer term time scales, with short term goals for concrete policy actions (Jaeger *et al.*, 1997). Unfortunately, there is not yet a sound scientific method how to do this, thus so far only heuristic methods have been used.

The trend in current IA modelling is to move toward greater and greater disaggregation, assuming that yields better models. In general, it is difficult to know when to stop building more detail into an IA model. With each incremented level of sophistication from the interlinkages of more processes in more detail, comes new insight. However, past decades of model building have shown that small and transparent models are often superior in that they provide similar results to large models faster and offer ease of use. In this respect, it is useful to distinguish between *complicated* and *complex* models. Complicated models are models that include a variety of processes, many of which may be interlinked. If incremental changes in these processes generally lead to incremental changes in model output, the dynamics of the model is almost linear and not complex at all. The more complicated the model, the higher the possibility of errors and bugs. It requires thorough testing to pick up most if not all errors and bugs, an activity, which is, unfortunately, heavily underrated. Complex models, however, may contain relatively few processes, but incremental changes in these processes may result in considerable changes in the results of the overall model. This non-linear behaviour may be due to the inclusion of feedbacks, adaptation, self-learning, and chaotic behaviour, and is often unpredictable.

Practically speaking, this means that disaggregation of IA models has profound consequences for the dynamics of the model. Breaking down a global model into various regions, implies that the regional dynamics should be dealt with in an adequate manner. Regional IA models use grid cells or classes for representing geographical differences and heterogeneities in regional IA models. They don't capture, however, the regional dynamics with regard to population growth, economic development, resource use and environmental degradation, let alone regional interactions through migration and trade.

7.4.2 Treatment of uncertainty

Any exploration of future developments inevitably involves a considerable degree of uncertainty. Because of the cross-disciplinary character of IA modelling, it includes many different types and sources of uncertainty. Because IA-models are end-to-end approaches, they also contain an accumulation of uncertainties. Uncertainties may arise from incomplete knowledge of key physiological, chemical and biological processes. Many uncertainties are of a socio-economic nature – related to people's behaviour – and reflect inadequate knowledge with respect to the driving forces of human behaviour.

Various attempts have been made to classify the different types and sources of uncertainty. Morgan and Henrion (1990) distinguish uncertainty about empirical quantities and uncertainty about the functional form of models, which may have arisen from: subjective judgement; disagreement among experts; systematic errors; approximation; and inherent randomness. Funtowicz and Ravetz (1989, 1990) classified uncertainties in three categories: i) technical uncertainties (concerning observations versus measurements); ii) methodological uncertainties (concerning the right choice of analytical tools); and iii) epistemological uncertainties (concerning the conception of a phenomenon).

Whatever classification is chosen, the various types and sources of uncertainties in IA-models need to be addressed in an adequate manner. Unfortunately, current techniques to analyse uncertainties, such as Monte Carlo sampling, and the usage of probability density functions, are merely useful to address technical uncertainties. These techniques are not suitable for analysing methodological and epistemological uncertainties, which primarily arise from subjective judgements and fundamental disagreement among experts. Another problem is that classical uncertainty analysis methods only address uncertainties in model inputs and neglect the interactions among multiple, simultaneous uncertainties which are crucial in IA. The resulting estimates of minimum, maximum and best guess values are therefore often erroneous and misleading.

A simple way of presenting uncertainties is by specifying a set of future scenarios, where scenarios then span a range of plausible, representative futures. This, however, doesn't give an indication of the cumulative uncertainty as well as the origin and meaning of the uncertainty range.

In IA models uncertainties are often reduced to technical artefacts. By attaching deterministic intervals or stochastic probability distribution functions to uncertain model parameters, it is suggested that variations in parameter values do yield estimates of the uncertainty in the model outcome. However, that may be true in the mathematical sense (although only partly), but it doesn't reflect the nature and source of the real world uncertainties. You may compare one type of uncertainty with the other in mathematical terms, but in physical terms that could lead to comparing apples and pears. It is not allowed to simply compare the uncertainty of the climate sensitivity (representing uncertainties in geophysical feedbacks) with the uncertainty of the fertility rate (representing uncertainties in triggering factors behind fertility behaviour). While the geophysical uncertainty might be reduced by future research, the demographic uncertainty might be structural in the sense that it cannot be reduced in the longer term.

In sum, current methods to analyse uncertainties in IAs are unable to: i) produce an order of magnitude of the cumulative uncertainty; ii) render various sources and types of uncertainties explicit; iii) provide systematic and coherent clusters of uncertainties; and iv) explain and clarify uncertainties in a manner understandable to decision-makers. In order to meet the above requirements, we therefore need new methods for uncertainty analysis (see Chapter 8).

7.4.3 Blending qualitative and quantitative knowledge

In most frameworks for IA, quantitative and qualitative knowledge are considered and treated as mutually exclusive. For instance, usually those aspects of a problem under concern that are not well known, or about which there is only vague and qualitative knowledge, are left out in the modelling process. This means, however, that we miss crucial links in the causal chains that form archetypal patterns of human-environment interactions. Quantitative rigor therefore prevents IA from being comprehensive, in the sense of studying all relevant aspects of a complex problem. It is therefore illusory to think that the full complexity of human-environment interactions could be integrated into a formal, quantitative modelling framework.

IA needs modelling frameworks, which are combinations of quantitative and qualitative approaches, from the perspective that they complement each other. A promising way to blend quantitative and qualitative information is to incorporate vague and qualitative knowledge in IA models, by using fuzzy logic. Contrary to Boolean logic, fuzzy logic makes use of a continuum of values, which represent elements of fuzzy knowledge. This technique is, amongst others, applied in the Syndromes project, where syndromes are the symptoms of global change as proposed by the German Advisory Council of Global Change (Schellnhuber *et al.*, 1997).

7.5 Challenges

During the last decades, IA models have proven themselves as legitimate and powerful approaches to complex issues. In particular to the climate change debate, IA models have contributed by exploring impacts of climate change and evaluating mitigation and abatement strategies. IA models have also provided useful information on balancing the carbon budget, sulphate aerosols and on various integrated aspects of land use.

However, there is still a long way to go before IA models will be fully accepted by the scientific modelling community on the one hand, and by the decision-making community on the other. Therefore, an ultimate challenge for IA modelling is to build up scientific and political credibility. To improve both scientific and political credibility of IA models, IA scholars should enhance IA models, improve the communication with disciplinary scientific and policy-making communities, and enrich and augment communication techniques.

Much of the criticism against IA models in this context has to do with their lack of transparency and the rather technocratic abstraction of reality they display. The following suggestions might be helpful in overcoming this communication barrier:

- Models should be documented (most of them are now poorly documented), and descriptions and findings should be published regularly in the open scientific and policy literature.
- The communication with the fundamental research community and with the decision-making world can be enhanced by using established and well-defined research methods and goals. Presenting these research methods in an open and clear way make models both comprehensible and defensible.
- Models can be provided with interactive communication shells or user interfaces that enhance the communication between designers and users. In general, visualisation techniques facilitate the presentation of information in a manner that corresponds to the way people intuitively perceive images. Innovative visualisation techniques can help to make models as transparent as possible, to make the underlying theories as clear as possible, to provide easy-to-use interactive interfaces, and to display uncertainties and complexity of the systems behaviour comprehensibly. This all helps to make the model outcomes, as well as the complex messages they want to convey, more immediately available and broadly accessible to a wide range of users, who may vary from the general public to decision-makers. Examples of user-interfaces are the interactive scenario scanner built around the IMAGE 2.0 model (Berk and Janssen, 1997), and the interactive user-interface around the TARGETS model (Rotmans and de Vries, 1997), the latter of which has appeared on CD-ROM.
- Indicators can also be useful tools in improving the communication with decision-makers. Indicators can be defined as pieces of information that help measure changes and progress in decision-making. Decision-makers already have indicators in their lexicon, using them to formulate policies and resulting targets to be achieved. Models can be linked systematically to indicators (see Rotmans, 1997). In this way, indicators

serve as vehicles to communicate model results, and can be used as a basis for mapping response strategies.

In sum, current IA tools and methods can be relatively easily improved, through making them more transparent to a wide audience, varying from decision-makers to the general public. Although full transparency of current IA tools and methods is illusory, it can be enhanced through:

- better documentation and more frequent publication;
- using established and well-defined research methods and goals;
- developing interactive user-interfaces (communication shells);
- using indicators that are linked to IA tools and methods;

The majority of IA studies is 'supply-driven'. The most prominent example is climate change, which was dictated by the (multi-) disciplinary research agenda, rather than the policy agenda. In order to increase the number of '*demand-driven*' IA studies, amongst others advocated by the post-normal school of thought (Funtowicz and Ravetz, 1994), the strength of analytical IA methods should be combined with the vigour of participatory methods. There is a need to develop models in close conjunction with decision-makers. This is more easily said than done. For instance, having decision-makers as stakeholders in the IA model design and building process is desirable indeed, but may lead to high expectations that cannot be fulfilled by IA. If the state-of-scientific knowledge doesn't permit reliable estimates of geographically-explicit impacts of global climate change, scientists need to make clear to decision-makers the difficulty of attaining such an unrealistic goal.

This means in practice that, already in its conceptual phase, the IA model should be co-designed by modellers, scientists, and decision-makers (and possibly by other stakeholders as well). This means involvement of the community of stakeholders to establish the model's credibility and authority during its development. Involvement includes discussions on the model's inputs and outputs, the temporal and spatial scale level used, the level of aggregation (detail) needed for developing policy strategies, issues and processes to be included or left out, and presentation of the model set-up and model results in terms of transparency. This culminates in a so-called 'user model' (van Asselt, 1994), that helps tailor the model to the user's needs. Since this is a continuing process, an intermediary layer of project managers may be established between the group of modellers and decision-makers. Except for the RAINS model (Alcamo *et al.*, 1990; Hordijk, 1991), which was partly co-designed by scientists and potential users, such a demand-driven IA modelling process would be unprecedented.

Enhancing IA models is a continuous process, and various strategies for further improving IA models can be followed. First, realising that an IA framework is as good as its weakest part of the whole model chain, it is much more effective to improve the rather weak or poorly defined model parts, rather than refining and disaggregating the already adequate submodels. Second, a no-regrets modelling strategy is to bring in, where necessary and where possible, region-specific information for submodels, but maintain as overall modelling strategy the ability to run the model at a fairly high level of aggregation. Third, building generic submodels (models independent of regional or temporal differences) has many disadvantages. This means that the theories and assumptions must be applicable at different levels of spatial aggregation and for different regions in different periods (Rotmans and Dowlatabadi, 1998). And finally, empirical testing is a crucial element in enhancing the quality of IA models. To this end, various model components could be tested against disciplinary expert models, not only to compare the outcomes of both types of models, but also to validate the scientific hypotheses, process formulation, boundary conditions, and model structure with each other.

For brevity's sake, we won't discuss here all IA modelling strategies, but instead focus on improving the weak parts of IA models. Let's consider some of the weaker parts of IA models. In general, knowledge of key dynamics in both social and natural systems is limited. But there is no doubt that IA models are unbalanced with regard to the representation of both interacting systems.

While in many macroeconomic-oriented IA models the socio-economic aspects are adequately represented, but the biological-physical aspects poorly, in most biosphere-oriented IA models the natural system is reasonably well represented, but the socio-economic system only caricaturally. For example, the dynamics of the climate-atmosphere-biosphere system continues to be far from well understood, mainly because of a dearth of knowledge and data about the internal dynamics of this complex system. This is typified by a continuing stream of surprise findings (e.g. missing carbon and nitrogen sinks, aerosols, and solar spots). Still, there is a solid scientific knowledge base for disciplinary modelling of many biogeophysical processes. Disciplinary expert models exist, from which knowledge can be extracted and represented in reduced form.

With regard to the socio-economic system, the situation differs materially. The social scientific knowledge base is smaller, resulting in less disciplinary models to be represented in reduced form in IA models. In addition, the modelling of socio-economic aspects suffers from a dearth of basic data, in particular outside industrialised countries. Therefore, the models we have developed to describe social dynamics have not been calibrated and validated adequately. There is, however, still sufficient knowledge of vital socio-economic processes to use in the IA modelling context. So far, this knowledge available has hardly been used in IA modelling. Among the more pressing questions in social dynamics are the following:

- what brings about the demographic and epidemiological transition?
- what drives environmentally-related *human behaviour*?
- what are the roots and dynamics of technological innovation and diffusion?
- what drives processes such as *urbanisation and migration*?

Although the above questions cannot be answered unambiguously, sound scientific hypotheses do exist. Using these hypotheses in combination with data and available expert models, forms the basis upon which these socio-economic processes can be represented in IA models. Here, we will only present IA modelling examples in the fields of demography and health, and human behaviour. These cases are examples of generic model building strategies, i.e. models that can be applied to different scale levels. As far as the remaining socio-economic issues are concerned, the reader is referred to other sources (e.g. Schneider, 1997; Dowlatabadi, 1998).

7.5.1 IA Modelling of population and health

Fertility modelling approaches are widely accepted and, only recently, existing mathematical techniques have been introduced in the health area (Weinstein *et al.*, 1987; WHO, 1994). Usually, regression techniques are used to explore the relations between broad health determinants, like literacy, income status, nutritional status, water supply and sanitation, education and medical services, and the health status measured in healthy life expectancy. However, these regression techniques only can give some suggestive evidence on the causes of population and health changes. Statistical models to estimate future fertility and health levels are based on extrapolation of past and current data. They operate on a short time horizon, and are static in terms of specifying the dynamics behind changing fertility and health patterns. Therefore, there is a need for integrated approaches that take account of the simultaneous occurrence of multiple risk factors and diseases, as well as cause-effect relationships. Such an integrated approach cannot be used in the clinical area on an individual basis, but is appropriate at the population level.

In Rotmans and de Vries (1997), an integrated systems approach to population and health is presented. A generic model has been designed, that simulates the driving forces (socio-economic and environmental factors), the fertility behaviour and disease-specific mortality, the burden of disease and life expectancy as well as the size and structure of the population, and a number of fertility and health policies. The major objective of the population and health model is to simulate changes in morbidity and mortality levels under varying social, economic and environmental conditions.

A second but related example is the MIASMA framework, which is designed to describe the major cause and effects relationships between atmospheric changes and human population health (Martens, 1997, 1998). The model focuses on climate change (in terms of changes in temperature and precipitation) and ozone depletion (in terms of changes in UV-B radiation). Under varying climate and ozone regimes, changes in the dynamics and distribution of vector-borne diseases are simulated (malaria, schistosomiasis and dengue), changes in cardiovascular diseases, changing patterns of skin cancer incidences, and changing mortality levels as a result of thermal stress. Although the model operates on the global level, it is being validated for specific developing regions in Kenya and Brazil.

It should be noted that this model does not take into account change in the health care system. A further integration step would then imply estimates of future human health conditions as a result of both environmental and socio-economic changes.

7.5.2 IA Modelling of consumption behaviour

Human consumption of goods and services is a major factor in the process of global change. So far, IA models only implicitly incorporate the behaviour of consumers. Two main reasons for modelling consumer behaviour more explicitly are: (i) to enhance and validate future projections generated by IA models; (ii) to offer the possibility to simulate effective and efficient behavioural strategies in the context of IA.

Social scientists have developed various models to conceptualise the processes underlying environmentally detrimental behaviour. However, none of these models is suitable for IA modelling for several reasons: (i) these models do not account for the full range of processes involved in 'real-life' consumer behaviours; (ii) the model relationships are rarely quantified; (iii) the models that do quantify such relationships use regression techniques, resulting in non-causal static models, in which processes of feedback, adaptation and learning cannot be accounted for; and (iv) the models are domain-specific, e.g. they model domestic energy consumption.

Box 7.2
A conceptual model of consumer behaviour

The conceptual model of human behaviour presented here is described in detail in Jager et al. (1997, and 1999). The model comprises consumer motivation, the opportunities (products and services) they might consume, and consumer abilities. Moreover, the conceptual model provides a taxonomy of major theories on behavioural processes, differentiating between reasoned versus automatic behaviour, and between individual versus social information processing.

As a basis for the model, an integrated systems-based approach is chosen, using the Pressure-State-Impact-Response framework. The Pressure system describes the driving forces behind consumer behaviour, while distinguishing between consumer motivation, consumption opportunities and consumer abilities (MOA-model). The State system simulates the cognitive and behavioural processes preceding consumption, using a comprehensive theoretical framework concerning behaviour. The Impact system describes the level of consumption, changes in abilities and opportunities, and the satisfaction of various needs in relation to quality of life. In the Response system four types of policy strategies to influence consumer behaviour are adopted: (i) provision of physical alternatives; (ii) regulation; (iii) financial stimulation; and (iv) social and cognitive stimulation.

In translating the conceptual model into a behavioural simulation model, consumers can be represented as computerised agents, so-called "consumats". The basic design of every consumat comprises a set of rules, which are rooted in the used theories of human behaviour. However, the consumats differ with regard to their motivations and personal abilities, thus representing various types of consumers. By including interactions and feedbacks between different consumats, various social processes can be simulated: e.g. status-related consumption, the dispersion of new consumer opportunities ("trickling down") and the manifestation of scarcity.

In the field of economics, modelling consumer behaviour has a long history. Demand theory is well developed, and allows for specification of demand as a function of, e.g. prices and income. Models based on this theory suppose that there is a general equilibrium between demand and supply (which is also price-driven). However, the problems with the neo-classical equilibrium approach are that it does not allow for multiple equilibria, and that it is based on the rational actor paradigm (Jaeger *et al.*, 1998). This means that actors are considered as individuals with fixed preferences, who maximise their utility over a set of alternatives, and ignore the preferences of other actors. In practice, however, people do not always behave according to rational actor principles because of a variety of reasons. People have imperfect knowledge, and, as a result of that, they are confronted with uncertainty and surprises, they often tend towards habitual behaviour, and under certain circumstances they simply imitate the behaviour of other people.

An alternative for the rational actor paradigm is based on psychological theory, which contains many theories on human behaviour. The art is now to specify under what conditions which theory-based rules will guide the agent's behaviour. Conceptual models to integrate various relevant theories on consumer behaviour are in development (Jager *et al.*, 1999). Box 7.2 presents a prototype of such a generic behavioural model.

7.5.3 Multi-agent modelling

The introduction of multiple actors (agents) allows to model self-organisation phenomena in complex systems. Agent-based models focus on the micro-level interactions in a system that may lead to unexpected emergent behaviour of the system. These models specify rules that determine agent behaviour (individual organisms, people or firms), where evolution of agents is feasible. The advantages of multiple-agent modelling are that: (i) the agents to be modelled are not identical and homogenous; (ii) the low level interactions between agents are important; (iii) spatial location is important; and (iv) it is easier to specify rules for the actions of individual agents than it is to specify the system behaviour.

7.5.4 Regional IA modelling

Integrated Assessment modelling is still on the way upwards of building up authority. Thus far, many IA models had a fairly abstract character, which is partly due to the global scale level at which IA models often operate. Notwithstanding the dominant orientation towards global modelling, there is a growing need and interest in regional IA models. Case studies at the local or regional level can convincingly demonstrate the added value of IA models. At the regional (or local) level the tensions between different forms of development can be made spatially-explicit. Regional development is characterised by increasing competition for space, resources and finances between a variety of stakeholders. An IA model of regional development then involves the estimation of trade-offs between space and resources claimed by economic development (industry, services, public transport), social development (houses, recreation, and private transport), and environmental development (nature preservation, quality of water, soil, air, noise pollution). So far, spatially-explicit modelling has been dominated by so-called GIS approaches. GIS models describe the spatial changes in a landscape in a detailed way, but they are static and thus don't give insights into the dynamic mechanisms behind past and future changes. On the other hand, models that endeavour to describe the dynamics of a number of relevant social, economic and ecological processes, system-dynamic models, do not adequately (or not at all) describe spatial changes. Recently, however, the technique of cellular automata has been developed, derived from Artificial Intelligence, that enables the linkage between system-dynamic and GIS models. Cellular automata are cells (automata) in a grid-field, where all cells in the grid-field communicate with each other through simple decision-rules or algorithms. In this way these protocols determine the dynamic spatial behaviour of all cells.

The basic procedure is then as follows: *macroscopic patterns* indicate changes on the macro-scale for a region with regard to: physical, demographic and socio-economic changes, which are generated by scenarios in combination with IA models. These macroscopic trends will be passed on to a region-specific cellular automata model, which will allocate the detailed localisation of these changes on the basis of its own spatial dynamics. Using site-specific information in combination with GIS-information, cellular automata allow for representation of the *microscopic dynamic patterns*. Then the cellular automata model will feedback local-specific information to the macroscopic trends to see what kind of consequences this may have. An example of dynamic spatially-explicit IA modelling is presented in Box 7.3.

Box 7.3
Region-specific modelling using cellular automata

Let's consider a nature area that is surrounded by a number of urban agglomerations. These cities have as a common characteristic that they want to expand, which brings the nature area under intensifying pressure because of macroscopic socio-economic trends. There is an increasing need for new houses, for drastic infra-structural changes. On the other hand, there is increasing opposition against those interventions in favour of preservation of the nature- and recreational function. According to the different perspectives of the various stakeholders, the nature area could be filled in differently. The anthropocentric perspective would lay the emphasis on the extension of social and economic activities, clearing the way for far-reaching infra-structural changes (road transport, air traffic, harbour function and housing). The ecocentric perspective will emphasise the important ecological and recreational function the area has. Whereas the

Box 7.3 Continued.

holistic perspective will favour the dynamic equilibrium between social, economic and ecological needs and interests (more houses with space for infra-structural changes in combination with a nature- and recreational function).

Macroscopic trends could indicate that in the nature area 50,000 new houses would be needed. The cellular automata model will allocate these houses on the basis of availability of land, suitability of land, accessibility of the location, the immediate vicinity of economic and industrial activities, etc. Next, the cellular automata will indicate what and where the amount of available land is, what the price and suitability is, etc. In addition, information can be provided on how to fill up space in the sense of: How will land use change? What kind of ecological valuable land will be lost? What is the risk of surface water pollution? What type of activity clusters will arise? Is there a tendency towards over-concentration or ghetto formation? Will economically weak activities be threatened by economically strong ones? Based on this information macroscopic trends could be adjusted, using simulation models, to reconsider the suitability and attractiveness of the specific region for new houses.

An interesting experiment is to supplement a cellular automata model by a multi-agent approach, which allows to model self-organisation phenomena in complex systems. Linking the decision-rules underlying cellular automata models to an agent-based model, enables to model the regional implications of human behaviour in a spatially-explicit way.

An interesting experiment is to supplement a cellular automata model by a multi-agent approach. Linking the decision rules underlying cellular automata models to an agent-based model, enables to model the regional implications of human behaviour in a spatially-explicit way.

Let's consider the same case study as discussed above. However, we now place agents in the form of decision-makers in control of the nature area that may evolve in any kind of direction. The decision-makers are responsible for infrastructure, land use, energy services and taxation. Based on the outline given by the decision-makers, the regional population, representing the second type of agents, decide whether or not to build, where to live (so long as there are houses for them), what routes to take to travel to work, etc. The fiscal environment encourages business to locate or move out, affecting employment. The result could be a rich and absorbing simulation of a developing nature area, where the decision-makers take "top-down" decisions, and the regional population makes their "bottom-up" responses.

7.6 The next generations of IA models

The next generation of IA models should meet the demands of scientists and decision-makers adequately, probably resulting in an even greater variety of IA tools and methods addressing the scientific issues and the policy debate at different levels. A guiding principle for the development of new tools and methods for IA could be a more realistic representation of the social and natural system. Still, the majority of the tools and methods we use follow the research paradigm that is based on equilibria, linearity, and determinism. The result is that incremental changes in parts of the subsystems we analyse, will cause gradual and incremental changes in the system as a whole. Unfortunately, the world does not function in such a simple, linear way. The real world shows strongly non-linear, stochastic, complex, and chaotic behaviour. This implies that incremental changes in conditions of subsystems may result in considerable changes in the results of the overall system, which may not be predicted beforehand.

Neo-classical economy, for instance, is based on the prevailing theory of general equilibrium. It supposes that the free market brings about an efficient resource allocation, except for a few "market failures" which can not be remedied by corrective taxes or subsidies. But without taking into account the possibility of multiple equilibria, the complex dynamics of price adjustments, and the role of positive feedbacks such as "learning by doing", which makes established technologies considerably cheaper over time, complex issues such as the role of induced technological change (ITC) can hardly be discussed. An alternative, more realistic way of describing the contemporary world economy would be to consider multiple equlibria, in the form of a transition of one economic equilibria to another. New approaches are underway that take into account multiple equilibria and positive feedbacks. However, they do not yet provide a coherent, comprehensive picture of the economy, still producing equivocal results.

A new and promising research paradigm that has emerged during the last decade is that of the *complex, adaptive systems approach*. Complex adaptive systems are composed of many agents who interact with their environment and can adapt to changes. These systems organise themselves, learn and remember, evolve and adapt. Evolutionary models, for instance, focus on social and natural systems as dynamically interwoven, where impacts on natural systems feedback to the social system, and vice versa. Both systems continuously adapt to human-induced perturbations of the system as a whole. This evolutionary approach is applied in various disciplines and uses a whole range of methods and tools, for instance, genetic algorithms. A genetic algorithm is a robust problem-solving approach based on the mechanics of the survival of the fittest. After reproduction, crossover and mutation, the fittest solutions pass on elements of themselves to later generations in an effort to find the most successful solutions. A crucial assumption is that the degree to which the agents' expectations meet measurements determines the fitness of the agents' perspective.

Some preliminary exercises are already underway, testing various elements of this new research paradigm (Rotmans and Dowlatabadi, 1998). It would be extremely useful to expand these experiments, and apply these experimental methods to a limited number of concrete case-studies. Obviously, in these experiments, the agents' representations used are quite abstract images of decision-makers and stakeholders. Only a weak isomorphism may exist between the real-world adaptation, and the way in which simulated agents adapt to their changing environment. Therefore, it may take a while before the application of complex, adaptive systems theory has matured to real world problems. Nevertheless, the first preliminary experiments, which deal with the notion of surprises and flexible strategies to respond to them, are encouraging and already produce interesting results.

Therefore the overall challenge for the next generation of IA models is to incorporate insights emerging from new scientific streams such as the areas of complex systems, adaptive behaviour, bifurcations, and ignorance about systems. Incorporation of these insights will broaden the scope of IA models and considerably improve their quality. In addition, the incorporation of cultural values, biases and preferences, which determine the manner in which complex problems are perceived and formalised, will allow the IA modelling community to inform the decision-making process in a more balanced way.

References

Alcamo, J. (1994). "IMAGE 2.0: Integrated Modeling of Global Climate Change.", Kluwer Academic Publishers, Dordrecht, The Netherlands.

Alcamo, J., Kreileman, E., and Leemans, R. (1996). "Global models meet global policy." *Global Environmental Change*, 6(4).

Alcamo, J., Shaw, R., and Hordijk, L. (1990). "The RAINS Model of Acidification: Science and Strategies in Europe.", Kluwer Academic Publishers, Dordrecht, The Netherlands.

Baybrooke, D., and Lindblom, C. E. (1963). A strategy of decision: Policy evaluation as a social process, London, UK.
Bertalanffy, L. v. (1968). General Systems Theory: Foundations, Development, Applications, Braziller, New York, USA.

Berk, M.M., and Jansen, M.A. (1997). The interactive scenario scanner: a tool to support the dialogue between science and policy on scenario development. RIVM Report No. 481508005, RIVM, Bilthoven, The Netherlands.

Checkland, P. (1981). *Systems Practice, Systems Thinking*, John Wiley and Sons, New York, USA.

Costanza, R., and Tognetti, S. S. (1998 (in press)). "Ecological economics and integrated assessment: Participatory process for including equity, efficiency and scale in decision-making for sustainability." , SCOPE, Paris, France.

Dinkelman, G. (1995). The interaction between problems and solutions in Dutch air pollution policy, Jan van Arkel, Utrecht, the Netherlands.

Dowlatabadi, H. (1998). "Sensitivity of climate change mitigation estimates to assumptions about technical change." *Energy Economics (accepted)*.

Dowlatabadi, H., and Morgan, M. G. (1993a). "Integrated Assessment of Climate Change." *Science*, 259, 1813-1814.

Dowlatabadi, H., and Morgan, M. G. (1993b). "A Model Framework for Integrated Studies of the Climate Problem." *Energy Policy*(March), 209-221.

Edmonds, J., Pitcher, H., Rosenberg, N., and Wigley, T. "Design for the Global Change Assessment Model." *Integrative Assessment of Mitigation, Impacts and Adaptation to Climate Change*, Laxenburg, Austria.

Fisher-Vanden, K. (1997), "International Policy Instrument Prominence in the Climate Change Debate: a case-study of the United States", IR-97-033, Kennedy School of Government, Harvard University and IIASA, Cambridge, USA and Laxenburg, Austria.

Forrester, J. W. (1961). *Industrial Dynamics*, MIT Press, Cambridge, USA.

Forrester, J. W. (1968). *Principles of Systems*, Wright-Allen Press Inc., Cambridge, USA.

Franz, W. (1997), "The Development of an International Agenda for Climate Change: Connecting Science to Policy", IR-97-034, Kennedy School of Government, Harvard University and IIASA, Cambridge, USA and Laxenburg, Austria.

Funtowicz, S. O., and Ravetz, J. R. "Managing Uncertainty in Policy-related Research." *Les experts sont formels: Controverse scientifiques et Decisions Politiques dans le domaine de l'environment*, Arc et Sanas, France.

Funtowicz, S. O., and Ravetz, J. R. (1990). *Uncertainty and quality in science for policy*, Kluwer, Dordrecht, the Netherlands.

Funtowicz, S. O., and Ravetz, J. R. (1994). "The Worth of a Songbird: Ecological Economics as a Post-Normal Science." *Ecological Economics*(10), 197-207.

Gibbens, J.H. (1993), "Statement of John H. Gibbons before the committee on energy and natural resources, US Senate", Office of Science and Technology, Washington, USA.

Goodman, M. R. (1974). *Study Notes in System Dynamics*, Wright-Allen Press Inc., Cambridge, USA.

Hordijk, L. (1991a). "An Integrated Assessment Model for Acidification in Europe," , Free University of Amsterdam, Amsterdam, The Netherlands.

Hordijk, L. (1991b). "Use of the RAINS Model in Acid Rain Negotiations in Europe." *Environmental Science and Technology*, 25(4), 596-603.

Jacoby, H. D. "Six Keys to Success in Interdisciplinary Work." Prospects for Integrated Assessment: Lessons learned from the case of climate change, Toulouse, France.

Jaeger, C. C., Barker, T., Edenhofer, O., Faucheux, S., Hourcade, J.-C., Kasemir, B., O'Connor, M., Parry, M., Peters, I., Ravetz, J., and Rotmans, J. (1997). "Procedural leadership in Climate Policy: a European Task." *Global Environmental Change*, 7(September).

Jager, W., van Asselt, M. B. A., Rotmans, J., Vlek, C. A. J., and Boodt, C. (1997). "Consumer behaviour: A modelling perspective in the context of integrated assessment of global change." *461502C17*, RIVM, Bilthoven, the Netherlands.

Kandlikar, M. and Sagar, A. (1997), "Climate Change Science and Policy: Lessons from India", IR-97-035, Kennedy School of Government, Harvard University and IIASA, Cambridge, USA and Laxenburg, Austria.

Lashof, D. A., and Tirpak, D. A. (1989). "Policy Options for Stabilising Global Climate." , US Environmental Protection Agency, Washington, USA.

Laszlo, E. (1972). The Systems View of the World: The Natural Philosophy of the New Development in Sciences, Braziller, New York, USA.

Long, M. and Iles, A. (1997), 'Assessing Climate Change: Co-evolution of Knowledge, Communities and Methodologies", IR-97-036, Kennedy School of Government, Harvard University and IIASA, Cambridge, USA and Laxenburg, Austria.

Martens, W. J. M. (1997). "Health impacts of climate change and ozone depletion: An eco-epidemiological approach," , Maastricht University, Maastricht, the Netherlands.

Martens, W. J. M. (1998). Health and climate change: Modelling the impacts of global warming and ozone depletion, Earthscan Publications, London, UK.

Meadows, D. H., Meadows, D. L., Randers, J., and Behrens, W. W. (1972). *The Limits to Growth*, Universe Books, New York, USA.

Morgan, G. M., and Henrion, M. (1990). Uncertainty - A Guide to Dealing with Uncertainty in Quantitative Risk and Policy Analysis, Cambridge University Press, New York, USA.

Nordhaus, W. D. (1979). *The Efficient Use of Energy Resources*, Yale University Press, New Haven, USA.

Nordhaus, W. D. (1992). "The DICE Model: Background and Structure of a Dynamic Integrated Climate Economy." , Yale University, USA.

Nordhaus, W. D. (1994). Managing the Global Commons: The Economics of Climate Change, MIT Press, Cambridge, USA.

Parson, E. A. (1996). "Three Dilemmas in the Integrated Assessment of Climate Change." *Climatic Change*(34), 315-326.

Parson, E. A., and Fisher-Vanden, K. (1997). "Integrated Assessment of Global Climate Change." *Annual Review of Energy and the Environment*, 22.

Prinn, R.H.J., Sokolov, A., Wand, C., Xiao, X., Yang, Z., Eckaus, R., Stone, P., Ellerman, D., Melillo, J., Fitzmaurice, J., Kicklighter, D. and Liu, Y. (1996), 'Integrated global system model for climate policy analysis: model framework and sensitivity studies', Global Change Center, Massachusetts Institute of Technology, USA.

Ravetz, J. R. (1997). "Integrated Environmental Assessment Forum: Developing Guidelines for Good Practise." *ULYSSES WP-97-1*, Darmstadt University of Technology, Darmstadt, Germany.

Root, and Schneider. (1995). "Ecology and climate: Research strategies and implications." *Science*, 269(52), 334-341.

Rotmans, J. (1990). *IMAGE: An Integrated Model to Assess the Greenhouse Effect*, Kluwer Academics, Dordrecht, The Netherlands.

Rotmans, J. (1997). "Indicators for Sustainable Development." Perspectives on Global Change: The TARGETS approach, J. Rotmans and H. J. M. de Vries, eds., Cambridge University Press, Cambridge, UK.

Rotmans, J., and de Vries, H. J. M. (1997). "Perspectives on Global Change: The TARGETS approach." , Cambridge University Press, Cambridge, UK.

Rotmans, J., and Dowlatabadi, H. (1998). "Integrated Assessment of Climate Change: Evaluation of Methods and Strategies." Human Choice and Climate Change: An International Social Science Assessment, S. Rayner and E. Malone, eds., Battelle Press, Washington, USA.

Rotmans, J. (1998), 'Methods for IA: the challenges and opportunities ahead', Environmental Modeling and Assessment 3, no. 3, 155-179.

Schellnhuber, H.-J., Block, A., Cassel-Gintz, M., Kropp, J., Lammel, G., Lass, W., Lienenkamp, R., Loose, C., Lüdeke, M. K. B., Moldenhauer, O., Petschel-Held, G., Plöchl, M., and Reusswig, F. (1997). "Syndromes of global change." *Gaia*, 6(1), 19-34.

Schneider, S. (1997). "Integrated assessment modelling of climate change: Transparent rational tool for policy making or opaque screen hiding value-laden assumptions?" *Environmental Modelling and Assessment*, 2(4), 229-250.

van Asselt, M. B. A. (1994). "Global Integrated Assessment Models as Policy Support Tools: A Triangular Approach." , University of Twente, Enschede, the Netherlands.

van der Sluijs, J. P. (1997). "Anchoring amid uncertainty," , Utrecht University, Utrecht, the Netherlands.

Weinstein, M. C., Coxson, P. G., Williams, L. W., Pass, T. M., Stason, W. B., and Goldman, L. (1987). "Forecasting coronary heart disease incidence, mortality and cost: the coronary heart policy model." *American journal of public health*, 77, 1417-1426.

Weyant, J., Davidson, O., Dowlatabadi, H., Edmonds, J., Grubb, M., Parson, E. A., Richels, R., Rotmans, J., Shukla, P., Tol, R. S. J., Cline, W., and Frankhauser, S. (1996). "Integrated Assessment of Climate Change: An overview and comparison of approaches and results." Economic and Social Dimensions of Climate Change, J. P. Bruce, H. Lee, and E. F. Haites, eds., IPCC, Cambridge University Press, Cambridge, UK.

WHO. (1994). "Health future research." *World health statistics quarterly*, 47(3/4).

Chapter 8

PERSPECTIVES AND THE SUBJECTIVE DIMENSION IN MODELLING

M. van Asselt and J. Rotmans

8.1 Introduction

Any exploration of future developments inevitably involves a considerable degree of uncertainty and Integrated Assessment (IA) modelling is no exception. As described in the previous chapters the uncertainties pertaining to the climate issue are fundamental. These uncertainties are inherent to the study of climate change, because of the following characteristics:

a) climate change is a global issue with long term impacts
b) the available data is lamentably inadequate, and
c) the underlying phenomena, being novel, complex and variable, are themselves not well understood.

Nevertheless, people in general, and decision-makers in particular, are interested in exploring the climate change issue in order to prepare climate-related policy measures. One particular role of science is to assist decision-makers in this effort by sketching images of the future climate and the likely consequences for humankind and the environment. As discussed in the previous chapters, one way to fulfil this role is to build an IA model and to use this model to explore possible pathways into the future and to assess appropriate policy responses. However, due to the fundamental uncertainties associated with climate change, such IA modelling is not a purely objective exercise. As argued by many scholars, no modelling is possible without a long sequence of decisions based on the modeller's subjective judgement

(Keepin and Wynne, 1984; Funtowicz and Ravetz, 1990, 1993; Morgan and Henrion, 1990; Robinson, 1991; Lave and Dowlatabadi, 1993; Morgan and Keith, 1995; Shackley and Wynne, 1995).

Subjectivity already enters the stage in the conceptual phase in deciding which elements will be included in the model, and which will be left out. Decisions in the modelling phase involving subjectivity range from which value is used for uncertain parameters to the choice of algorithms, from the treatment of problematic data sets to definitions of functional forms. Subjective judgements thus affect both uncertainties in model inputs and the model formulations themselves.

However, these subjective choices, even if they have been made in a very careful manner, are usually hidden in IA models. As soon as the uncertain parameter or relationship is covered by a quantitative value or a mathematical formulation, it is for a non-modeller impossible to grasp the difference between model parts based on facts and model parts that are more uncertain. Hitherto, there have been no adequate methods available to IA modellers for addressing the issue of subjectivity in a systematic way. Current uncertainty analysis techniques used in IA, e.g. Monte Carlo sampling and probability distribution functions, merely aim at data-uncertainty, but are not suitable for addressing subjectivity. Furthermore, classical methods suffer from the fact that they only address uncertainties in model inputs and neglect the structure of the model itself. Estimates of minimum, maximum and best guess values used in classical uncertainty analysis are often erroneous or misleading. The fact is that such estimates ignore the interactions among multiple, simultaneously occurring uncertainties, precisely the relationships that cannot be ignored in IA. Finally, because IA models of climate change are intended to capture as much as possible of the cause-effect relationships of climate change, such models are prone to an accumulation of uncertainties.

Current methods do not give decision-makers an indication regarding sources of the underlying uncertainties. As is stated in Chapter 7, IA models often reduce uncertainties to technical artefacts. By attaching probability distribution functions to uncertain model parameters, it is suggested that variations in parameter values do yield estimates of the uncertainty in the model outcome. However, that may be (partly) true in the mathematical sense, but it does not reflect the nature and source of the real world uncertainties. Furthermore, the kind of outputs produced by such uncertainty analysis (see Box 8.1) does not enable decision-makers to understand the magnitude of uncertainty. Quite often the reaction of decision-makers towards such results is either that they conclude that climate change is so uncertain and thus implausible, or they use the statistical mean value as the most likely trajectory on which they base their policy. The drawback of the latter attitude is that the resulting policy is likely to be inadequate if the actual development deviates from the mean value, which is highly plausible (because mean does not imply 'the most likely development in reality'). On the other hand, the first attitude may result in the situation in which waiting for certainty implies that preventive measures that can be taken today are ignored and that mitigation measures (i.e. measures that will be carried out as climate change has manifested itself) are superfluous, because the associated impacts are irreversible.

Box 8.1
Results of standard uncertainty analysis

Standard uncertainty analysis involves performing a significant set of model experiments in which the uncertain model inputs and parameters are varied according to pre-set distribution intervals. Intelligent clustering methods are used to experiment in an efficient manner. The different outcomes are collected and aggregated in terms of 95-percentiles (see Figure 8.1 for an example). This range has just statistical meaning, namely it indicates the range that comprises the trajectories of 95% of the outcomes. The remaining 2.5% beyond both the upper and lower bound are considered to be outliers that should not be taken into account. Apart from this statistical meaning the upper and lower bound as well as the range in between do not explain the adopted interpretation of the underlying uncertainties.

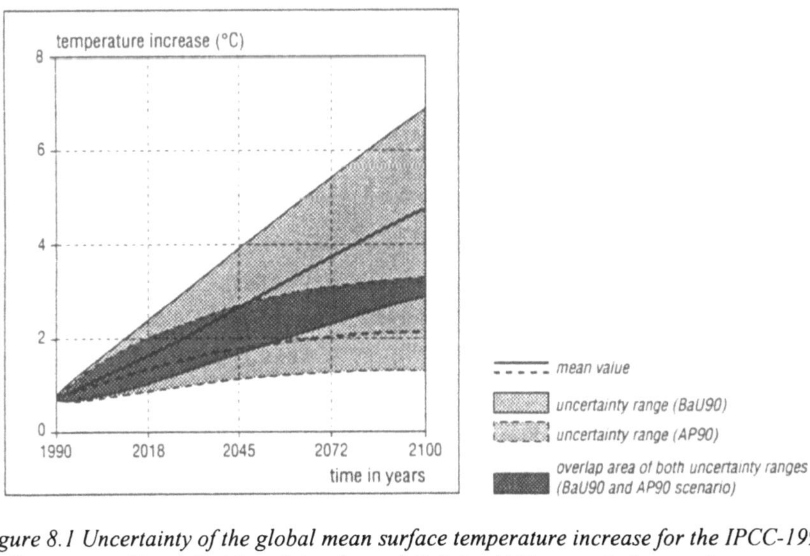

Figure 8.1 Uncertainty of the global mean surface temperature increase for the IPCC-1990 Business-as-Usual (BaU) and Accelerated Policies (AP) scenario (uncertainty range represents the '95% confidence band' range between the 2.5% and 97.5% percentiles)

In sum, uncertainty legitimates different ways of modelling climate change. Which interpretation of blind spots in scientific knowledge is considered the most appropriate, is a subjective decision of the modeller or the modelling team. It is expert judgement, but judgement nevertheless. Both modellers and users of models have to be aware of the subjective feature of IA models of climate change. To raise this awareness in a constructive manner, methods to address uncertainty and subjectivity in a systematic way are needed. Since recently, a multiple perspective approach is advocated in the IA community (e.g. Lave and Dowlatabadi, 1993; van Asselt and Rotmans, 1995, 1996; Rothman *et al.*, 1999; Biggs *et al.*, 1998; Kasemir *et al.*, 1998). In this approach the IA model does not merely include one perspective, but comprises plural perspectives. The present Chapter discusses a specific methodology that belongs to this class of methods, namely the methodology of 'multiple perspective-based model routes'. The method is developed by van Asselt and Rotmans (for further reading see van Asselt and Rotmans, 1996; van Asselt *et al.*, 1996) and is applied to the IA model TARGETS (see Chapter 7). This Chapter describes the main features of this methodology and illustrates by way of example how this method can be actually applied in a modelling exercise.

The aim of this Chapter is to give the reader a better understanding of the major role of uncertainty in climate change modelling, and to demonstrate why the subjective dimension should not be neglected in this modelling field. Furthermore, it aims to offer the reader but one way to deal with subjectivity in a systematic manner. After having studied the chapter, the reader should understand how such a multiple-perspective approach could be applied in other modelling efforts.

8.2 From subjectivity to plurality

Social psychology (e.g. Tversky and Kahneman, 1974, 1980, 1981)) indicates that subjective interpretations of uncertainty are seldom fully arbitrary. These interpretations tend to relate to value systems or so-called 'perspectives'. Social sciences teach us that there are different perspectives to which (groups of) people adhere.

Uncertainty thus provides room for different interpretations. The subjective interpretation in turn is related to the adopted perspective. This implies that models describing issues that involve significant uncertainty are, implicitly or explicitly, the reflection of merely one specific perspective. Because of the uncertainty and complexity, different perspectives that do not contradict the facts are all equally valid. There is no way to discriminate between the legitimacy of different plausible interpretations of what is (yet) unknown. In other words, in case of complexity we cannot escape pluralism.

One way to approach the issue of subjectivity in modelling exercises is to incorporate multiple perspectives into the model. In such an effort, uncertainty is "marked" by different interpretations according to different perspectives. Different interpretations are reflected by different choices concerning model inputs, parameter choices, model structure and equations. In this way, experimenting with the model implies choosing among perspective-dependent options. In this way, subjectivity is not hidden, but it is openly on the table.

Social sciences teach us that it is not necessary to involve the endless number of individual preferences to account for variety in perspectives (e.g. Rayner, 1987). The challenge then is to find a typology of perspectives that sufficiently covers the pluralism in value-systems. The advantage of a limited number of perspectives is that it enables modellers to prepare different coherent 'routes' through the model (see Box 8.2). The identified uncertainties in the model can be compared to man-made 'crossings' in a landscape. Model routes signify different choices at the various crossings. In technical terms, a perspective is used to invest such a model route with coherence. A perspective-based model route is thus a chain of interpretations of the crucial uncertainties coloured with the bias and preferences of a certain perspective.

Box 8.2
Model routes as walks through an uncertain landscape

A group of hikers embark on a walk from the same point in a landscape. They look at the landscape surrounding them, but they interpret what they see in different ways, so that they do not all march off in the same direction (Figure 8.2). For example, if they see a mountain ahead of them, one hiker will see it as a challenge and he wants to climb it, while a companion prefers a route that avoids the mountain for fear of falling off. A third hiker would like to enjoy the panorama at the top, but thinks climbing is too risky, and thus ascends by established footpaths. At every juncture the hikers are obliged to choose how they will cross the landscape. Furthermore, they cannot always sum up the situation in advance. The hikers' assessments of the landscape behind the mountain may range from envisaging another, even higher mountain to a ravine or a plateau. Their different conceptions of the unknown landscape make them inclined to choose different pathways.

Box 8.2 Continued.

Moreover, the landscape might change as a result of human actions. As an example we might consider the building of a campfire, the impacts of which may range from scorching the surface beneath the fire itself, to unexpectedly great and probably irreversible damage caused by a forest-fire. The risk assessments and thereby the attitude toward any specific action will certainly differ among the group of hikers: from abandoning the idea of a fire, or taking preventive measures such as building trenches filled with water round the site, to simply building a large campfire. Apart from being affected by human activities, the landscape can also be changed by natural factors like the weather, or events like earthquakes, which may create unexpected obstacles for the hikers.

Our hikers start out without much knowledge of the landscape and the processes within it, and similarly lack understanding of the consequences of both natural and human-induced changes on the features of the landscape. Nevertheless, they learn from experience how to survive. If, for example, they had once been washed out in a storm, because they had erected their tent on a hill, the hikers would subsequently pay more attention to their choice of campsite.

Any IA model designed to describe (aspects of) the functioning of the real world can be thought of as a 'map' of such a landscape. Multiple model routes can then be thought of as representations of the different pathways followed by the hikers, reflecting different assessments of the landscape's uncertain features.

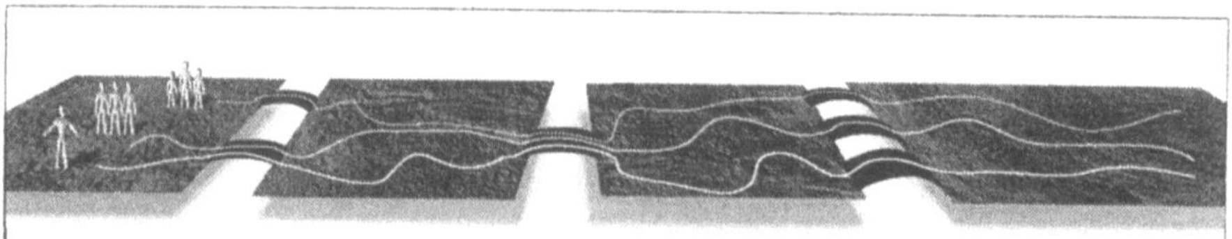

Figure 8.2 Metaphor of multiple routes

8.3 Framework of perspectives

The approach of multiple model routes clearly requires a general framework of perspectives. The first is step is to define the concept 'perspective'. The definition of perspective adopted in this chapter holds:

A **perspective** is a *coherent* and *consistent* description of the *perceptual screen* through which (groups of) people *interpret* or make sense to the *world* and its *social dimensions*, and which guide them in *acting*.

What is needed to operationalise the concept of multiple perspective-based model routes is a typology of perspectives that satisfies the following conditions:

- It should be social scientifically credible (i.e. in line with social scientific empirical insights and theoretical reasoning);
- It should be structured in a systematic manner;
- It should be generic, i.e. applicable to different temporal, geographical and aggregational scales;
- Each perspective should comprise both a 'worldview' (i.e. how people interpret the world) and a 'management style (i.e. how they act upon it).

Unfortunately, the social sciences do not provide a ready to hand, generally accepted typology of perspectives, which is independent of time and scale. To this end, van Asselt and Rotmans (1995, 1996, 1997) have chosen to adopt a top-down approach in designing a framework of perspectives that lends itself for implementing model routes. In the top-down approach social scientific insights and arguments are used to arrive at an aggregated typology.

Cultural Theory, as developed by anthropologists (Douglas and Wildavsky, 1982; Thompson *et al.*, 1990; Schwartz and Thompson, 1990; Rayner, 1984, 1991, 1992; O'Riordan and Rayner, 1991) and used in political science, has been a basic source of inspiration. Cultural Theory does not represent social science as a whole. Its schematism is rigid and cannot fully take account of the real world variety of perspectives. The typology associated with Cultural Theory is nothing more, but also nothing less, than an attempt to systematically address the complex issue of different perspectives at a high level of aggregation. In this effort the developers have aimed to take account of theoretical and empirical work done in this field (see e.g. Thompson *et al.*, 1990). As any model, it is merely a limited and defective reflection of reality. However, in spite of the lacunae and inconsistencies, we did not find a typology that better satisfied the criteria mentioned above. In the context of their aims, it therefore seems legitimate and reasonable to use the types put forward in Cultural Theory to characterise the spectrum of perspectives and to use the associated typology to implement multiple perspective-based model routes in IA models. Some alternative typologies found in the social scientific literature are presented in Box 8.3, to provide the reader with some food for comparison.

Reasoning from the adopted definition of a perspective, perspectives can be characterised by two dimensions: i) worldview, i.e. a coherent conception of how the world functions, and ii) management style, i.e. policy preferences and strategies. A 'worldview' can be defined as a coherent description of: i) a way of structuring reality, and ii) an accompanying vision of the relationship between human and the environment, see Figure 8.3 (Zweers and Boersema, 1994). A view on the structure of reality comprises a myth of nature, which addresses issues such as ecosystem vulnerability and a view on humanity. We will first describe worldview and management style consecutively, and then connect them by using Cultural Theory in order to provide a limited set of coherent perspectives. In this approach, Cultural Theory thus merely serves as an organising principle to connect worldview and management style.

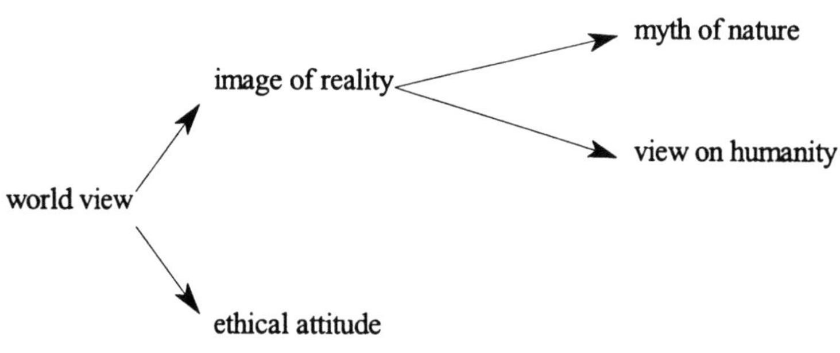

Figure 8.3 Multiple dimensions of worldview concept

Box 8.3
Examples of alternative typologies of perspectives

Acting perspectives

The Netherlands Scientific Council for Government Policy (WRR) distinguishes four 'acting perspectives' in their report 'Sustained risks: a lasting phenomenon' (WRR, 1994), which are applied to sectors, as energy, water and food supply. The WRR defines these acting perspectives reasoning from the question: 'are levers for a sustainable strategy especially sought in the consumption needs or in the production activities?', which yields the following perspectives:

	Changes in consumption needs	**No changes in consumption needs**
Changes in production activities	*Guard*	*Manage*
No changes in production activities	*Save*	*Utilise*

Utilise

In this acting perspective, only minimal adaptations are necessary to cope with the environmental and social risks. Both the present level of consumption and the production technology can, with minor adaptations, be continued without creating grave and structural problems. In the light of the actual knowledge on climate change, an increase of concentrations of greenhouse gases in the atmosphere is allowable. Waste of resources should be prevented. Reduction of greenhouse gases emissions might be a concomitant effect of waste reduction measurements, but is not an aim in itself.

Save

*Searching levers for sustainable strategies primarily in the volume of consumption and the consumption patterns, without affecting production activities, is characteristic for the **Save** acting perspective. The risk of scarcity for future generations dominates the attitude towards energy supply. Reduction of energy consumption is necessary. Strategies towards such a transformation include both energy-efficient technologies and an austerity of the energy-intensive Western life-style.*

Box 8.3 Continued.

Manage

*The **Manage** perspective can be described as changing the production activities, for example by technological shifts or by using alternative resources, while persisting at the present level of consumption. In this perspective, the environmental risks associated with the supply and use of energy are not acceptable, which holds for the risks of global warming in particular. Future energy consumption should be reduced by transformations of the production activities.*

Guard

*In the perspective **Guard,** changes in both consumption needs and production activities are needed for a sustainable future. In the very short term, energy supplying activities need to be transformed into a on recurring energy-based production system. Together with a significant increase of energy efficiency, austerity of the Western life-style is needed.*

Ethical positions

Coward and Hurka (1993) and Dotto (1993) studied the spectra of policy preferences from an ethical point of view. Central to their analysis is the issue of consequences: 'if an act has good consequences, this counts ethically in its favour; if it has bad, or especially, disastrous consequences, this counts ethically against it' (Hurka as cited in Dotto, 1993). But this raises yet another complex question: 'good and bad for whom?' They argue that the evaluation of consequences depends on whom 'ethical standing' is accorded to. Four different categories are distinguished: 1) humans here and now; 2) humans everywhere and now; 3) humans everywhere at all times; and 4) the environment valued for itself. Generally speaking the wider one throws the net of ethical standing, the more avoidance strategies are favoured over adaptation.

Humans here and now

In this ethical position merely people living in one's own family, ethnic group or country deserve ethical standing.. This approach favours adaptation over avoidance. Due to the time lag between the release of greenhouse gases now and its consequences, the negative effects will be felt by humans in the future, whom are accorded no ethical standing. On the other hand, if avoidance measures were to be implemented, there would be immediate economic and social costs for humans here and now, who do have ethical standing. This ethical position does not absolutely exclude avoidance measures, but such measures should have here-and-now benefits.

Box 8.3 Continued.

Humans everywhere and now

Including the rest of the world -and notably the developing countries- in one's ethical considerations moves even more strongly in the direction of adaptation strategies. If people presently living in developed countries did not want a dramatic drop in their standard of living and if people in the developing countries wanted a significant increase in their standard of living, strategies that allow increased greenhouse gas emissions were to be favoured.

Humans everywhere at all times

In the ethical position that grants ethical standing to humans at all times, good and bad consequences that occur in the future will be just as real as those that occur today. Therefore such future consequences matter ethically. Assuming that future effects of climate change will be largely negative, extending ethical standing to future generations pushes towards avoidance strategies. The same is true if ethical standing is extended not just to future generations, but to present and future generations all over the world. However, a concern for all humans at all times increases the ethical costs of avoidance measures: an avoidance strategy that would deny increased industrialisation might protect future generations, but it would also considerably harm present generations by perpetuating their low standard of living. In case the burden of avoidance falls entirely on the developed countries, this would undoubtedly require huge reductions in greenhouse gas emissions, leading to structural changes in the way of living. Achieving a level of well being above an acceptable threshold for all present and future humans demands for a climate policy of avoidance that includes measures to reduce population growth.

The environment

The idea of extending ethical standing to (components of) the natural environment -valuing them for themselves and not just for their utility to humans- marks a radical break with much traditional Western ethics, which have emphasised the anthropocentric point of view. The less radical form of this eco-centric view extends ethical standing to individual non-human organisms, particularly higher animals, which are thought to be capable of feeling pleasure and pain as humans are. The more radical position, which can be called 'holistic environmental ethics', extends ethical standing to species, ecosystems and even the biosphere as a whole. Individuals have ethical significance only through their contribution to the whole of which they are part. Both ecocentric positions favour avoidance strategies: no

Box 8.3 Continued.

significant additional costs in avoidance are seen, since the bad effects (if any) of this policy would fall almost entirely on humans. The negative consequences of global warming, however, will fall on the environment, i.e. affect a natural world that has ethical standing in its own right.

Five views on sustainable development

De Vries (1989) introduced five views on sustainable development, which typology Van de Poel (1991) applied to the issue of climate change: i.e. the Technocrat-Adventurer, the Manager-Engineer, the Steward, the Partner and the Humanist. Metaphors are used to characterise these views.

Technocrat-Adventurer

The prevailing image of this view is Spaceship Earth in its potential abundance. It is expansionistic in its outlook. Human resources in terms of skills, potential for adaptation and ingenuity are the means to overcome probable (physical) constraints. Technology and economic growth are the means to progress. Concerning the climate issue, the Technocrat-Adventurer emphasises the adaptability of the technological and economic system.

Manager-Engineer

The prevailing image of this view is Spaceship Earth in its physical finiteness: human activities are constrained by physical boundaries. In essence, the future is a management problem, focussing primarily on physical means and economic ends. Concerning climate change, the Manager-Engineer insists that measurements have to be taken to prevent passing the limits of the physical system, through technical solutions, based on international agreements.

Steward

An appropriate image for this view is the Garden Earth. A garden needs design and maintenance, but nature has to play her role as well. The essence is care. Applied onto the climate issue, the Steward adheres to what is commonly called the 'precautionary principle'. Preventive measurements that have no short-term disadvantages should be taken. Emphasis is put on international solidarity and the need of austerity in addition to technical and economic measures.

Box 8.3 Continued.

Partner

The metaphor assigned to this view is Gaia as put forward by Lovelock's hypothesis that life on Earth controls atmospheric conditions for contemporary biosphere (Lovelock, 1988). The relation between humans and nature is characterised by interdependence, harmony and partnership, and not by exploitation and utility. Concerning environmental problems in general, and the climate issue in particular, the emphasis is put on transformation of the present life style to an ecological responsible way of living.

Humanist

This view is expressed in the metaphor of humanity as 'enfolding organism'. In this view emphasis is put on 'higher aims' and the 'quality of life', which means that the Humanist does not focus on the establishment of rules and laws, but on a continuously reflexive (re)adjustment of choices. With regard to the climate issue, the Humanist plausibly advocates structural institutional and social transformations.

Cultural Theory

Cultural Theory (Douglas and Wildavsky, 1982; Thompson *et al.,* 1990; Schwarz and Thompson, 1990; Rayner, 1984, 1991, 1992; O' Riordan and Rayner, 1991) claims that distinctive sets of values, beliefs and habits with regard to risk are reducible to only a few ideal types. Cultural Theory generally distinguishes four perspectives from which people perceive the world and behave in it, namely: the ***hierarchist***, the ***egalitarian***, the ***individualist*** and the ***fatalist***. Hierarchism, egalitarianism, individualism and fatalism are not novel concepts. In some sense, Cultural Theory 'duplicates' conventional science. Hierarchism and individualism, for example, reflect a conventional duality expressed variously as 'collectivism versus individualism' and 'state versus market' (Grendstad, 1994).

In the following just the so-called 'active perspectives', i.e. hierarchist, egalitarian and individualist, are taken into account. The fatalist 'survivalists', for whom everything is a lottery, are excluded from our framework of perspectives, because they cannot systematically be described by any characteristic function and are frequently excluded from, or uninterested in, active participation in debates (Rayner, 1984, 1994). The hierarchist, egalitarian and individualist are considered as the extremes. The resulting spectrum that these extreme stereotypes define, comprises a variety of less extreme, or rather hybrid, perspectives on the structure of reality and the relationship between people and nature.

Grid-group scheme

Point of departure in Cultural theory are the group and grid dimensions as proposed by Douglas *et al.* (1982), and represented in Figure 8.4. The term *group* refers to the extent to which an individual is incorporated into bounded units. The greater the degree of incorporation, the greater the extent to which individual choice is subject to group determination. This dimension is comparable to what Hofstede (1994) describes as *collectivism versus individualism*. A group is individualistic, as the internal relationships between individuals are loose: everyone is expected to care for oneself. A group is collectivistic, as individuals are part of a close-knit group, which provide protection in exchange for unconditional loyalty. The term *grid* denotes the degree to which an individual's life is circumscribed by externally imposed prescriptions. The more binding the prescriptions are, and the more extensive their scope, the fewer the facets of social life that are open to individual negotiations. This dimension is comparable to the concept of *power distance* (Hofstede, 1994), defined as 'the degree in which less powerful members of a group expect and accept that power is distributed unequally'.

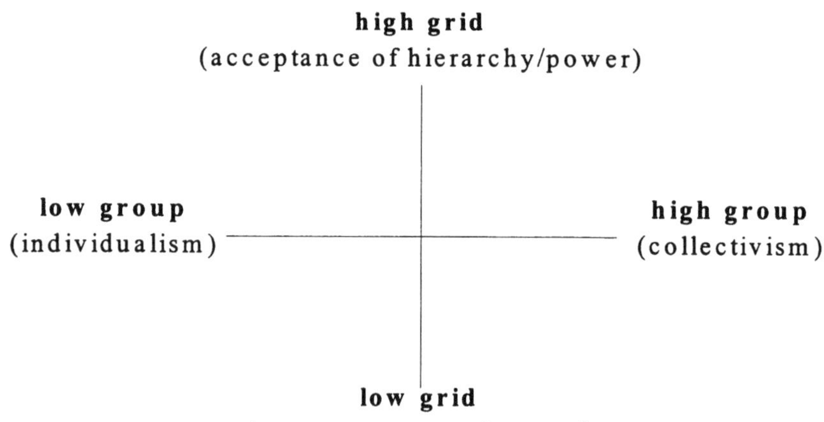

Figure 8.4 Group-grid frame

Along the grid dimension, it is the element of social control that sets the various perspectives apart from each other (Figure 8.5). Strong group boundaries coupled with minimal prescriptions produce strong relations that are **egalitarian**. Because such groups lack internal role differentiation, no individual is granted the authority to exercise control over another. When an individual's social environment is characterised by strong group boundaries and binding prescriptions, the resulting social relations are **hierarchical**. Individuals in this social context are subject to the control exerted by other members in the group as well as to the demands of socially imposed roles. The exercise of control (and more generally the very existence of inequality) is justified on the grounds that different roles for different people enable people to live together more harmoniously than alternative arrangements would. Individuals, who are neither bound by group incorporation nor by prescribed roles, inhabit an **individualistic** social context. In such an environment, all boundaries are provisional and subject to negotiation. Although the individualist is relatively free from control by others, this does not imply abstention from exerting control over others; the individualist's success is often measured in terms of the size of the following the person can command.

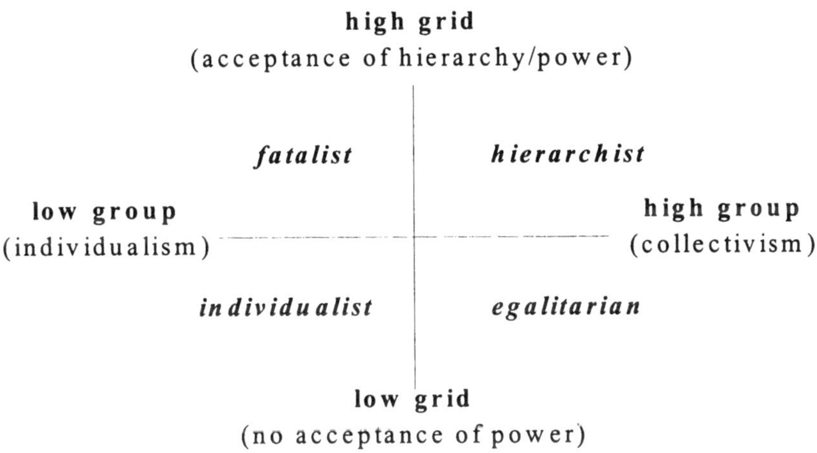

Figure 8.5 Perspectives derived from grid-group dimensions

Characteristics of the cultural perspectives

The view of nature of the three active perspectives can be characterised in terms of the various myths of nature identified by ecologists (e.g. Holling, 1979, 1986; Timmerman, 1986, 1989). There are four possible myths of nature: Nature Benign, Nature Ephemeral, Nature Perverse/Tolerant and Nature Capricious. Each myth can be represented graphically by reference to the metaphor of a ball rolling in a landscape (Figure 8.6).

A view on humanity involves addressing the question: who are we, what special qualities do we possess and how do we relate to other human beings? In the individualistic perspective, human nature is extraordinarily stable. Human nature is described as essentially *self-seeking*, i.e. human beings are considered to be rationally self-conscious agents seeking to fulfil their ever-increasing material needs. Such conception of human nature denies that individuals can be motivated by pursuing the collective good. This view on humanity is very individual-oriented. The view on human nature tied to the egalitarian perspective holds that humans are *born good, but are highly malleable*. Just as human nature can be corrupted by "evil" institutions (e.g. markets and hierarchies), so can it be rendered virtuous by an intimate and equal relationship with nature and other humans. Human satisfaction or self-realisation lies in spiritual growth and maturity, rather than in the consumption of goods. Egalitarians have it that individuals are members of groups, and that the collective is the more important dimension. The hierarchists consider humans to be born *sinful*, but who can and should be redeemed by good institutions for the sake of man and environment. This is again a collective view, but one in which certain groups have more power than others, who should obey the rules by those higher up in the hierarchy.

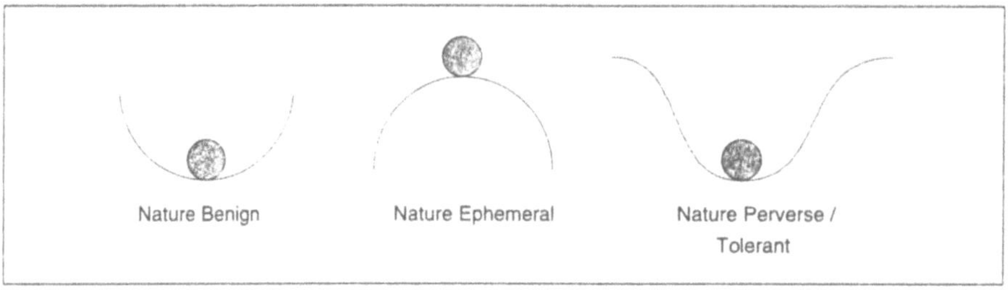

Figure 8.6 Myths of nature

With regard to the variety of ethical attitudes towards the relationship between man and nature two mainstreams of thought are usually recognised, namely: anthropocentrism and ecocentrism. Admitting that such a dichotomy is far too simple to account for the observed variation in ethical standpoints, we would like to involve 'partnership' as a third alternative. A schematic representation of the core of each stream of thought is set out in Figure 8.7.

Anthropocentrism is the view in which the development of humankind is considered as the ultimate goal of the universe, while nature is seen merely as providing resources, which are there to be exploited. Humans are the only beings that have value in themselves, and therefore everything else is valued in terms of benefit for humans, an attitude which Zweers and Boersema (1994) describe as 'humanity's monopoly on values'. Nature is seen as a purely factual, material entity. The fundamental attitude towards nature associated with anthropocentrism is 'supremacy' (Zweers and Boersema, 1994). The anthropocentric worldview is held by the individualistic perspective, that has an unlimited confidence in technological possibilities, and for whom there are no limits to growth.

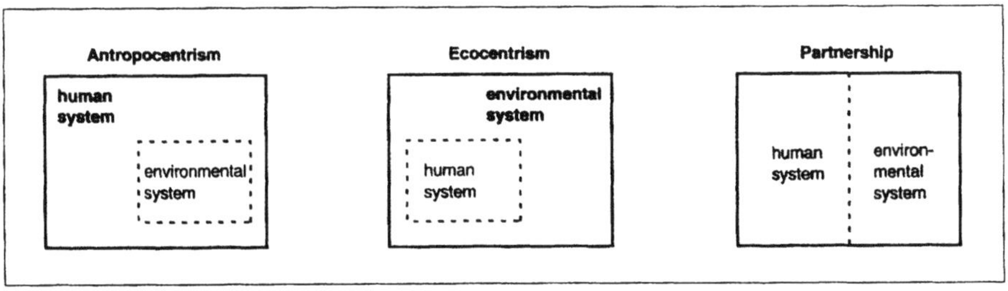

Figure 8.7 Streams of thought with regard to the relationship between man and nature

In discussions on environmental issues, ecocentrism is generally considered to be the opposite of anthropocentrism. Ecocentrism imputes intrinsic value to nature: which can have its own 'aims'. Aims, which will continue to be present, even if humans are not involved in any way. Instead of applying human criteria to value nature, each animal, each organism should be looked at according to its own nature. Nature is defined as a complex whole which organises and maintains itself and which includes everything, including humans. In other words, humans are "plain citizens" (Leopold, 1949) of the biotic community. The fundamental attitude can therefore be described as 'participation' (Zweers and Boersema, 1994), i.e. subordination of humanity to nature. The ecocentric worldview is associated with the egalitarian perspective.

While anthropocentrism and ecocentrism are common notions, partnership is usually less prominent in discussing attitudes towards the relationship between people and the environment. Following Zweers and Boersema (1994) we define partnership as the ethical attitude, which views the Earth as a totality, where humans and nature are valued equally. In this view the mutual dependency between humans and nature is stressed. For example, humans are dependent for their survival on nature because of resources, while nature's diversity and quality have to be protected against overexploitation. Thus, a balance between environmental and human values has to be ensured. The hierarchist perspective is associated with this ethical attitude.

Management styles are defined in terms of approaches towards response strategies, and thus include preferences regarding policy instruments. Cultural Theory enables us to ascribe stereotypes of management styles to the various perspectives. Hierarchism is associated with a preference for bureaucratic management, while the stereotypes of management styles to the various perspectives. egalitarian preference can be described as communal anti-managerialism. The egalitarian favours bottom-up collective initiatives. Individualism in its extreme manifestations advocates an anti-intervention laissez-faire attitude; let the market do the work. Following Cultural Theory (e.g. Schwarz and Thompson, 1990; Rayner, 1991), the hierarchistic management style can be characterised as 'control', while the egalitarian management style is characterised as 'preventive'. The management style associated with the individualistic worldview is characterised as 'adaptation'.

Hierarchists maintain that is possible to prevent serious global problems by careful stewardship of the opportunities that nature provides for controlled economic growth. Assessment of the bounds of human perturbation of the environment is an essential feature of this approach. The 'control' management style can be associated with the risk-accepting attitude shown by hierarchists. The salient value here is system maintenance. Therefore, rational allocation of resources is the principal mandate given to managing institutions. To this end regulation and financial incentives are the preferred policy instruments.

The egalitarian management style implies prudence and prevention, and therefore can be characterised as risk-aversive. The managing institutions are obliged to approach the environmental system with great care. Anticipation of negative consequences of human activities is a central element in this management style. Activities that are likely to harm the environment should be abandoned. With regard to the capitalist economic system, drastic structural social, cultural and institutional changes are advocated in order to arrive at a sustainable socio-economic structure· regardless of short-term economic costs. For example, policy measures should aim to radically reduce western consumption styles· The preventive management style stresses communication programmes, enhancement of education, and research, development and demonstration incentives oriented towards appropriate technology.

Individualists hold that changes provide new opportunities for human ingenuity that will be revealed through the workings of the marketplace. Negative consequences of human activities will be resolved by technological solutions. Economic development should not be curtailed by policy measures, so as to ensure a high level of technological progress and increasing welfare. The adaptive management style can easily be related to the risk-seeking attitude associated with individualism. The managing institutions must allow the market system to operate freely, and should therefore refrain from any regulation or prohibition. Preferred policy instruments are financial incentives, and research and development programmes.

8.4 Methodology of multiple model routes

Model routes consist of chains of alternative formulations of model relationships and model quantities. How to implement such model routes in an IA model? And who are involved in such an exercise? The answer to the latter question first. Implementing model routes in an IA model of climate change should be a collaborative effort involving modellers, analysts and Cultural Theory experts. The question of implementation is addressed by providing a process flow in terms of implementation steps. For your help, the multiple model route methodology is summarised in a flow chart (Figure 8.8), where Box 8.4 illustrates what building model routes implies in terms of the metaphor of a map of an uncertain landscape.

Step 1 - Controversy

The first step involves analysis of the relevant scientific debates in order to identify the major controversy within the scientific community. The reason for this entrance point is that value differences are at the core of such disputes (Colglazier, 1991). In other words, analysis in terms of controversy enables to identify disagreement among experts. To give an example of how such a controversy may look like: the controversy that is prevailing within the climate community can for example be summarised as:

Is the global climate disturbed in a serious and irreversible way; and if so to what extent, at what rate of change and with what regional pattern, and what are the human and environmental consequences? (Den Elzen et al., 1997)

The next step is to unravel this controversy in terms of facts, hypotheses and uncertainties. To this end comprehensive study of the scientific literature is needed. This search then leads to an inventory of the scientific uncertainties underlying the debate in the scholarly community.

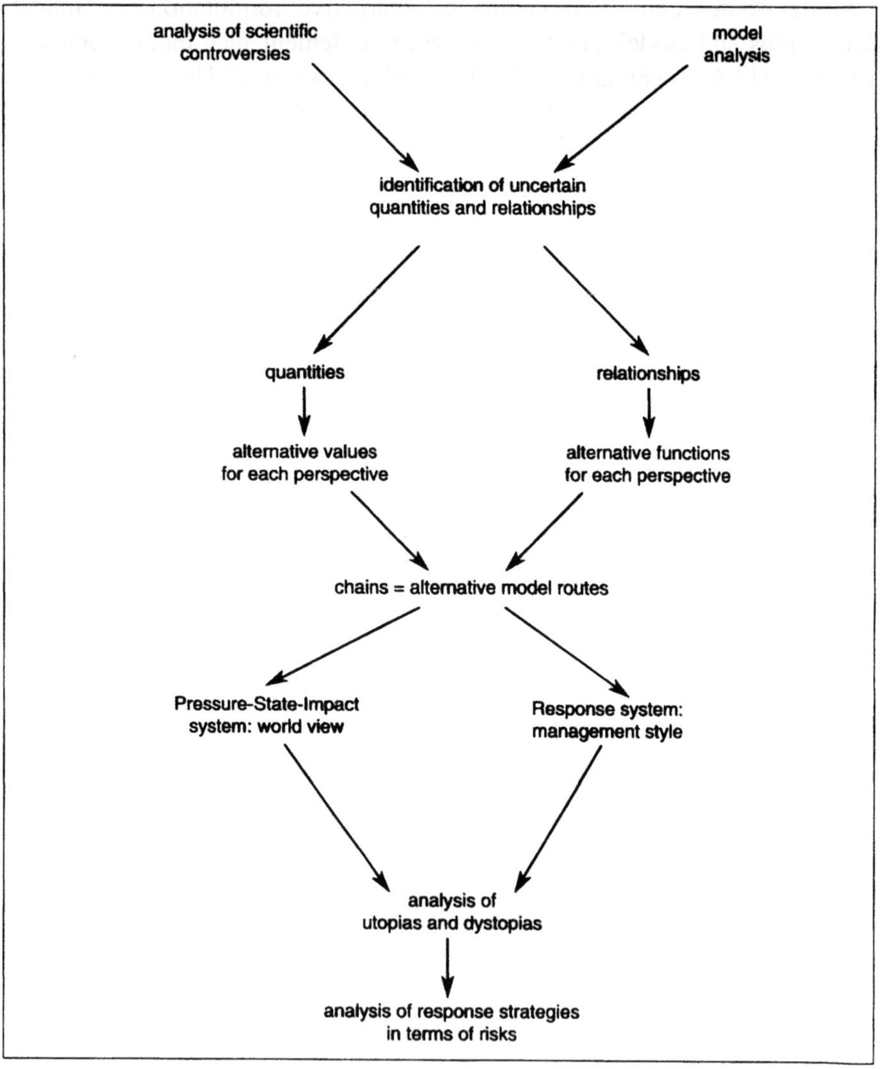

Figure 8.8 Flow chart of the multiple model route methodology.

Box 8.4
Building model routes in terms of mapping

In terms of the map-metaphor, it is the modellers' role to erect signposts in the landscape and to indicate them on the map to provide routes that reliably lead to a specific end. This requires the modellers to wander through the landscape on successive reconnaissances, during which they successively adopt another perspective. In other words, modellers are obliged to depart from their own preferred route, and have to acknowledge that other routes are likewise possible. Instead of providing merely one description of the functioning of (aspects of) the real world, thereby adopting a cavalier attitude to the uncertainties and leaving aside the wide variety of alternative perspectives, modellers need to illuminate the major uncertainties and to provide reasonable alternative interpretations. "Erecting signposts" is thus equivalent to identifying crucial uncertainties in state-of-the-art knowledge, indicating them in the model, and then providing mathematical translations consistent with the bias and preferences of a particular perspective. Less experienced walkers, notably, users of the model, can then choose to follow one of these routes to explore the landscape, instead of blindly hoping to stumble across a suitable path.

Step 2 – Model analysis

The inventory of the uncertainties enables to identify which model components are necessarily influenced by subjective judgement. All model choices that are related to the scientific uncertainties are by definition representing a particular interpretation of uncertainty. Such model choices comprise both the model formulations, i.e. the structure of the model, and the estimations of model inputs, i.e. quantification of parameters. By going back and forth between the scientific literature and the model, a list of crucial model components that involve subjectivity has to be produced. Sensitivity analysis is useful to decide what model components are crucial.

Step 3 – Qualitative interpretations

The next step involves defining alternative interpretations of the uncertain model components reasoning from the above framework of perspectives. Qualitative descriptions of the perspective-based interpretations should be the result of intensive and iterative dialogues between the participants in the implementation process. In case of the climate change issue, Table 8.1 summarises the main features of the qualitative descriptions of climate uncertainties in terms of the hierarchist, individualist and egalitarian perspective, as was used as starting point in the implementation of multiple model routes in the TARGETS model as described in Chapter 7, see also Rotmans and de Vries (1997).

Table 8.1 Qualitative descriptions of multiple perspectives on climate change (Based on: den Elzen *et al.*, 1997)

Crucial uncertainties	**Hierarchist**	**Egalitarian**	**Individualist**
Perception of climate change	Serious threat	Catastrophic threat	Environmental system can adapt to climate change
Overall effect of terrestrial feedbacks	Moderate CO_2 fertilisation	Amplifying temperature feedbacks	Strong CO_2 fertilisation
Geophysical feedbacks	Amplifying effect	Strong amplifying effect	Minor amplifying effect
Radiative effect	Moderate cooling of aerosols	Slightly cooling effect of aerosols	Strong cooling effect of aerosols

Step 4 – Implementation of model routes

The crucial next step involves translation of the qualitative descriptions of the perspectives into model terms. In the case of model quantities, this means that alternative values have to be determined. This is achieved by choosing different parameter values, which to a more or less extent represent the perspective's interpretation. When seeking more sophisticated implementation, subjective probability functions (Morgan and Henrion, 1990; Morgan and Keith, 1994) or alternatively the concept of 'fuzzy sets' may be considered. In case the scientific literature provides a range of quantitative estimates for the parameter/input under concern, this range can be used to arrive at reasonable model choices. In case no specific quantitative estimates are available, the team has to decide upon representative quantitative values.

In the case of uncertain relationships, functional forms need to be reformulated. In the simplest manifestation, this means either changing the function, or deleting, adding or changing the function's arguments. Notwithstanding their mathematical simplicity, such changes are fundamental in a conceptual sense. Ideally, different equations are provided by the scientific literature, which can be assigned to the perspective-based interpretations of the underlying uncertainty. These equations found in the literature can then easily be translated into model terms. In practise, it is rather likely that alternative interpretations are not spelled out in the literature in explicit mathematical terms. In this case, modellers have to study the perspective-based interpretation of the uncertain issue and search for an adequate translation into model terms.

Step 5 – Calibration

The freedom of interpretation is limited by facts. In model terms, this implies that the multiple model routes have to represent historic trends and should match available data. In other words, each model route should be calibrated against the observational data available.

MANAGEMENT STYLE	WORLD VIEW: **egalitarian**	**hierarchist**	**individualist**
egalitarian	***utopia***	dystopia	dystopia
hierarchist	dystopia	***utopia***	dystopia
individualist	dystopia	dystopia	***utopia***

Figure 8.9 Utopias and dystopias

Step 6 – Model experiments

The implemented and calibrated model routes allow for systematic experimentation with the IA model. To this end, the distinction between worldview and management style becomes relevant. Matching each perspective's management style to its respective worldview is a technique used to assess the utopias. A utopia is a world in which a certain perspective is dominant and the world functions according to the reigning worldview.

Dystopias describe either what would happen to the world if reality proved not to resemble the adopted worldview following adoption of the favoured strategy, or vice versa, i.e. where reality functions in line with one's favoured worldview, but opposite strategies are applied. Thus dystopias are mismatches between worldviews and management styles (Figure 8.9).

The result of this step is a flow of outputs representing various pathways into the future. A first inspection of these results involves evaluating whether the outcomes differ significantly from previous qualitative or model scenario studies. The next step then involves explanation of these differences. In case no failures in the model account for these differences, the experiment is likely to provide new insights into the future. Box 8.5 provides an example of such an assessment. In this example population projections generated with multiple perspective-based model routes are compared with the authoritative population projections of the United Nations (UN).

Step 7 – Risk assessment

The challenge is to go beyond the multiple perspective-based 'sprinkler' of "would-be-worlds" (Casti, 1997). Assessment implies providing general insights relevant for decision-makers that are valid independent of the preference for a certain perspective. To this end, van Asselt and Rotmans (1996) propose to indicate future risks by evaluating the whole range of utopian and dystopian outcomes and to assess whether the majority of the resulting images of the future shows undesirable trends. In this respect, dystopias are especially interesting. Dystopias indicate the risks of decision-making in uncertain conditions by showing to what kind of future the chosen strategies may lead, in the event that the adopted worldview fails to describe reality adequately.

One way to identify the risks for the future from a set of utopian and dystopian model experiments is to select two key indicators for the issue under concern. For example, if we consider the climate change issue, key indicators are global mean temperature increase and the rate of change. These two indicators enable to define low risk, moderate risk and high risk areas for the resulting two-dimensional space. The next step is to plot the outcomes of the set of utopian and dystopian experiments in this two-dimensional space. The resulting picture indicates whether from a comprehensive point of view, the assessed issue should be a key priority in decision-making. A further analysis of the experiments that are on the safe side may indicate which policy options seem to be interesting to consider. On the other hand, thorough analysis of the experiments that landed in the high-risk area may provide insights in which policy strategies may cause an undesirable future.

Box 8.5
Comparison of perspective-based population scenarios with UN projections *(based on Hilderink and Van Asselt, 1997)*

The spectrum emanating from the three utopias, generated with the population sub-model of the IA model TARGETS, runs from 7.9 to 13.0 billion people in 2100 for population, while the life expectancy projected for 2100 varies between 76 and 86 years. Comparison of our projections with the current UN projections (UN, 1993, 1995), of which the 2100 values range from 6.4 billion in the low projection to 17.6 in the high projection, shows a smaller range provided with perspective-based model experiments. Why do the perspective-based projections significantly differ from the ones provided by the UN? First of all, the utopian projections all assume that the fertility transition succeeds, it is only the onset and the transition rate that differ among the three perspectives. By the end of the 21st century the utopian projections show a fertility level approaching replacement level (2.1 children per woman), which eventually causes stabilisation of the population under the condition that the mortality level also stabilises. In designing their scenarios, the UN choose fertility levels which are taken constant for the next 50 to 100 years. Their assumptions with regard to fertility levels for the second half of the next century vary between 1.6 in the low-fertility case to 2.6 in the high-fertility case. The high UN projection thus comprises futures in which the fertility level is far above replacement level, and thus yields a higher population projection then our utopian projections reveal. Secondly, the population and health sub-model in TARGETS accounts for the recognised mutual relationships between fertility and mortality. The assumptions with respect to food, water and environmental circumstances underlying the utopian population projections reason from a secure food supply, access to clean fresh water for everyone and moderate environmental changes. The result is that the major health determinants, i.e. food security, safe drinking-water supply and health services, evoke in any perspective an epidemiological transition towards low mortality levels. A low mortality level implies a high life expectancy, which, in turn, increases the level of human development, stimulating a decrease in fertility level (see Niessen and Hilderink, 1997). On the other hand, a lower fertility level causes a smaller increase in the population, thereby enhancing the availability of resources per capita, which in turn results in better health conditions. These causal relationships imply that a high life expectancy excludes a high fertility level for the world at large. The perspective-based projections therefore do not comprise a scenario, which describes a large

Box 8.5 Continued.

excess of births over deaths for a healthy population. The high UN projection seems to presuppose such an implausible development.

The low UN projection provides a picture in which the population declines very fast, i.e. a decline of 1.4 billion in 50 years during the second half of the 21st century, which corresponds to an average decline of 30 million persons per year. None of the utopian projections shows a decline that can be compared to such a fast decrease. The rapid and huge decline in population as described in the low UN projection presupposes an excess of deaths over births. None of the stages of the demographic transition features a situation in which the crude death rate exceeds the crude birth rate. In other words, the low UN projection assumes a very low fertility level (i.e. a global average of 1.6 children per woman as early as 2050, which is supposed to remain constant for the next 50 years) and/or an extraordinary situation featuring very high mortality levels. One can probably imagine a large number of deaths due to famine and severe water shortages. As said before, the utopian projections do not account for grave lack of food and water. Other potential factors, which influence mortality, such as wars and natural disasters, are ignored in our model due to them only accounting for a negligible part of the total mortality. When, for example, the Gulf War, and wars in Rwanda, Sudan, Uganda, Angola and Liberia raged in 1990, only a mere 0.6% of the total mortality could be attributed to wars (WorldBank, 1993). What do the utopian projections tell us with respect to plausible futures in the light of state-of-the-art knowledge? They tell us that population projections exceeding 15 billion inhabitants on Earth by 2100 are implausible, even if we take account of a variety of perspectives on demographic and epidemiological dynamics. Globally speaking, a reverse fertility transition, resulting in an enduring situation of very high fertility rates and low mortality rates, would be necessary to obtain such high population numbers. Even countries presently featuring a high fertility rate show a rapid decline in fertility levels and an increase in life expectancy. In Kenya, for example, the number of children per woman fell from 7.4 to 5.2 in the last decade, while the life expectancy increased by three years. A situation in which, globally speaking, fertility levels remain high or even increase is therefore highly unlikely. The three images of the future suggest that an improvement in life expectancy in the course of the next century is probable. An extension of longevity is generally associated with a better health status. Summarising, the utopian experiments suggest that a decline in fertility and an improvement in health are reasonable pathways into the 21st century.

Step 8 – Communication of insights

Last but not least, the insights derived from perspective-based IA have to be communicated in an understandable manner to decision-makers. In this effort, it may be wise to translate concepts that are relevant for the analysis (hierarchist, egalitarian, individualist, utopia, dystopia) into terms that match closer to the decision-makers vocabulary.

8.5 Application of multiple model routes[1]

In the context of the present book, an example of the multiple model routes methodology in the field of climate change is highly relevant. To this end, this section discusses the application of the method to the CYCLES-sub-model of the TARGETS model. In this sub-model the climate-related global cycles of the basic elements C, N, P and S are described.

Reasoning from the controversy on climate change, as presented above, and the inventory of crucial uncertainties, as described in the previous Chapters (see Table 8.2 for a summary of the major uncertainties pertaining to climate change), we start the description of this particular application with step 3, which involves a more comprehensive description of the perspective-based interpretation of the crucial uncertainties in model terms. While the CYCLES-model of TARGETS primarily focuses on the biogeochemical cause-effects-chains (see Chapter 4), and in which emissions are inputs from other sub-modules of TARGETS on energy, economy and land, no explicit decision variables are to be distinguished in this particular sub-model. In our description we therefore focus on the worldview dimension.

The hierarchist interpretation of climate change uncertainties

The hierarchist perspective has it that global climate might be seriously and irreversibly disturbed by human activities. Consistent with the tendency to control, hierarchists tend to ignore speculations about possible processes and feedbacks, and confine themselves to the ones of which the probability of occurrence and magnitude are to a certain extent known. They interpret uncertainties in line with the estimates of prominent scientific experts and institutions, for example the Intergovernmental Panel on Climate Change (IPCC). Taking into account current IPCC estimates, hierarchists expect that CO_2- and N-fertilisation form important dampening mechanisms of the future atmospheric CO_2-concentration, see also Chapter 4. The overall future effect of the temperature feedbacks is a slightly positive one (IPCC, 1996). Geophysical feedbacks, especially water vapour, are assumed to have an amplifying effect on the initial warming process induced by anthropogenic

[1] This Section is based on den Elzen *et al.* (1997)

emissions of greenhouse gases. The radiative forcing related to sulphate aerosols is believed to explain most of the difference between the observed global temperature increase, and the global mean temperature increase as simulated by the previous generation of climate models. In sum, hierarchists will follow the conclusion of the IPCC that climate change is likely to occur in the long term if anthropogenic emissions of greenhouse gases continue to increase.

Table 8.2 Crucial uncertainties pertaining to climate change

Process	**Crucial uncertainty**
Primary processes in terrestrial biosphere	Terrestrial CO_2 uptake
Secondary processes in terrestrial biosphere	CO_2 fertilisation Temperature feedbacks Soil moisture changes
Radiation	Direct effects of aerosols Indirect effect of aerosols
Geophysical feedbacks	Water vapour Clouds

The egalitarian interpretation of climate change uncertainties

According to the egalitarian myth, minor human-induced changes already have a major influence on the behaviour of the environmental system. Egalitarians fear that nature is increasingly disrupted by the growing pressures on the environment and they have it that adaptation of natural systems to the new situations is not or only partly possible. Egalitarians expect environmental problems to be aggravated by amplifying feedbacks, although dampening feedbacks may delay serious disasters. Speculative positive feedbacks with possibly catastrophic impacts are considered, even it they are controversial within the scientific community. On the other hand, potential negative feedbacks tend to be ignored in this risk-aversive worldview. Egalitarians argue that global climate is already seriously and irreversibly disturbed, and believe that the net effect on terrestrial feedbacks is dominated by the positive temperature feedback on soil respiration. The temperature feedbacks affecting net primary production will be positive within a small range: slight temperature changes will result in an increase in the primary production, but radical changes will strongly reduce it. This results in a situation in which substantial climate change forces a decrease in the terrestrial uptake of CO_2, thereby amplifying the initial warming process. Egalitarians argue that present and anticipated future pressures are likely to result in major, probably catastrophic climate change. Positive geophysical feedbacks due to water vapour and clouds are expected to dominate the response of the climate system to initial disturbances. Cooling effects from aerosols are considered to be of minor importance. An increased concentration of aerosols is anyhow undesirable, because its sources, the SO_2-emissions from fossil fuel combustion, are associated with acidification.

The individualistic interpretation of climate change uncertainties

Consistent with the individualist's perception of nature, the environmental system is robust in reacting to disturbances and has the ability to adapt and evolve. Individualists do not believe that human activities result in irreversible impacts. The environmental system is either able to cope with the fluctuations, or humans are ingenious enough to find solutions. This attitude makes that individualists emphasise dampening feedbacks, even if they are speculative. On the other hand, amplifying feedbacks, if they occur, are considered negligible. Disturbance of the climate system due to human activities is thought to be minor: negative feedbacks as CO_2- and N-fertilisation will dominate the overall effect of terrestrial feedbacks. Consequently, the net primary production of terrestrial ecosystems, and also the food supply, may significantly increase. The climate-system is considered to be resilient and self-regulating, providing fairly stable mean climatic conditions. Individualists deny the possibility of persistent transformations of the climate. It is argued, for example, that radiative effects of aerosols might offset an eventual global warming, and that geophysical feedbacks will only slightly amplify any initial climate change.

These qualitative descriptions are used to implement multiple model routes in the CYCLES sub-model of TARGETS. Reasoning from estimates and equations found in the scholarly literature, the model routes involve perspective-based choices, for example, with respect to:

- CO_2-fertilisation factor β;
- The Q_{10}^{res} factor that indicates the rate of intensification of respiration processes given a 10°C temperature increase;
- The $Q_{sox,direct}$ factor that indicates the direct radiative effect of aerosols;
- The $Q_{sox,indirect}$ factor that indicates the indirect radiative effect of aerosols;
- The climate sensitivity factor λ that represents the geophysical feedbacks;

- The functional form of the relationship between temperature and net primary production.

Two examples illustrate the variety of interpretations among the three perspectives: i) The values for CO_2-fertilisation factor β range from 0 (i.e. no effect) in the egalitarian model route to 0.7 (i.e. substantial effect) in the individualistic model route, and ii) the relationship between net primary production and global mean temperature is a parabolic function in both the hierarchist and egalitarian model route, where it is a continuously increasing function in the model route representing the individualistic perspective.

Model experiments

With the three model routes various experiments have been carried out. The outputs of the utopian experiments in terms of atmospheric CO_2-concentration and global mean temperature increase are presented in Figure 8.10.

Inspection of these graphs leads to the interesting conclusion that a high global mean temperature increase (to about 3.5°C by the end of the next century) can be explained by a relatively low CO_2-concentration (i.e. the egalitarian utopia). On the other hand, a concentration of CO_2 in the atmosphere to about twofold the present level (i.e. the individualistic projection) does not necessarily yield high temperature increases. The individualistic utopia features a global mean temperature increase of about 1 °C in the course of the 21st century. Standard climate scenarios always show the concentration-temperature combinations low-low, middle-middle and high-high. How to explain the counter-intuitive results produced with multiple perspective-based model routes? Such an analysis of the climate controversy indicates by way of narratives how large the uncertainties are and how to understand outcomes in terms of interpretations of the underlying uncertainties. Comparison of the egalitarian and individualistic utopia teaches us how crucial the uncertainties concerning amplifying and dampening feedbacks are in projections of future climate conditions. The outcomes yield that if amplifying feedbacks dominate the biosphere-atmosphere response to initial warming signals, the absolute global temperature increase will be significant, even if the CO_2-emissions do not increase dramatically. On the other hand, if dampening feedbacks dominate, the future average climate will not be affected much, even by high CO_2-emissions.

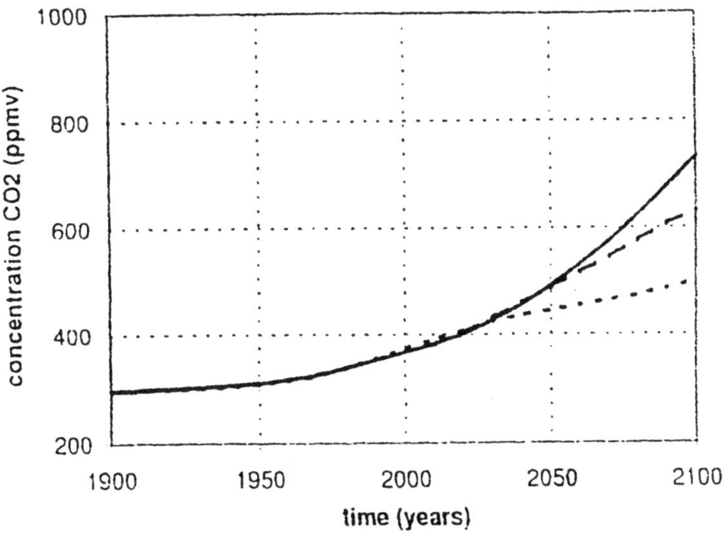

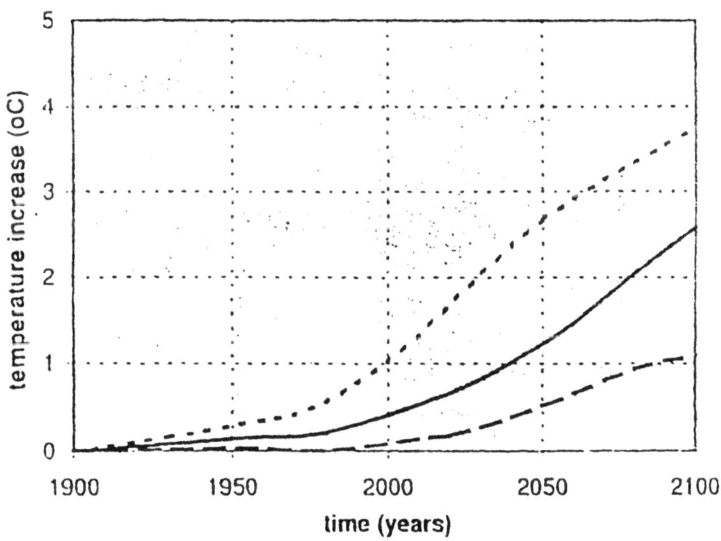

Figure 8.10: CO_2 concentrations and temperature increase over the period 1900-2100 for the three utopias
(—: hierarchist; ---: egalitarian; - - - : individualist (unpublished TARGETS runs)).

Another interesting observation is that it is not appropriate to consider the hierarchist as a middle-ground perspective. The utopian results with the

CYCLES model show that the hierarchist is not always in the middle. The hierarchistic utopia shows the highest atmospheric concentration of CO_2, which can be explained as follows: the hierarchistic perspective is more pessimistic than the individualist concerning the development and penetration of energy-efficient technologies and the market-mechanism with regard to alternative energy sources. On the other hand, the hierarchist does not expect severe life-style changes as the egalitarian perspective does. As a result the CO_2 emissions, and thus the atmospheric CO_2 concentration, in the hierarchistic utopia are significantly higher than in the two other utopian images of the future.

The outcomes of the CYCLES model in terms of atmospheric CO_2-concentrations have been compared with the scenarios designed by the Intergovernmental Panel on Climate Change (IPCC, 1996). The IS92a scenario is the principal one in the set of IPCC scenarios. In particular comparison between the hierarchistic scenario and the IS92a scenario is interesting, because the CO_2-emissions and land use changes as well as assumptions concerning the main parameters are comparable. The future atmospheric CO_2-concentration as simulated by the hierarchistic model route is about 5% higher than the central IPCC-1995 estimate. This is due to the differences in the carbon balancing mechanisms (den Elzen *et al.*, 1997) for which there are two possible explanations: (i) the N-fertilisation effect does not increase in the future, whereas the CO_2-fertilisation does; and (ii) the temperature-related feedbacks can lead to additional atmospheric CO_2-releases due to increased soil respiration.

Risk assessment

Dystopian experiments with the CYCLES-model have been performed in the following way: for each perspective a coherent set of inputs was available, signifying anticipated trends with regard to population, economy, energy, water and land. These trends account for policy strategies in line with the management styles of the three perspectives. In this way six dystopian combinations are possible, namely:

- hierarchistic inputs – egalitarian model route in CYCLES
- hierarchistic inputs – individualistic model route in CYCLES
- egalitarian inputs – hierarchistic model route in CYCLES
- egalitarian inputs – individualistic model route in CYCLES
- individualistic inputs – hierarchistic model route in CYCLES
- individualistic inputs – egalitarian model route in CYCLES

The projections associated with these experiments are presented in Figure 8.11, where global mean temperature increase and rate of global mean temperature increase are taken as the two key indicators that define risk levels.

As critical threshold, the levels proposed by Alcamo et al. (1996) are taken, namely an absolute global mean temperature increase of 1°C and a perspective-dependent level for the rate of global mean temperature increase, i.e. 0.1°C per decade for egalitarian assessments and 0.2°C per decade for hierarchistic and individualistic evaluation. In this way three risk areas can be defined:

I – high climate risk area: both climate risk levels are exceeded
II – moderate climate risk area: one climate risk level is exceeded
III – low climate risk area: no climate risk level is exceeded

Inspection of Figure 8.11 teaches us that in most cases (i.e. 7 out of 9) the climate risk levels are exceeded. To prevent a pathway into the future that leads to climate risks, reductions in CO_2-emissions in line with the egalitarian set of inputs are necessary. This implies that CO_2-emissions associated with energy use have to decline to a level of 3-4 GtC per year in 2100 (de Vries *et al.*, 1997), which is a reduction of about 50% compared to the present 6 GtC per year. However, if the global climate system functions according to the egalitarian worldview, climate risk levels are still exceeded. In other words, climate risks are inescapable in case amplifying feedback mechanisms dominate the biogeochemical process. Such a transition to lower emission levels is supposed to be triggered by the following portfolio of policy measures:

- energy policies that involve energy taxes, R&D programs on renewable energy sources;
- population and health policies aimed at human development and health care improvement;
- economic policies that advocate modest economic growth.

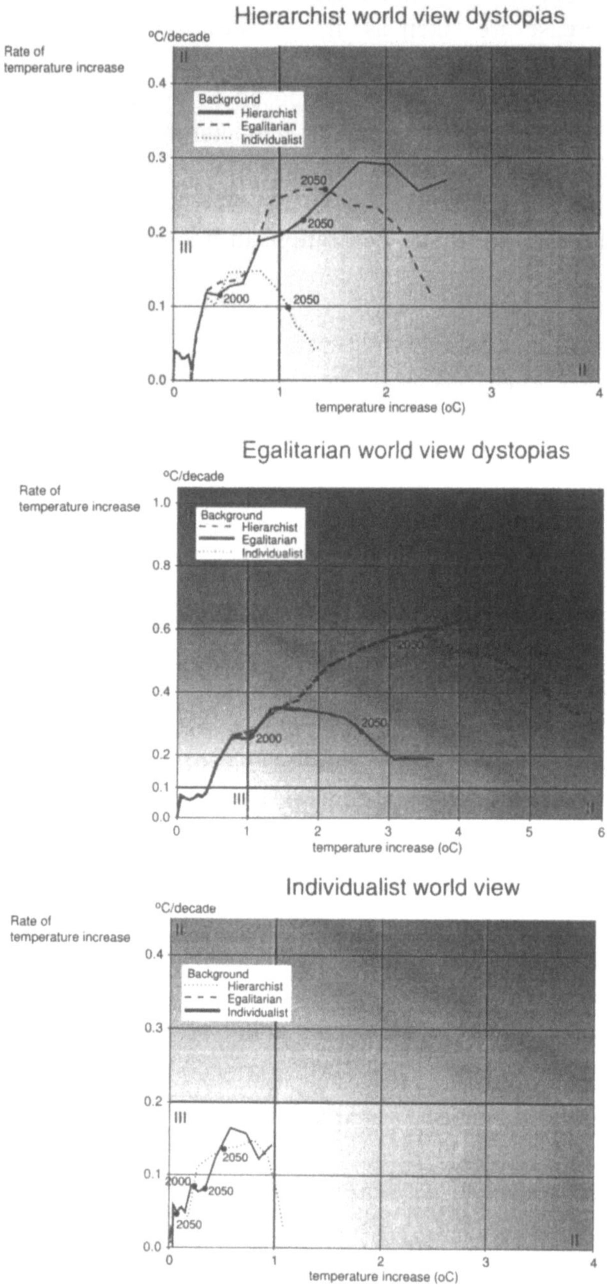

Figure 8.11 The absolute and relative global mean surface temperature increase for the three dystopias related to absolute and relative temperature target level (solid lines refer to utopias): I maximum climate risk area: both climate targets are exceeded; II high climate risk area: one climate target is exceeded; III low climate risk area: no climate target is exceeded

8.6 Conclusions

The future will remain uncertain, and images of the future will therefore remain subjective, and thereby perspective-dependent. Therefore, scientists engaged in exploring future developments should think in terms of 'management' of uncertainty, rather than its definitive resolution. Addressing a wide variety of perspectives is then one of the central tasks. In this Chapter a methodology is described that allows IA modellers to do so: to take account of a variety of perspectives in relation to uncertainty, and to 'grasp' uncertainty by expressing it in terms of risk.

The multiple model routes methodology may prove to provide a complementary methodology useful to uncertainty analysis in IA models. Using the perspective-based model routes as ways to envisage coherent clusters of various interpretations of the uncertainties in the model, differences in future projections can be motivated and explained, instead of merely arriving at minimum, maximum and 'best-guess' values. The advantage of the proposed approach is that it allows consideration of plurality in a rather consistent way. For the moment we recognise three major benefits: i) it enhances reflexivity of integrated assessors; ii) it enables us to consider more than one (hidden) perspective, and iii) it offers a framework for discussion on subjectivity in IA models.

Limitations of the methodology reside in the fact that it allows us to do no more than taking account of three rather deterministic perspectives, of which each is by definition a stereotype. Therefore, model routes can never be more than heuristic tools to explore consequences of value differences. Ambiguity in interpretation within the terms of reference of the various perspectives seems rather unavoidable. This reveals the need to view the implementation of model routes as an iterative process, going back and forth between interpreting qualitative descriptions and implementing model choices.

We see the model routes as building blocks in fashioning a new methodology for constructing scenarios that take account of a (greater) variety of perspectives. If one starts with a response scenario purely based on a single perspective, evaluation of the utopia and dystopias might lead to adjustment of the initial strategy. The resulting new set of policy options can again be analysed and evaluated in terms of dystopias and utopias, leading to insights useful in further refinement of the strategy. In such iterative and interactive experiments with IA models equipped with multiple model routes, uncertainties are no longer hidden, but rather used in a constructive manner in scenario exercises.

Current uncertainty analysis techniques do not give decision-makers a good insight into all uncertain dimensions relevant for decision-making. In particular the role of multi-interpretability of uncertainties and, as a consequence, the role of subjectivity are generally overlooked in such traditional uncertainty analysis. The perspective-based uncertainty analysis of climate change, as described in this Chapter, illustrates that considering multiple perspectives leads to insights that may be of interest for the global decision-making process on climate change.

References

Biggs, B., Robinson, J. B., Tague, C., and Walsh, M. (1998). "Tools for Linking Choices and Consequences." Sustainability Issues and Choices in the Lower Fraser Basin: Resolving the Dissonance, M. Healey, ed., UBC Press, Vancouver, Canada.

Casti, J. L. (1997). Would-be worlds: How simulation is changing the frontiers of science, John Wiley & Sons, New York, USA.

Colglazier, E. W. (1991). "Scientific Uncertainties, Public Policy and Global Warming: How Sure is Sure Enough?" *Policy Studies Journal*, 19(2), 61-72.

Coward, H., and Hurka, T. (1993). "Ethics and Climate Change: The greenhouse effect.", Wilfrid Laurier University Press, Waterloo, Canada.

de Vries, H. J. M. (1989). "Sustainable resource use: An inquiry into modelling and planning," , University of Groningen, Groningen, the Netherlands.

de Vries, H. J. M., Buesen, A. H. W., and Jansen, M. A. (1997). "Energy systems in transition." Perspectives on Global Change: The TARGETS approach, J. Rotmans and B. de Vries, eds., Cambridge University Press, Cambridge, UK.

den Elzen, M. G. J., Beusen, A. H. W., Rotmans, J., and van Asselt, M. B. A. (1997). "Human disturbance of the global biogeochemical cycles." Perspectives on Global Change: The TARGETS approach, J. Rotmans and B. de Vries, eds., Cambridge University Press, Cambridge, UK.

Dotto, L. (1993). *Ethical choices and global greenhouse warming*, Wilfrid Laurier University Press, Waterloo, Canada.

Douglas, M., and Wildavsky, A. (1982). *Risk and Culture: Essays on the Selection of Technical and Environmental Dangers*, University of California Press, Berkley, USA.

Funtowicz, S. O., and Ravetz, J. R. (1990). *Uncertainty and quality in science for policy*, Kluwer, Dordrecht, the Netherlands.

Funtowicz, S. O., and Ravetz, J. R. (1993). "Science for the Post-Normal Age." *Futures*, 25(7), 739-755.

Grendstad, G. (1994). "Classifying Cultures." *9502*, Norwegian Research Center on Organization and Management and Department of Comparative Politics, Bergen, Norway.

Hilderink, H. B. M., and van Asselt, M. B. A. (1997). "Population and Health in Perspective." Perspectives on Global Change: The TARGETS approach, J. Rotmans and H. J. M. de Vries, eds., Cambridge University Press, Cambridge, UK.

Holling, C. S. (1979). "Myths of Ecological Stability." Studies in Crisis Management, G. Smart and W. Stansbury, eds., Butterworth, Montreal, Canada.

Holling, C. S. (1986). "The Resilience of Terrestrial Ecosystems." Sustainable Development of the Biosphere, W. C. Clark and R. E. Munn, eds., Cambridge University Press, Cambridge, UK.

IPCC. (1996). *Climate Change 1995: The Science of Climate Change*, Cambridge University Press, Cambridge, UK.

Kasemir, B., van Asselt, M. B. A., Dürrenberger, G., and Jaeger, C. C. (1998). "Integrated Assessment: Multiple Perspectives in Interaction." *International Journal of Environment and Pollution (accepted)*(special issue 'Methods and Models for Decision Support').

Keepin, B., and Wynne, B. (1984). "Technical Analysis of the IIASA energy scenarios." *Nature*, 312, 691-695.

Lave, L. B., and Dowlatabadi, H. (1993). "Climate Change: The effects of personal beliefs and scientific uncertainty." *Environmental Science and Technology*, 27(10), 1962-1972.

Leopold, A. (1949). *The Land Ethic*, Oxford University Press, Oxford, UK.

Lovelock, J. E. (1988). *The Ages of Gaia: A Biography of Our Living Earth*, Norton and Company, New York, USA.

Morgan, G. M., and Henrion, M. (1990). Uncertainty - A Guide to Dealing with Uncertainty in Quantitative Risk and Policy Analysis, Cambridge University Press, New York, USA.

Morgan, M. G., and Keith, D. W. (1995). "Subjective Judgments by Climate Experts." *Environmental Science and Technology*, 29(4/5), 468-476.

Niessen, L. W., and Hilderink, H. (1997). "The population and health submodel." Perspectives on Global Change: The TARGETS approach, J. Rotmans and B. de Vries, eds., Cambridge University Press, Cambridge, UK.

O' Riordan, T., and Rayner, S. (1991). "Risk Management for Global Environmental Change." *Global Environmental Change*, 1(2), 91-108.

Rayner, S. (1984). "Disagreeing about Risk: The Institutional Cultures of Risk Management and Planning for Future Generations." Risk Analysis, Institution and Public Policy, S. G. Hadden, ed., Associated Faculty Press, Port Washington, USA.

Rayner, S. (1991). "A Cultural Perspective on the Structure and Implementation of Global Environmental Agreements." *Evaluation Review*, 15(1), 75-102.

Rayner, S. (1992). "Cultural Theory and Risk Analysis." Social Theory of Risk, G. D. Preagor, ed., Westport, USA.

Rayner, S. (1994). "Governance and the Global Commons." *Discussion Paper 8*, The Centre for the Study of Global Governance, London School of Economics, London, UK.

Rayner, S. F. (1987). "Risk and relativism in science for policy." The social and cultural construction of risk, B. B. Johnson and V. T. Covello, eds., Reidel, Dordrecht, the Netherlands.

Robinson, J. B. (1991). "Modelling the interactions between human and natural systems." *International Science Journal*(130), 629-647.

Rotmans, J. and de Vries, H.J.M. (1997), "Perspectives on Global Change: the TARGETS approach", Cambridge University Press, Cambridge, U.K.

Rothman, D., Robinson, J., and Biggs, D. (1999). "Signs of Life: Linking Indicators and Models in the Context of QUEST." Ecological Economics and Integrated Assessment: A participatory process for including equity, efficiency and scale in decision-making for sustainability, R. Costanza and S. Tognetti, eds., SCOPE, Paris, France.

Schwarz, M., and Thompson, M. (1990). *Divided We Stand: Redefining Politics, Technology and Social Choice*, Harvester Wheatsheaf, New York, USA.

Shackley, S., and Wynne, B. (1995). "Integrating Knowledges for Climate Change: Pyramids, Nets and Uncertainties." *Global Environmental Change*, 5(2), 113-126.

Thompson, M., Ellis, R., and Wildavsky, A. (1990). *Cultural Theory*, Westview Press, Boulder, USA.

Timmerman, P. (1986). "Myths and Paradigms of Intersections Between Development and Environment." Sustainable Development of the Biosphere, W. C. Clark and R. E. Munn, eds., Cambridge University Press, Cambridge, UK.

Timmerman, P. (1989). "The Human Dimensions of Global Change: An International Programme on Human Interactions with the Earth.", HDGCP Secretariat, Toronto, Canada.

Tversky, A., and Kahneman, D. (1974). "Judgement under Uncertainty: Heuristics and Biases." *Science*, 185, 1124-1131.

Tversky, A., and Kahneman, D. (1980). "Causal Schemes in Judgements under Uncertainty." Progress in Social Psychology, M. Fishbein, ed., Lawrence Erlbaum Associates, Hillsdale, USA.

Tversky, A., and Kahneman, D. (1981). "The Framing of Decisions and the Psychology of Choice." *Science*, 211, 453-458.

UN (1993). "World Population Prospects: The 1992 revision." *ST/ESA/SER.A/135*, United Nations, Department for Economic and Social Information and Policy Analysis, New York, USA.

UN (1995). "World Population Prospects: The 1994 revision." , New York, USA.

van Asselt, M. B. A., Beusen, A. H. W., and Hilderink, H. B. M. (1996). "Uncertainty in Integrated Assessment: A Social Scientific Approach." *Environmental Modelling and Assessment*, 1(1/2), 71-90.

van Asselt, M. B. A., and Rotmans, J. (1995). "Uncertainty in Integrated Assessment Modelling: A Cultural Perspective-based Approach." *RIVM-report no. 461502009*, National Institute of Public Health and the Environment (RIVM), the Netherlands, Bilthoven.

van Asselt, M. B. A., and Rotmans, J. (1996). "Uncertainty in Perspective." *Global Environmental Change*, 6(2), 121-157.

van Asselt, M. B. A., and Rotmans, J. (1997). "Uncertainties in perspective." Perspectives on Global Change: the TARGETS approach, J. Rotmans and B. de Vries, eds., Cambridge University Press, Cambridge, UK.

van de Poel, I. (1991). "Environment and Morality." (in Dutch) M.Sc. thesis, University of Twente, Enschede, the Netherlands.

WorldBank. (1993). *World Development Report 1993 - Investing in Health*, Oxford University Press, New York, USA.

WRR. (1994). *Sustained Risks: A Lasting Phenomenon*, WRR: Netherlands Scientific Council for Government Policy, SDU Uitgeverij, The Hague, The Netherlands.

Zweers, W., and Boersema, J. J. (1994). "Ecology, Technology and Culture: Essays in Environmental Philosophy." White Horse, Cambridge, UK.

Chapter 9

GLOBAL DECISION MAKING: CLIMATE CHANGE POLITICS

J. Gupta

9.1 Introduction[1]

Science can provide the basis for political action. However, as Shiva and Bandyopadhyay (1986) argue, scientific evidence can only generate the social authority for political action on complex environmental problems when there is consensus within a society about the nature of the problem. Scientific information is not automatically translated into political decisions, but goes through a complex process in which stakeholders and decision-makers interpret the information and negotiate with each other.

This chapter examines the science policy relationship with respect to the climate change problem. It first looks at the role of scientific uncertainty and controversy on the climate change issue (9.2.1), the different types of scientific reasoning (9.2.2), and then explains different theories on how science influences the policymaking process (9.2.3 and 9.2.4). It develops an integrated model on the relationship between science and policymaking (9.2.5). Sections 9.3-9.6 review the formation and development of the climate change regime in terms of the different stages identified in the integrated model.

[1] This Chapter uses and further develops ideas that have been published in Gupta 1997. It is based on on-going research being undertaken within the post-doctoral research work on 'Climate Change: Regime Development in the Context of Unequal Power Relations' which is being financially supported by the Netherlands' Organization for Scientific Research

9.2 From scientific description to problem definition

9.2.1 Scientific uncertainty and controversy

From the previous chapters it is clear that climate change science is beset with *uncertainties* in relation to the predictions of future climate change, because of the incomplete knowledge on the interrelationships between the causal factors and complex feedback effects. Till recently, governments did not take action on problems that may never occur, i.e. uncertain problems. However, uncertain problems such as climate change may have huge irreversible impacts. While countries hope that they will be "winners", the very uncertainty of the nature of the change may in itself make countries "losers". Hence, at the Second World Climate Conference (SWCC, 1990) scientists and policymakers adopted the *precautionary principle*, which was later also adopted in the Rio Declaration of the United Nations Conference on Environment and Development (UNCED, 1992) and is framed as follows in the United Nations Framework Convention on Climate Change (FCCC, 1992): "The Parties should take precautionary measures to anticipate, prevent or minimise the causes of climate change and mitigate its adverse effects. Where there are threats of serious or irreversible damage, lack of full scientific research shall not be used as a reason for postponing such measures, .." (Article 3 (3)).

The precautionary principle assumes that there is scientific consensus that the problem is serious, albeit uncertain. While policymakers and scientists associated with the Intergovernmental Panel on Climate Change (IPCC) conclude on the basis of all the available data and research that "the balance of evidence suggests a discernible human influence on global climate" (Houghton *et al.*, 1996: 4), there are others who are sceptical about the way in which scientific results are generated and presented (Singer, 1996) and argue that not only are specific results uncertain and questionable (Idso, 1996; Corbyn, 1996; Segalstad, 1996; Barrett, 1996), but that the scientific consensus is in reality "political consensus" and thus *controversial* (Priem 1995; Boehmer-Christiansen 1995; 1996).

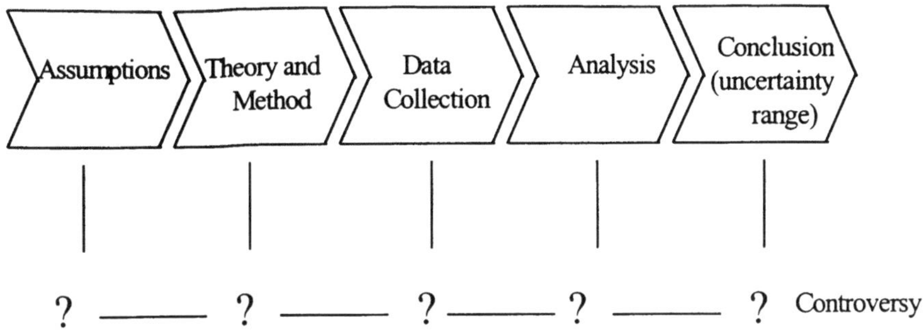

Figure 9.1 Uncertainty and controversy

The uncertainty of the problem makes decision-making complex; but the controversy surrounding the uncertainty makes it even more complex and may lead to the precautionary principle being interpreted symbolically.

9.2.2 Types of science and problems

Types of science

The 'uncertainty' and 'controversy' relate to different types of science and scientific practice. The philosophy of the dominant method of science is that scientific research does not embody any specific values or ethics and consists of different types of experiments which aim to explain reality. In general, scientific research implies the adoption of certain explicit assumptions, the formulation of certain research hypotheses, the development of a methodological and theoretical framework, collection of data, analysis and the testing of the hypotheses and conclusions. The research results aim to be accurate and objective, given the assumptions, theoretical framework and methodology. The bulk of modern research uses such scientific techniques. Such science could be referred to as '*normal science*' (cf. Kuhn, 1962).

There is, however, an alternative school of scientific thought that argues that no scientific research is truly 'objective'. This school argues that, first, politicians and policymakers influence the scientific methods developed through selective financing and promotion of projects, through institutional structures where such methodologies are to be developed and through other more subtle ways. Second, science is also not entirely value-free, since scientific research is influenced by the dominant ideology of science, i.e. rationality, dominant political and economic ideologies such as individualism, capitalism, liberalism, etc. and sub-conscious cultural views and values of the researchers. Critics argue in favour of systemic approaches to science in which social aspects are taken into account and an integrated, holistic view is developed. "Modern scientific knowledge identifies development with sectoral growth, ignoring the underdevelopment introduced in related sectors through negative externalities and the related undermining of the productivity of the ecosystem; it identifies economic value merely with exchange value of marketable resources, ignoring use values of more vital resources and ecological processes and it identifies utilisation merely with extraction, ignoring the productive and economic functions of conserved resources" (Bandhyopadhyay and Shiva, 1985: 199). They argue further that the introduction of resource and energy-intensive production technologies leads to the creation of wealth for a small minority, while at the same time undermining the material basis for the survival for the large majority. Dominant scientific methods tend to reduce data (reductionism) and remove it from its context (decontextualisation Miller 1993: 566) and tend to lack 'social responsibility'. Hence, Myrdal (1944: 1043) argues that "there is no other device for excluding biases in social science than to face the valuations and to introduce them as explicitly stated.." "Science is used as a final arbiter in all resource conflicts. The 'scientific' is taken as synonymous to public interest. However, since dominant science is partisan, decisions based on it will serve the special interest groups. Public interest science is a tool which makes explicit the political nature of partisan science and makes it a factor located within environmental conflicts, not a source of independent and neutral judgements about conflicts" (Shiva and Bandyopadhyay, 1986: 87). In *public interest science*, scientists, policymakers and target groups, representing different interests and ideologies are invited to define and address the problem in an interactive process in order to generate the sort of policy measures that are likely to succeed at the implementation level. Such science does not exclude the knowledge of lay persons, but instead considers them experts since "an expert is not a special kind of person, but each person is a special kind of expert, especially with respect to his or her own problems" (Mitroff *et al.*, 1983: 125). Hence, some scientists argue in favour of *post-normal science* in relation to problems where facts are uncertain, values in dispute, stakes high and decisions urgent. In such post-normal science, a dialogue between stakeholders is considered necessary in which values, perspectives and policy considerations are explicitly taken into account in the scientific process. Such science is especially necessary in order to take decisions on how the access to products and services of the environment are to be shared within the domestic or the North-South context (cf. Functowicz *et al.*, 1996).

Third, since scientific research tends to be undertaken within a theoretical context, it is seen as less relevant for analysing complex real-life problems. Thus, climate change sceptics argue that the models used to predict climate change are merely theoretical models and cannot capture the complexity of reality (Priem, 1995: 1-2; Barrett, 1996: 60-70).

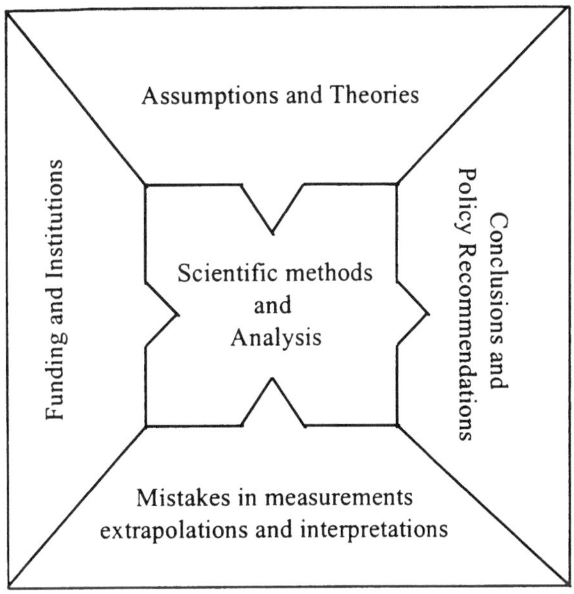

Figure 9.2 Sources of bias in science

Fourth, scientific methods of data collection, extrapolation and interpretation of data can be controversial, since such methods may be context specific and may lead to mistakes when applied at a global level. In relation to the climate change issue, there have been some controversies about scientific methodologies used to calculate the emission levels of greenhouse gases and the building of emissions scenarios (see for a summary of these controversies, Gupta, 1997: 150-165). Agarwal and Narain (1991) argue that even the so-called objective calculation of data is based on subjective decisions about which methods should be adopted and how such measurements should be taken. Figure 2 provides a diagrammatic representation of the key sources of bias that may creep into scientific methods.

Having said that, no scientific method is truly holistic, and the methods and tools of the post-normal/public interest school will also be based on its own choice of assumptions, etc. It is in the confrontation and dialogue between the two schools of thought on science that scientific knowledge can be fruitfully generated.

Types of problems

While scientific tools can be used to find solutions to specific problems, 'problems' do not necessarily have unique definitions. Problems tend to be socially constructed as Hoppe and v.d. Graaf (1991:5) put it. People tend to perceive problems and related uncertainties differently and in relation to the social organisation they belong to and in relation to their own context and situation. According to Hisschemöller (1993), problems can be classified into four types depending on the nature of the consensus on science and values. Problems are *structured* when there is societal consensus on the science and on the values at stake. Problems are considered as *unstructured* when there is neither consensus on the science nor on the values at stake. There are also problems that are *moderately structured* in terms of *consensus* on science or on *values*.

Problem structuring can also be applied to the international level (Gupta, 1997). Structured problems are problems in which there is consensus between negotiators from different countries at international meetings and between them and the people they represent. Moderately structured problems (horizontal) are problems in which there is consensus between negotiators, but there is less consensus between the negotiators and the people they represent. This is specially the case when the people in some countries are relatively unaware of the problem. In such cases, although international consensus in generating solutions to the problem on the basis of the available science is possible, implementing the consensus views in the domestic context may be difficult as there is limited domestic support. This may lead to vertical implementation bottlenecks. Moderately structured problems (vertical) are problems in which there is consensus between negotiators and the people they represent but there is less consensus between negotiators from different countries. This implies that the problem is perceived differently by countries and international consensus on potential solutions is difficult to generate. This may lead to horizontal negotiation bottlenecks. Unstructured problems are problems in which neither international negotiators can agree upon the problem, nor can the domestic public agree with their negotiators (Gupta, 1997: 37-39).

Vertical Bottlenecks	Horizontal Bottlenecks: Low	Horizontal Bottlenecks: High
Low	Structured problem Normal science adequate	Moderately structured problem (vertical)
High	Moderately structured problem (horizontal)	Unstructured problem Post-normal/public interest science to supplement normal science

Figure 9.3 Structured and unstructured problems

Structured problems are generally easier to address, in that there is clarity on the contours of the problem. This implies that the problem can be clearly defined. Scientific methods can also be used to determine the appropriate solutions to the problem. Moderately structured and unstructured problems are complex. They cannot be clearly defined and it is difficult to find solutions to such problems. The application of normal scientific techniques has limited value in such problems since these theoretical solutions cannot take into account the complexity of the real life problem.

A small-scale issue facing a small community with common values may tend to be structured and be more amenable to scientific solutions. However, a more large-scale, uncertain problem facing a non-homogenous global community is likely to be moderately structured or unstructured. Such problems cannot be easily addressed by using normal scientific techniques. The climate change problem is a long-term, large-scale problem. As sections 3-6 will show it is not a structured problem.

9.2.3 The use of science by policymakers

There are different theories about how scientific information is used in policymaking. Most theories on democracy and law around the turn of the century emphasised that policymakers should use knowledge and science to develop policies and laws in the *public interest* (cf. Rousseau, cited in Held,

1987: 74). These were normative theories. In more recent years empirical and behavioural theories have gained in importance.

The *two-cultures theory* states that policymakers are often unable to make effective use of science since the contexts in which scientific research is undertaken and policy is made have different cultural rules, values, reward systems, language and conceptions of time (Caplan, 1979; Rich, 1991). Scientists prefer to undertake research that aim at generating and testing hypotheses, or at formulating and investigating certain research questions within a theoretical context. They would like to publish their research in scientific journals where the work is evaluated on the basis of its theoretical and methodological contributions. They are reluctant to formulate policy recommendations for policymakers, which take into account the political, economic and practical problems faced by the policymakers. At the same time, policymakers are not interested in all the details of scientific research and the complexity of scientific understanding. They need information that they can translate into decisions that are implementable. In fact for policymakers, "the right decision is not necessarily compatible with the scientific truth" (Hisschemöller and, quite different and there is frequently a big communication gap between the cultures. This is the reason why scientists are often asked to generate 'usable knowledge' or 'functional knowledge', i.e. knowledge which can be understood and applied by the policymaker.

On the other hand, supporters of the *rational actor theory* argue instead that the gap between policymakers and scientists is caused by the process by which policymakers adopt some elements of the available scientific information on the basis of what is perceived to be:

- in their own interests;
- in line with their own ideas and world views;
- confirming their own intuitive expectations;
- consistent with the policies of their organisations;
- and politically feasible (Rich, 1981; 1991; Lindblom *et al.*, 1979; Van de Vall, 1988).

From the rational actor model it follows that the use of information for policies is not always in the public interest. *Public choice theorists* (Bate, 1996) argue that the notion that policymakers work in the public interest is naive, and that private interests provide the sustained influence on policymaking. Such selective use of knowledge is also referred to as 'half-knowledge' (Lazarsfeld, 1967; Marin, 1981). These theories of knowledge use in the policymaking process have been summarised in figure 4.

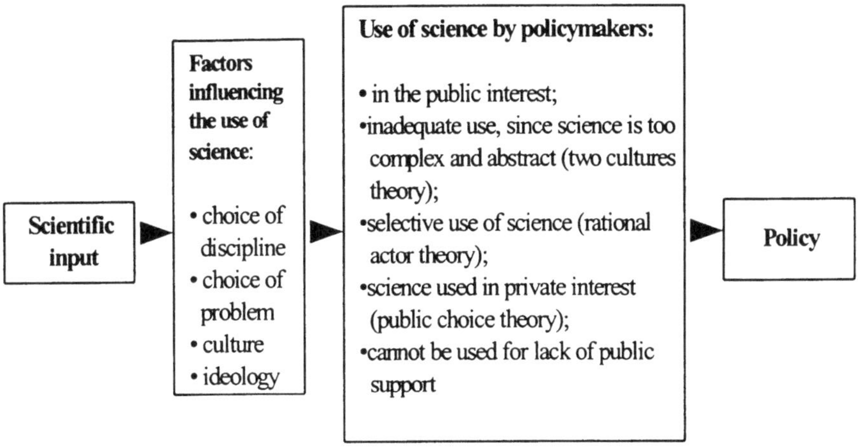

Figure 9.4 Theories on knowledge use

Different models can be developed on how science influences the policymaking process (Hisschemöller and Olsthoorn, 1996; Weiss, 1977). Building on these models, I argued (Gupta, 1997: 40-41) that there were different scientific models that were consistent with different phases of the policy process.

The *technocracy model* refers to the situation in which scientists define the problem and identify specific solutions to the problem and policymakers use the available science and knowledge as best they can, subject to the limitations predicted by the two cultures theory.

The *adhocracy model* refers to the situation in which policymakers use the available scientific knowledge to justify their existing goals, expectations and policies in line with the rational actor model.

The *multistakeholder dialogue model* refers to the situation in which the complexity of the problem makes it difficult for the policymakers to use the available knowledge and here the policy and scientific process needs to be better coordinated and also simultaneously needs to involve different stakeholders in society.

These models can be related to the type of problem to be addressed and the science necessary. For *structured* problems a (normal) scientific approach may be adequate and the technocratic model could apply. For *unstructured problems* public interest and post-normal science in relation to the multistakeholder dialogue are relevant.

Finally, an issue that is often overlooked is summed up in the *structural imbalance in knowledge generation theory* which argues that since scientific development in some countries occurs more quickly than in others, policymakers may have to use 'foreign science' to develop their policies. Such foreign science may be irrelevant to the domestic situation, having been developed in another context, and may also be inaccurate being based on different types of assumptions. This can lead to mutual distrust and defensive policymaking at the international level (Gupta, 1997: 164).

9.2.4 From scientific issue to political agenda item

There are different theories that explain the process of policymaking. Normally, for an issue to be solved at a political level, it must come on to the political agenda from the scientific or public agenda. In the process of *agenda building*, four stages are distinguished: the want stage (when a problem is sensed), the demand stage (when a problem is defined and policy measures are asked for), the issue stage in which groups and agencies recognise the problem, and the political item stage, in which the problem has reached the political agenda (Cobb *et al.*, 1976).

Winsemius (1986) looks at the policy process in a slightly different way. In his *policy life cycle* theory he argues that in the first phase there is disagreement over the definition of the problem. This is followed by a phase in which the disagreement decreases and the political importance of the issue increases. In the third phase solutions are devised, and in the fourth phase the management and self-regulation of the problem begins. Downs (1972) argues in his *issue-attention cycle* theory that problems do not always get solved. This depends on the attention that the public gives to the issue and the relationship between public concern and the policymaking process. According to Downs there are five phases, a pre-problem phase, an alarmed discovery phase, a third phase in which the costs are identified, a fourth phase in which there is a decline in the public interest and a fifth post-problem phase. In the first pre-problem phase, although the condition/ phenomenon comes into existence, the public is as yet unaware of the problem. In the second alarmed discovery and euphoric enthusiasm phase, public awareness on the problem increases and the public and the government are enthusiastic about being able to solve the problem. "This outlook is rooted in the great American tradition of optimistically viewing most obstacles to social progress as *external* to the structure of society itself, the implication is that every obstacle can be eliminated and every problem solved *without any fundamental reordering of society itself*, if only we devote sufficient effort to it" (Downs, 1972: 39). In the third phase, information is generated about the costs and benefits of the measures to deal with the problem. In the fourth phase, the public becomes gradually discouraged by the problem, or it feels threatened by it or it becomes bored by it especially if the problem is complex and expensive. In the post problem phase, the issue enters into a limbo phase. Only if adequate problem-solving measures are taken in phase 2 or if the costs of taking measures as identified in phase 3 are quite low, the problem may be addressed.

In the policymaking process, decisions are taken via the filtering of knowledge as explained earlier. Such a process also leads to *non-decisions* in which the selective use of options leads to premature closure of discussions on other options. Non-decisions occur, when parties that exercise power try to narrow the scope of decisions to protect certain interests at the cost of other interests. A non-decision is a "decision that results in suppression or thwarting of a latent or manifest challenge to the values or interests of the decision-maker" (Bachrach and Baratz, 1970: 39-52). Non-decisions are especially relevant in moderately structured and unstructured problems, since such problems may be addressed at the cost of certain stakeholders.

In the international sphere, decision-making occurs generally within the framework of the United Nations. Countries may make agreements with each other via bilateral or multilateral treaties. The preparatory work for such treaties may take place within specific United Nations Organisations or ad hoc bodies established by the international community of nations. Each country is a sovereign entity and decision-making within the United Nations tends to follow the rule of consensus. All countries need to agree to the decisions being adopted and the rules of procedure for adopting further decisions on the specific issue at hand.

9.2.5 An integrated science-policy model

The above theories on types of science, problems, science-policy models, and policymaking processes are integrated into a model with four stages.

Stage 1: Technocratic stage

In the technocratic stage of a problem-solving model, scientists signal the existence of the problem and indicate certain essential features of such a problem. There is gradual public awareness on the problem and the public demands policy measures by the government. In ensuing discussions on the problem, the degree of disagreement among policymakers and the public on the nature of the problem gradually becomes small, the issue is defined, and the problem gradually comes on to the (international) agenda (the outside mobilisation model of Cobb). Alternatively, the government takes note of the problem and decides to create public awareness on the issue (the mobilisation model of Cobb). Or the government takes note of the problem and decides that it can take action to address the issue without involving the public (the inside mobilisation model). The technocratic model of science determines the contours of the problem and the potential solutions to it. If the problem is simple and structured, adequate policies to solve the problem may be devised and the policy process may reach the management and self-regulation stage. If the problem is moderately structured or unstructured, then in the technocratic stage the policy process may be limited to identifying the different aspects of the problem, and to developing a framework approach to dealing with the problem and to listing possible principles and policy options.

Stage 2: Adhocratic stage

Simple and structured problems that can be easily solved through simple technologies may never reach such a stage. However, some problems may appear to become more complex as more information becomes available. There may appear to be relatively high consensus on the science but less consensus on the values that need to be embodied in the policy. If the problem is perceived as moderately structured or unstructured, then the policymaking process may pass through a second stage. In this stage, the degree of disagreement between policymakers and/or the public increases as new information emerges. The political difficulties in taking far-reaching measures may lead countries to adopt no-regrets measures as a first step (i.e. measures that can be justified for other reasons as well). Countries may re-define the problem and selectively accept the recommendations of the scientists so that the adopted policies fit their existing frame of policies, and they may use such recommendations and advice to justify measures in other policy fields. It is theoretically possible that the no-regrets measures may be adequate to address the problem and the problem may reach the management and self-regulation stage. On the other hand, such measures may fall short of dealing with the issue.

Stage 3: Stakeholder stage

Where the no-regrets measures are inadequate to solve the problem, and where the problem concerns a large number of stakeholders, the adhocratic model of science-policy dialogue may be inadequate. In such cases, because of the range of values and perceptions involved, the government may not be in a position to take far-reaching measures without actively involving the domestic population. In this stage, I believe that public interest and post-normal science will be very important in identifying the critical issues and in identifying solutions that may have a high compliance pull. At the same time, the policymaking process will need to actively involve the domestic stakeholders. The ideal situation would be a three-way dialogue between researchers, policymakers and the domestic stakeholders. If the problem is international in scope, it is also conceivable that issue-based coalitions between like-minded countries and transnational actors will be formed. The structural imbalance in knowledge will remain, and may influence the good faith underlying the negotiations. If stakeholders can address the national elements, and international coalitions can resolve the international components then the policy process may reach the management or self-regulatory phase.

Stage 4: Government as mediator model

The above models assume that policymaking aims at solving problems, and that the problems can be addressed within a national scope. However, for unstructured, global problems on 'eco-space', these three models may not be adequate. In order to preserve the global environment, global pollution should not exceed the carrying capacity of the Earth. Economic growth is generally accompanied by pollution. If we are to reduce global pollution, this implies that pollution entitlements will have to be shared between countries. This raises the question: how should the quantity of permissible pollution be shared among countries? Since, there is no global government that can impose a distribution of permissible rights, countries have to negotiate these with each other. However, since domestic stakeholders are unlikely to be willing to sacrifice their existing or potential rights to eco-space, negotiators will have limited mandates to negotiate with in international fora and this may lead to a (bottom-up) stalemate in the discussions. Alternatively, negotiators may negotiate a deal which although legal may not have domestic legitimacy (as many developing country negotiators have done in the climate negotiations; see Gupta *et al.*, 1996: 45-46). This may lead to symbolic policies at national level, in an attempt to meet international requirements and still satisfy the domestic public, i.e. a top-down stalemate. The need for governance to balance the self-interests of the stakeholders may become stronger calling for the government to mediate between what domestic actors support and what is internationally seen as necessary. Thus, I envisage that the role of government will evolve in the future to a stage in which the government is empowered to mediate between domestic and international priorities; when the legitimacy of the government depends not just on its domestic population but also on the way in which it presents international concerns to the domestic population. The policymaking process may systematically follow the technocratic, adhocratic, stakeholder models and then move on to the fourth stage where the government will have to mediate between what is perceived as internationally necessary independent of domestic constraints and what the domestic public is willing to accept. The role of international relations science will be to help governments understand the international and domestic constraints and to find a synergy between the two. Such science should make use not only of normal but also of public interest and /or post normal science.

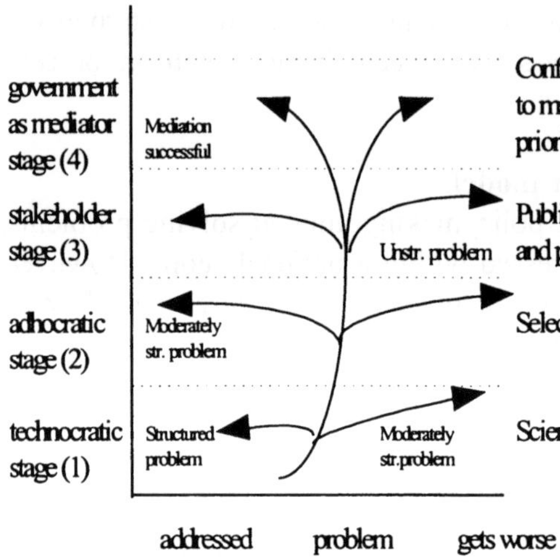

Figure 9.5 The possible paths of the life-cycle of a science-policy process

The above sections have provided a brief history of the relationship between science and the policymaking process. The following sections will explain the history of the climate change regime in terms of the four stages outlined above.

9.3 The technocratic stage - I

9.3.1 Regime formation: a brief history

Since the early half of the nineteenth century, there has been some research on and off on the issue of greenhouse gases and climate change, notably the work of Fourier in 1827 and Arrhenius in 1896. The process of research continued until the issue reached the global scientific agenda in 1979 when scientists attending the (first) World Climate Conference organised by the World Meteorological Organisation (WMO) declared that "it is now urgently necessary for the nations of the world: [...] to foresee and to prevent potential man-made changes in climate that might be adverse to the well-being of humanity". This Conference led to the establishment of the World Climate Programme (WCP) in 1979 by WMO and the United Nations Environment Programme (UNEP). Six years later, a major scientific meeting on climate change was organised at Villach by WMO, UNEP and the International Council for Scientific Union (ICSU). This meeting was followed by the establishment of the Advisory Group on Greenhouse Gases (AGGG) to provide scientific advice on climate change. This Group established three ad hoc working groups to study various aspects of the climate change problem. In 1988, 300 scientists and policymakers from 46 countries concluded at the Toronto Conference on the Changing Atmosphere that climate change could be a dangerous experiment and that countries should: "educe CO_2 emissions by approximately 20% of 1988 levels by the year 2005 as an initial global goal. The Advisory Group was a relatively small body influenced considerably by the views of environmental NGOs and institutes and was not dependent on government approval. Although it could generate consensus on certain scientific and policy aspects, it did not have the necessary influence on governments. Hence, later that year, UNEP and WMO set up the Intergovernmental Panel on Climate Change (IPCC), consisting of a small secretariat and a world-wide network of scientists. These scientists assess the latest scientific information and its implications for policymakers. The policymakers' summaries are approved by governments.

In January 1989, a resolution on climate change was adopted by the United Nations General Assembly. Two months later, heads of 24 States met at the Hague and adopted the Hague Declaration on Climate Change. In May, heads of Francophone countries met at Dakar and endorsed the Hague Declaration. In June, the Council of the European Community decided that member states had to play their full part in addressing the climate change issue without further delay. One month later, the G7 countries met and declared the need to limit the emissions of greenhouse gases (GHGs) and increase energy efficiency. The heads of Non Aligned Countries met in September and called for a global multilateral approach to the problem of climate change. In October, the Commonwealth Heads of Government met in Langkawi and declared that they were deeply concerned about the deterioration of the Earth's Environment. In November 1989, ministers from 67 countries discussed the issue and adopted the Noordwijk Declaration on Climate Change with three provisional targets (to reduce global greenhouse gas emissions by more than 50%; that industrialised countries should stabilise their emissions at 1990 levels by the year 2000 and to increase afforestation), based on the results of background scientific reports. In the same month, the small island states organised the Small States Conference on Sea Level Rise and adopted the Male Declaration on Global Warming and Sea Level Rise urging actions from all states to reduce the emissions of greenhouse gases. In December 1989, the Cairo Climate Conference attended by 400 participants from all six continents adopted the Cairo

Compact. Following the trend set in 1989, there were several conferences all over the world in 1990 that discussed the climate change issue.

In 1990, IPCC published its reports on Scientific Assessment, Impacts Assessment and Response Strategies. The Scientific Assessment predicted that in a business as usual scenario, the global mean temperature will rise by about 0.3 degrees Centigrade per decade (with an uncertainty range of 0.2 to 0.5 degrees Centigrade per decade) and the sea-level is likely to rise by about 6 cm per decade during the next century. The Impacts Assessment Report concluded that these changes would lead to stress in the agricultural, natural terrestrial and hydrological systems, human settlements and the oceans and coastal zones. The Response Strategies concluded that policies to limit GHG emissions need to be undertaken and that adaptation strategies should include coastal zone management and resource use and management. Response strategies include public education and information, technology development and transfer, economic, financial, legal and institutional mechanisms.

The AGGG concluded that given the characteristics of the global system, the maximum permissible rate of sea-level rise should be between 20-50 mm per decade, with a maximum of 0.2-0.5 m above the 1990 global mean sea level. Furthermore, a maximum rate of change in temperature of 0.1 degree Centigrade per decade was allowable. Global CO_2 equivalent concentrations should not go beyond 330-400 ppm (Rijsberman *et al.*, 1990 viii-ix).

The scientific input provided the basis for ministers from 137 countries to adopt a policy Declaration at the Second World Climate Conference in 1990. In December, the United Nations General Assembly adopted a resolution to initiate international negotiations on a framework convention on climate change. By 1990 the climate change issue was firmly on the international political agenda. In the meanwhile, the global community was also preparing for the United Nations Conference on Environment and Development. The preparatory meetings for this Conference were focused on preparing an Integrated Agenda for Action on all major environment and development issues, and the climate change problem was integrated into chapters concerning, for example, oceans and the atmosphere.

9.3.2 Different country positions

While the issue came on to the domestic agenda of all countries, it had different characteristics in different countries. For many industrialised countries (ICs) domestic science, freak climatic events and public awareness motivated domestic governments and helped to define the domestic position. By 1992, all OECD countries, except Turkey and the US, had adopted unilateral or group targets to stabilise their GHG emissions. In 1990, the

Council of the European Community adopted a Communiqué stating that the member states "are willing to take actions aiming at reaching stabilisation of the total CO_2 emissions by 2000 at 1990 level in the Community as a whole". The European Free Trade Association of seven countries also announced that it would stabilise its CO_2 emissions in 2000 at 1990 levels. During the Bush presidency, the USA was reluctant to adopt any national targets and argued that scientific uncertainty surrounding the issue justified further research before taking such far-reaching measures. In April 1993, the Clinton administration adopted a political target to stabilise its GHG emissions by the year 2000 at 1990 levels. The OECD countries have policies to promote efficiency standards, energy conservation, cogeneration, fuel switching, renewable energy and improved coal combustion technology; and to focus on energy demand management and pricing schemes. Some countries in transition (CITs) from a socialist to a market economy adopted targets for stabilising their emissions in 2000 at 1987-90 levels for the privilege of joining the group of developed countries.

Most developing countries (DCs) were influenced greatly by the concern on the problem expressed by developed country governments and their enthusiasm to address the problem. The issue was imported on to the foreign policy agenda of these countries from international initiatives of the developed countries. However, within the domestic context, there was little domestic debate and discussion on the issue. Climate change featured low on the domestic priority list and few researchers were concerned about it. Since the issue was imported, the representatives of these countries tended to see the issue more in terms of other foreign policy issues. Their common position was to demand that the developed countries should reduce their emission levels and to lobby for technology transfer and financial assistance to deal with the problem (Gupta, 1997: 46-98).

While these countries presented a common front during the negotiations, small groupings within them have specific interests. China, India, Brazil, Mexico and Argentina, large countries with high potential future emissions, have defended their right to grow and increase their emissions. The newly industrialising countries have been wary of making strong demands on the developed countries because of their fear of being subject to those very same demands if they became members of the industrialised world. The oil producers led by Saudi Arabia and Kuwait, have opposed specific measures to reduce GHG emissions in the climate negotiations and the reference to renewable energies as environmentally sound energies in the United Nations Conference on Environment and Development (UNCED) negotiations. They have opposed carbon taxes on fossil fuels in order to ensure the growing market of rich purchasers of their oil. The small island states are particularly vulnerable to sea level rise, and negotiated forcefully for strong emission reduction measures.

9.3.3 The North-South angle

The climate change problem has a very strong North-South element because of the substantial differences in emission and development levels. Developing countries are, by definition, less developed, have lower emissions both at a gross and per capita level and are likely to be more vulnerable to the potential impacts of climate change. Hence, in the scientific and political meetings, countries agreed that developed and developing countries had 'common but differentiated' responsibilities.

According to IPCC, *if* GHG concentrations are to be stabilised at 1990 levels, then the global emissions of CO_2 need to be reduced by more than 60%, methane by 15-20%, nitrous oxide and CFCs by about 70-85% (Houghton *et al.*, 1990: xviii). In 1995, the IPCC report stated that "Stabilisation at any of the concentration levels studied (350 to 750 ppmv) is only possible if emissions are eventually reduced well below 1990 levels" (Houghton *et al.*, 1995: 22). In order to stabilise emissions at any level, an annual emission budget will have to be assessed and the emissions will have to be shared between countries. Although there are differences in the statistics about the relative contributions of the different groups of countries, the per capita (and total) contribution of the developed countries is substantial. Against this background it is clear that the critical problem is: How should the principle of common but differentiated responsibilities to be applied in sharing the carbon budget among countries?

The differentiated responsibilities imply that:

- developed countries should, in view of their emissions, take the lead in dealing with the problem.
- The right of developing countries to develop and increase their emissions should be guaranteed by legal principles. Developing countries argue that developed countries have used up more than their rightful share of the 'eco-space' in the past and should compensate for that. They lobbied unsuccessfully for the inclusion of the objective that emissions should be eventually shared on an equitable per capita basis. The implicit developed country argument is that the effort should be shared equally and that they should go back to only a fixed percentage of their current emissions, and developing countries should only rise to a fixed percentage of their existing emissions. Given the somewhat linear relationship between fossil fuel use and growth, this would keep the relative economic position of countries intact.

- Since climate change provides an additional burden to developing countries, developed countries should provide assistance/compensation to help them finance relevant technology to improve their economy while minimising their contribution to climate change. At the SWCC, scientists stated that, "Developing countries are being asked to participate in the alleviation of the legacy of environmental damage from prior industrialisation.... It is clear that developing countries must not go through the evolutionary process of previous industrialisation but rather must 'leap frog' ahead directly from a status of under-development through to efficient, environmentally benign, technologies" (Scientific Statement, 1990).
- Since developing countries are likely to be more vulnerable, they should be assisted in coping with the impacts. The critical North-South issue is focused on how equity (justice and fairness) considerations will be taken into account in the interpretation of 'common but differentiated responsibilities'; i.e. whether due account will be taken of who is responsible for the problem, and who is in a position to take action.
- Finally, the special position of the most vulnerable countries and the countries with economies in transition also needed to be taken into account.

9.3.4 The Consensus in the Climate Convention

Despite the underlying differences in interests, countries agreed that climate change, as a common concern of humankind, "should be confronted within a global framework" (UNGA, 1989 Resolution). The negotiations on the Convention began in earnest in 1991. The negotiations led to the adoption of the United Nations Framework Convention on Climate Change (FCCC) in May 1992.

The Convention translated the scientific research material provided by the IPCC into a treaty language with legally binding measures. It consists of 26 articles and two annexes. The first annex includes the names of developed country parties. Annex II includes the names of developed country parties but excludes the countries with economies in transition. The ultimate objective of the Convention is the "stabilisation of greenhouse gas concentrations in the atmosphere at a level that would prevent dangerous anthropogenic interference with the climate system. Such a level should be achieved within a time-frame sufficient to allow ecosystem to adapt naturally to climate change, to ensure that food production is not threatened and to enable economic development to proceed in a sustainable manner" (Article 2). Given the above-mentioned reluctance of the US to accept binding targets, the FCCC includes two paragraphs that include the vaguely worded aspiration of the developed countries to bring back their emissions of greenhouse gas emissions to 1990 levels by the year 2000, individually or jointly.

Five principles guide the implementation of the FCCC. The principles include the principle of sustainable development, the precautionary approach, responsibilities towards present and future generations on the basis of equity and in accordance with the differentiated responsibilities and respective capabilities of countries. The Annex I Parties are expected to take the lead in implementing the Convention.

All parties are obliged to prepare national inventories of GHG emissions, climate policies and coastal zone management plans; to integrate climate considerations in relevant policies; to cooperate in relation to science, and to promote education, and public awareness. Annex I parties have a legally vague obligation to return their emissions of GHGs to about the 1990 level by the end of the century. Under Article 4.7, the implementation of developing country obligations is dependent on assistance from the developed countries.

The FCCC has established the Conference of the Parties (COP), which consists of negotiators from ratifying countries and meets once a year to take decisions on the basis of the advice of the Subsidiary Body for Scientific and Technological Advice and the Subsidiary Body for Implementation. The Secretariat undertakes the daily tasks of implementing the Convention.

The Convention states that a mechanism for providing financial resources to developing countries, including for the transfer of technology has to be established. The Global Environment Facility (GEF) of the World Bank, UNEP and UNDP is the operating entity of the financial mechanism. The Convention also allows countries to jointly implement their obligations. Joint Implementation (JI) has been interpreted to mean that developed countries or their domestic industries may invest in emission reduction possibilities in other countries. In return for such investments, the investor will receive a 'climate credit'. This implies that the investing countries may achieve their domestic targets in other countries if they wish to. This allows countries to adopt cost-effective emission reduction options in other countries.

9.3.5 Information sufficient for euphoric negotiation

In the first stage of the problem, scientists signalled the problem, people were alarmed by it, the epistemic community presented usable knowledge to the policymakers via, inter alia, IPCC reports on how the problem could be defined and the technological and policy options for addressing the problem. The environmental NGOs played a major role in supporting the epistemic communities. The growing public awareness that accompanied the first global United Nations Conference on Environment and Development (UNCED) provided a very strong impetus to the issue. There was a relative absence of an 'anti-lobby'. This stage was influenced by the optimism of policymakers and their notions of solidarity with the rest of the world. The enthusiasm of the policymakers from the industrialised countries was infectious and developing country negotiators were deeply influenced by not only the enthusiasm but also their concern. Their awareness of their physical vulnerability to the potential impacts of climate change, and the political attractiveness of climate change in its potential for providing developing countries an accelerated process of growth were important motivating factors.

The accelerated public interest, and the growing international momentum provided governments the support they needed to negotiate and adopt the FCCC, which provides a framework, defines principles and lists policy options. In this stage, as Down predicts, the need to do something to address the problem is seen as more important than the difficulties in the process of actually doing something. The problem *appeared* to be reasonably structured in that there was some amount of global consensus that there was a serious and irreversible climate change problem, there was a generally shared common perception that the climate change problem called for the reduction of greenhouse gas emissions, there was also a common perception that countries should address the problem in accordance with their common but differentiated responsibilities!

9.4 The adhocracy stage: II

9.4.1 Underlying North-South conflicts: problem definition, science, values and solutions

While the principle of "common and differentiated responsibilities" helped to ensure consensus in the first stage, it also tended to gloss over the conflict of interests between countries on the definition of the problem, its science and its solutions. The following sections covers in brief the developments in science and national policies and shows how with increased information, the pressure to take action to implement commitments, some underlying conflicts have surfaced strongly in the second stage of the climate change negotiations.

The developments in relation to the FCCC

As mentioned earlier the Convention was adopted in 1992. For a Convention to enter into force, countries must ratify the Convention. Such a ratification normally occurs after the Parliament or equivalent domestic body agrees to accept the obligations under the Convention. A minimum number of ratifications are necessary for a Convention to enter into force. In the case of the FCCC, the number is 50. In the period leading to the entry into force of the Convention, the Intergovernmental Negotiating Committee remained active and organised two meetings a year to undertake preparatory work so that the Conference of the Parties could take informed decisions once the Convention entered into force. Following fifty ratifications, the Convention entered into force in 1994. Every year since then, the Conference of the Parties has met to discuss the critical issues mentioned in the Convention. At the first COP held in Berlin in 1995, 21 decisions were taken. The most important of these was the Berlin Mandate which set up a process to review the adequacy of the targets adopted and to develop policies and measures and Quantified Emission Limitation and Reduction Options for specified time-frames. Furthermore, after extensive negotiations a pilot phase on Activities Implemented Jointly was adopted. Measures initiated in the pilot phase were not to be seen as fulfilling current objectives and no credits were to be given to countries. The pilot phase was to be reviewed annually and a decision on Joint Implementation was to be taken no later than 2000. Furthermore, the interim arrangements in relation to the GEF were extended. The secretariat was asked to develop a report of activities undertaken by countries in fulfilment of their obligation to transfer technologies to developing countries.

At the next COP, held in 1996, 17 decisions were taken. At this meeting, the secretariat reported that it was not possible to provide a clear report of the technology transfer taking place and that a round table on the issue should be organised in COP 3. The decisions included a programme for the Subsidiary Body for Implementation, the work of ad hoc group on Article 13 of the Convention,2 and a decision noting the slow progress on technology co-operation.

At the third COP in Japan in December 1997, the Kyoto Protocol to the United Nations Framework Convention on Climate Change was adopted (KPFCCC). This Protocol lists Quantified Emission Limitation and Reduction Objectives (QUELROs) for all developed countries. The European Union countries are expected to reduce their emissions by 8% in the period 2008-2012 in relation to emission levels in 1990; the USA by 7%,

[2] The FCCC calls for the establishment of a multilateral consultative process, available on request, when Parties needs to resolve questions regarding its implementation (Article 13).

Japan by 6%. Australia, Iceland and Norway are allowed to increase their emissions by 8%, 10% and 1%, respectively. New Zealand, Russia and Ukraine are allowed to stabilise their emissions. The remaining Annex I countries are allowed varying levels of reduction. This Protocol allows developed country Parties to internally transfer or acquire emission reductions via trading among countries with emission allowances, joint implementation with countries in transition to a market economy, and adopts a Clean Development Mechanism which enables investment in developing countries in return for emission credits.

The Kyoto Protocol has been ratified by only 3 countries thus far. The US has announced that it will not ratify the agreement until key developing countries agree to meaningfully participate in the agreement. The European Union and other developed countries are unlikely to ratify until the US does so, and there is at present a deadlock in the process. In response to the pressure, Argentina and Kazakhstan announced at the fourth meeting of the parties in Buenos Aires in November 1998 that they would be willing to participate meaningfully, but the other developing countries were extremely upset by the ratification politics.

Scientific developments

In the mean while the IPCC published its Second Assessment Report in 1996. This report indicated that the projected emissions are likely to lead to an increase in global mean temperature by 1-3 degrees C by the end of the next century and a projected sea level rise by 0.5m by 2000 (Houghton *et al.*, 1996: xiv; 6). These predictions indicated lower rates of temperature increases and sea level rise than those made in the 1990 report. The potential impacts of climate change on developed countries was estimated at 1-2% of GNP and that on developing countries at 2-9% (Pearce *et al.*, 1996).

During the negotiating process, the climate change problem was seen as a technical and scientific problem caused by emissions of greenhouse gases and by reductions in the natural sinks. This implies that national inventories of emissions and removals by sinks need to be made. The science should point out ways of reducing emissions and increasing sinks. The 'common but differentiated responsibilities' approach should be used to determine which countries should take which measures and why. This line of reasoning is fairly logical and systematic.

However, interviews with negotiators and domestic actors from the developing countries indicate that they took a somewhat bottom-up approach to the problem. If emissions have to be reduced drastically and sinks have to be substantially increased, this implies, in the absence of a structural revolution, that at some point of time the available emissions will have to be divided among countries. Since the present industrial structure of countries is such that there is a strong correlation between emissions and economic growth, very few countries will be actually willing to give up their right to economic growth. This will imply a strong conflict of interest between countries and it is unlikely that countries will wish to solve the problem. The only way the problem can be solved, assuming that technological developments are inadequate, is if the problem of climate change is not seen as the problem itself, but is seen as the symptom of a bigger problem, i.e. as a symptom of the wrong development ideology; an ideology that may not be compatible with environmental protection and social justice. Since the justification for taking action to minimise climate change is that we should protect the environment for future generations, this implies that present generations must also be provided for.

In this stage, some scientific conflicts also emerged. At a more general level, there was more literature that was sceptical about the existence of the climate change problem. In this period several publications in the United States and in continental Europe were also made by scientists who stated that they were sceptical of the projections made by the IPCC. At a specific level, there were a few controversies on scientific results and inputs into the negotiation process. The knowledge imbalance theory is of particular relevance here, for since developing countries lag behind the developed countries in climate change science and in the relevant domestic implications, they have had to rely on foreign scientific input. While most of the foreign information is credible, there are some anecdotal examples that show that the underlying assumptions integral to some of the research undertaken had some biases which tended to favour the interests of the Northern countries. Prof. Parikh (1992: 507-508) argues that some IPCC GHG emissions scenarios have in-built assumptions that work against the long term interests of developing countries. Agarwal and Narain (1991: 4-5) argue that some statistics on deforestation and methane emissions are biased heavily against developing countries as they are based on inaccurate projections. As and when new research and political statements based on such research were being made at international meetings, negotiators and actors from the South felt that these statements and research documents tended to focus much more on the possible future emissions of the South, rather than the present and past and also potential future emissions of the North (Interviews, 1994- 1998). Prof. Mitra (1992: 202) argues that apart from the few large developing countries, most of the developing countries are likely to only suffer from the problem; and for these countries the crucial issue is adaptation and not responsibility for causing the problem. For the large countries, the per capita emissions are low and are likely to remain so in the near future.

In this stage, there is also conflict in relation to the values underlying the international consensus. How is the notion of 'common and differentiated responsibilities' to be interpreted? Why is the division of responsibilities under the Convention more on the basis of capability for taking action rather than responsibility for having caused the problem. It is also felt that technology transfer provisions in the Convention are being given less importance. The amount of money in the Global Environment Facility that focuses on four global issues is considered very nominal - a symbolic gesture. There are also several other controversies regarding the management and the location of the financial mechanism of the financial mechanism for the climate change problem (see for details Gupta, 1995 and Gupta, 1997: 99-115). The other mechanism for North-South co-operation - Joint Implementation was also seen much more as a means for the North to avoid its own responsibilities in relation to reducing its own emissions at home. Developing countries accepted a pilot phase on Joint Implementation despite their fears that this instrument is neo-colonialist in that it might permit developed countries to continue their current lifestyles, while exhausting cheap emission reduction options in the developing countries, thus making it more difficult for them to implement their own obligations at some later date (Gupta *et al.*, 1996: 43-61). Verbruggen (1995: 12) concludes that "Joint Implementation does indeed increase efficiency [relative to targets and timetables], but by a rather transparent ruse. First, the North refuses to distribute emission rights at international level; then having opted for a second-best system of individual reduction targets, comes up with an instrument of efficiency, which reduces the North's own emission reduction costs most of all" (see for details regarding the controversy on Joint Implementation, Gupta, 1997: 116-131). The 'Joint Implementation' controversy has been partially addressed through the establishment of the Clean Development Mechanism, but even here the majority of the developing countries see this as a new name for 'Joint Implementation', although they hope that this mechanism may be better suited to meet the specific concerns of the developing countries.

Finally, if the climate change problem is more a lifestyle and development paradigm problem, then reducing emissions addresses the issue at the margin and is a symptomatic approach. It does not address the issues of a faulty development paradigm, the distribution issues and the fact that meeting present needs is a pre-condition to addressing future needs. Hence, such a problem calls for a more serious look at the way the global community defines development.

9.4.2 Underlying domestic conflicts: Environment versus growth

The climate problem can also be seen as an environment versus economic problem. All countries aim for economic growth. Economic growth tends to be accompanied by some amount of pollution. The relationship between growth and environmental pollution is arguably in the form of an inverted 'U' (the Environment Kuznets Curve) curve. This implies that when a country is in an early phase of development, the emissions tend to increase with growth. However, as the society becomes more advanced, as technologies are developed further, as environmental measures are adopted, as the society develops the service sector more than the heavy industry sector, the economic growth may not necessarily be accompanied by corresponding increases in emissions (Malenbaum, 1978; Janicke, 1989; Grossman, 1995). This theory is optimistic, because it suggests that economic growth need not be linked to environmental pollution. On the other hand, there are no indications that such de-linking is likely to be permanent and although the de-linking in terms of economic growth is important, it does not imply that there is per capita delinking and the gross pollution may continue to increase. This implies that all countries face a quandary: how should they integrate environmental concerns in their developmental process?

Different countries are in different phases of development. For Japan, a highly efficient country, reducing emissions is not going to be easy. For CITs, where the efficiency is much lower, growth can be temporarily stimulated without an accompanying increase in emissions. For developing countries, although they are inefficient, their total growth per capita is so low, that efficiency increases may be compensated by the volume of economic growth and related emissions.

In the post 1992 phase, economists in the developed countries were demonstrating how costly it will be for countries to take domestic measures and were also showing the opportunities for cost-effective reductions elsewhere in the world. This led to increased reluctance to take domestic action and reduced the impetus in the policymaking process in the developed countries. Their arguments were strengthened by the growing scientific controversy that gained ground in this period. Further, unlike in the ozone layer regime and the Mediterranean Action Plan where the epistemic communities influenced the formation of the regime by shaping the international debate and also by contributing to the development of convergent domestic polices in compliance with the regime (Haas, 1989: 377-403), the international scientists (and the NGOs) who shaped the initial debate have not been adequately able to create the necessary domestic consensus needed to provide the regime its domestic backing. This is partly because climate change is much more complex as an issue and cannot be addressed without stakeholder involvement and commitment. Inevitably this means that the policy options listed in the first stage were filtered through a 'no regrets' criterion in which measures that are beneficial for other reasons than climate change are promoted because of their incidental benefits to climate change. However, the peer-review process in the IPCC is gradually increasing and there is potential for this body to develop the critical mass when it not only affects the international negotiation process but is also able to generate and create a high level of domestic support in the developing and the developed countries.

While developed countries have adopted a series of domestic measures an early review of domestic policies presents a sombre picture. Despite the huge number of domestic policies initiated, the high GNP, low oil prices, and inadequate federal funding has meant that there are early signals that the stabilisation target of the United States and many other developed countries may not be reached. However, there are some indications that the developed countries as a whole may be able to bring back their emission levels substantially by the year 2000 (Bolin, 1998; FCCC/CP, 1998/11); however, the reductions are attributable more to the economic decline in East Europe than due to a structural change in the developmental process. This indicates that reducing emissions in the developed countries is going to be very difficult. At the same time, the emission levels continue to increase in the developing countries, even though some of these countries are trying to encourage the development of renewable energy. Several multinationals have decided to develop their own initiatives to deal with the climate change.

9.4.3 A stage of slow-down?

In the second stage, i.e. the period between 1992-ca. 2000, there appears to be unprecedented speed in that a Kyoto Protocol with binding targets has been negotiated and adopted, and at the same time, a slow-down in the momentum. There appears to be a contradiction. While the hard and binding targets in the Kyoto Protocol fall far short of what environmentalists consider necessary, they go far beyond what most economists can see as politically and economically feasible. Though there is a legal agreement, the ratification politics shows that the initial enthusiasm has been tempered by the realisation of all the complexities of taking serious measures. Although, there is more scientific research confirming the climate change problem, there is also much more controversy. The economists have been more active in this period, focusing on the expenses. The realisation of the multivariate dimensions of the problem has led to no-regrets policymaking in the developed countries and symbolic support from developing countries.

The decline in political commitment also reflects the diminished public support in the post-UNCED stage, as the public and media have turned their attention to other international issues. The public support may have also diminished with the realisation that far-reaching measures may call for substantial sacrifices in the short-term.

While there is a slowing down in the momentum in relation to substantive measures, procedurally the regime has done quite well. The Convention entered into force in record time in 1994. More than 165 countries have ratified the Convention. This demonstrates that the bulk of the global community takes the issue seriously and are gradually consolidating the agreements they had made earlier. The various subsidiary bodies of the FCCC have been regularly meeting. The Kyoto Protocol was adopted in 1997. Thus at a procedural level, there appears to be no slow-down. In this stage, there appears to be considerable agreement between the negotiators about the problem, but the domestic difficulties in developing and implementing policies implies that the problem is seen more as a moderately structured (horizontal) problem.

It would appear that while in the formative stages of the regime, it was clear that the problem could not be addressed without the co-operation of governments, governments are beginning to realise that they themselves may not be able to address the problem without the assistance and co-operation of the people they represent. This brings us to the next stage in the problem.

9.5 Beyond adhocracy: Stages III and IV

9.5.1 Resolving domestic issues: the stakeholder model (III)

The political difficulties in implementing climate change targets, in generating resources to support technology transfer indicate that the problem is far more complex than initially visualised. Addressing the problem is also not as easy as initially anticipated. The political, if not economic, costs are high and the domestic support for far-reaching measures seems to be flagging. There is a very real risk that people will get either bored by the subject, discouraged or feel threatened, and the regime might slip into the fourth phase of Down's issue-attention cycle. The regime might become defunct. At the same time, the procedural momentum is there, and if the appropriate steps are taken the regime may move from strength to strength.

In the third stage of the regime, it is vital that the public is mobilised and motivated to take action. Without their support, governments will not be able to take far-reaching action. In this stage, inevitably the role of science must change. As mentioned earlier, there is a domestic conflict between economic growth and environmental protection. This conflict will hamper the development of appropriate policies. It is thus vital that ways and means to deal with this problem are developed. This can be done by developing some appropriate alternative development paradigms and/or by encouraging domestic actors to seriously examine the issue and to reach some conclusion about what is the best way to address the problem. Several research projects have been undertaken that focus on the positions of domestic stakeholders in several countries (Bernabo *et al.,* 1995; Gupta *et al.*, 1995; Gupta, 1997; Vellinga *et al.*, 1995; Hisschemöller *et al.*, 1995). It is clear that stakeholders have different views on whether climate change is important and whether action should be taken. Probably in the pre-1992 phase, there were more people who believed that climate change was a serious issue and that domestic action needed to be taken. In the post-1992 stage, although there has been a considerable growth in the number of people aware on the issue, there are more people that are reluctant about the need for domestic action. At the same time, there are opinion polls (Mellman Group, 1997) that show that the public is not as indifferent to the problem as governments appear to think.

Given the uncertainty, the controversy and the fact that the problem concerns all domestic actors, it is clear that the climate change issue is a mind boggling problem. In this stage, government leadership cannot survive on 'no-regrets' policy, which will probably be temporarily exhausted in the second stage. In this stage, governments may have to move towards a "a be prepared" policy. This section essentially speculates about the follow-up stages in the early part of the next century. Public support has to be actively lobbied for in this stage. This may even call for increasing the scientific literacy of the public so that they are able to make informed decisions on how they perceive the issue. In this stage governments may have to stimulate discussions between diverse stakeholders to see if these stakeholders can come to some solution for the problem, and can provide ideas about how governments need to address the issue.

In this stage, it is not inconceivable that the role of policy-makers will be to focus on mobilising social groups via public awareness campaigns; to focus on public interest and post normal science; to match global with local priorities; to ensure local commitment to global problems; to use issue-linkages as indirect policies: where stakeholders make issue-linkages with other issues, and such issue-linkages need to be exploited with a view to developing a set of indirect policies; and promote coalitions between like-minded actors (See Gupta, 1997).

In this stage, a minimum strategy for the developed countries will be:

- to ratify the Kyoto Protocol with urgency so that the right signal is sent to industry and they can begin to compete in earnest with each other and start to develop new technologies;
- to adopt policies that can provide the framework within which industries and consumers see the economic gains of engaging in a process of industrial transformation;
- to review their export packages and to see how the technology export package can be made more 'greening' with a view to maintaining the comparative advantage in that sector in the international arena;
- develop potential policy options that are kept in 'cold storage'; so that if events lead to increased public demand for action, there are measures that can be pushed through.

In this stage (2000-2015?), a minimum strategy for the developing countries will be to review the design of their future development process and trajectory and to engage in broad social discussion about how sustainable development needs to be defined and articulated.

9.5.2 International issues: Beyond the stakeholder approach (IV)

While raising scientific literacy, public awareness and stakeholder involvement is the key science-policy process in stage 3, there is a very real fear that the domestic stakeholder discussions may not be able to adequately address the North-South and international issues. These will gradually become more important as developing countries will also eventually have to undertake domestic policies and targets. It is possible that a technological revolution occurs and the global community is able to deal with the problems of access to such technology and the climate change problem will be resolved. On the other hand, in the absence of such a revolution, the issue of sharing emission levels between countries will still remain a sore point between countries. Citizens of different countries may be unwilling to give up their 'eco-space' for people from other countries and negotiators may not be given a mandate to negotiate on emission rights. This may lead to a situation in which the problem is side-tracked or when countries use force to manipulate others into action and thereby increase international stress.

In this stage, public support, in line with Down's prediction, may decrease as the public may be unwilling to accept the implications of reducing their right to 'eco-space'. At the same time, responsible governments may feel that the problem needs to be addressed seriously. In such a situation, I believe that it is very important that the role of governments in the international and domestic context should change. The feeling of global community and solidarity that has been growing especially in the last decade of this century, as demonstrated by a large number of global conferences, implies that the traditional approach to international decision-making where governments are accountable only to their own population; and the international community expects them to observe their international commitments in good faith will have to give way to a new order in which governments are not just accountable in the domestic context but also in the international context. Especially in issues where there is a strong conflict of interest, governments will have to mediate between the concerns of the domestic public and the more abstract concerns of the global community. If they do not, complex eco-space problems may not be resolved. "This calls for a re-examination of the role of government. Stage 3 was initiated on the realisation that governments cannot mobilise action on their own. In stage 4, it may become clear that the problem may not be addressable without governments" (Gupta, 1997: 202).

9.6 Conclusion

This chapter has shown that the initial signalling of the complex problem of climate change led to the first stage of euphoric decision-making leading to the large-scale ratification of the Climate Convention and some degree of euphoria continued into the second phase leading to the adoption of the Kyoto Protocol. This was followed by the stage of confronting political reality and the underlying conflicts that had been glossed over in the process of finding global consensus. These conflicts have become gradually more important. In this stage, what happens is guided by what is politically possible and climate change policy is being defined by no-regrets policy for developed countries and by symbolic commitment from developing countries. While thc dominant scientific communities were able to provide the initial impetus in the development of the regime they have been unable to provide the necessary objective answers needed for cultural consensus at a global level and have been thus far unable to generate thc local social authority for global decision making. The leadership of the developed countries has been, via ratification politics, dependent on the support from the developing countries. In the third stage, stakeholder involvement is necessary to generate social consensus within the domestic context, for governments may not be able to solve the problem on their own. While in this stage domestic conflict may be resolvable, the North-South issues may remain. In the fourth stage, the issue of climate change will either fade out of the international agenda, or with a technological revolution can be addressed technocratically. Failing that, this chapter argues in favour of a new role for government as mediator between domestic and international concerns.

References

Agarwal A. and Narain S. (1991). *Global Warming in an Unequal World: A Case of Environmental Colonialism*, Centre for Science and Environment, New Delhi.

Bachrach, Peter, and Mortons S. Baratz (1970). *Power and Poverty: Theory and Practice*, Oxford University Press, London.

Barrett, Jack (1996). Do CO_2 emissions pose a global threat? in John Emsley (ed.) *The Global Warming Debate: The Report of the European Science Forum*, Bournemouth Press Limited, Dorset. pp. 60-70.

Bate, R. (1996). An Economist's Foreword in John Emsley (ed.) *The Global Warming Debate: The Report of the European Science Forum*, Bournemouth Press Limited, Dorset. pp. 7-21.

Bernabo, C., S. Postle Hammond, T. Carter, C. Revenga, B. Moomaw, M. Hisschemöller, J.Gupta, P. Vellinga and J. Klabbers (1995). *Enhancing the effectiveness of research to assist international climate change policy development. Phase II Report.* S&PA, Washington DC and Amsterdam, IVM/VU.

Boehmer-Christiansen, S. (1995). Britain and the Intergovernmental Panel on Climate Change: The impacts of scientific advice on Global Warming Part 1: Integrated Policy Analysis and the Global Dimension, *Environmental Politics*, Vol 4, no. 1, pp. 1-19.

Boehmer-Christiansen (1996). Political pressure in the formation of scientific consensus, in John Emsley (ed.) *The Global Warming Debate: The Report of the European Science Forum*, Bournemouth Press Limited, Dorset. pp. 234-248.

Bolin, Bert (1998). The Kyoto negotiations on Climate Change: A Science Perspective, *Science*, Vol. 279, 330-31.

Caplan, N. Morrisson, A. and Stanbough R.J. (1979). *The use of social science knowledge in policy decisions on the national level*, Ann Arbor Institute for Social Science Research, University of Michigan.

Cobb, Roger W. JK Ross and MH Ross (1976). Agenda Building as a Comparative Political Process, in *American Political Science Review*, Vol. 70 pp. 126-138.

Corbyn, P. (1996). Carbon Dioxide Fluctuations Resulting from Climate Change in John Emsley (ed.) *The Global Warming Debate: The Report of the European Science Forum*, Bournemouth Press Limited, Dorset. pp. 71-78.

Downs, Anthony, (1972). Up and down with ecology- the issue-attention cycle, *The Public Interest*, Vol. 28, pp.38-50.

Functovicz, S., M O'Connor and Jerry Ravetz (1996). Emergent Complexity and Ecological Economics, in van der Straten and van den Bergh (eds.). *Economy and Ecosystems in Change*, Island Press.

Grossman, G.M. (1995). Pollution and growth: what do we know? In I. Goldin, L A. Winters (ed.) *The Economics of Sustainable Development*, Cambridge University Press, Cambridge, 19-42.

Gupta, J., R.v.d. Wurff and G. Junne (1995). *International Policies to Address the Greenhouse Effect: An evaluation of international mechanisms to encourage developing country participation in global greenhouse gas control strategies, especially through the formulation of national programmes*, W-95-07, Department of International Relations and Public International Law (University of Amsterdam) and Institute for Environmental Studies (Vrije Universiteit).

Gupta, J. (1995). 'The Global Environment Facility in its North-South Context', *Environmental Politics*, Volume, 4 no.1 Spring issue, pp. 19-43.

Gupta, J. R.S. Maya, O. Kuik and A.N. Churie (1996). A digest of regional JI issues: Overview of results from National Consultations on Africa and JI in Maya, RS and J. Gupta (eds.) *Joint Implementation: Carbon Colonies or Business Opportunities? Weighing the odds in an information vacuum*, Southern Centre for Energy and Environment, Harare, 43-61.

Gupta, J. (1997). *The climate change convention and developing countries - from conflict to consensus*?, Environment and Policy Series, Kluwer Academic Publishers, Dordrecht.

Haas, Peter M. (1989). Do regimes matter? Epistemic Communities and Mediterranean pollution control, *International Organization 43*, 3, summer 1989, p. 377-403.

Held, D. (1987). *Models of Democracy*, Polity Press, Cambridge.

Hisschemöller, M. (1993). *De Democratie van Problemen, De relatie tussen de inhoud van beleidsproblemen en methoden van politieke besluitvorming*, VU uitgeverij, Amsterdam.

Hisschemöller, M., J. Klabbers, M.M. Berk, R.J. Swart, A. van Ulden en P. Vellinga (1995), *Addressing the greenhouse effect: options for an agenda for policy oriented research*, Free University Press, Amsterdam.

Hisschemöller, M. and X. Olsthoorn (1996). Linking Science to Policy: Identifying barriers and Opportunities For Policy Response, in T. E. Downing, A.A. Olsthoorn and R.S.J. Tol (eds.) *Climate change and extreme events: Altered risk, socio-economic impacts and policy responses*, Institute for Environmental Studies R-96/4 ECU Research Report Number 12, pp. 273-295.

Houghton, J.T., Jenkins, G.J. and Ephraums, J.J. (1990). *Climate change: The IPCC Scientific Assessment*, Cambridge University Press.

Houghton, J.T.; L.G. Meira Filho, J. Bruce, H. Lee, B.A. Callander, E. Haites, N. Harris and K. Maskell (eds.) (1995). *Climate Change 1994: Radiative forcing of climate change and An evaluation of the IPCC IS92 Emission Scenarios*, Cambridge University Press.

Houghton, J.T.; L.G. Meira Filho, B.A. Callander, N. Harris, A. Kattenberg and K. Maskell (eds.) (1996). *Climate Change 1995: The science of climate change; Contribution of Working Group I to the Second Assessment Report of the Intergovernmental Panel on Climate Change*, Cambridge University Press.

Hoppe, Rob and Henk van der Graaf (1991). Readings on policy science and management, *Policy science and management*, Department of Public Administration, pp 2-4.

Kuhn, T.S. (1962). *The structure of scientific revolutions*, Chicago.

Idso, S. (1996). Plant responses to Rising Levels of Atmospheric Carbon Dioxide, in John Emsley (ed.) *The Global Warming Debate: The Report of the European Science Forum*, Bournemouth Press Limited, Dorset. pp. 28-33.

IPCC-III (1990). *Climate Change - The IPCC Response Strategies*, WMO/UNEP/IPCC.

Jänicke M. Monch, H. Ranneberg, T. Simonis, U.E. (1989). Economic Structure and Environmental Impacts: East West Comparisons. *The Environmentalist*, Vol. 19, pp. 171-182.

Lazarsfeld, P. (1967). Introduction. In Lazarsveld P. *et al.* (eds.). *The Uses of Sociology*, New York: Basic Books.

Lindblom, C.E. and Cohen, D.K. (1979). *Usable knowledge*. Yale University Press, New Haven.

Malenbaum, W. (1978). *World Demand for Raw Materials in 1985-2000*, Mcgraw Hill, New York.

Marin, B. (1981). What is half-knowledge sufficient for and when? Theoretical comment on policymakers uses of social science. *Knowledge: Creation, Diffusion, Utilization*, 3(1), pp. 43-60.

Mellman Group (1997). Memorandum dated 17 September 1997 from the Mellman Group to the World Wildlife Fund containing the Summary of Public Opinion Research Findings.

Miller, Alan (1993). The Role of Analytical Science in Natural Resource Decision Making, *Environmental Management*, Vol. 17, No. 5, pp. 563-574.

Mitra, A.P. (1992). *Scientific Basis for Response of Developing Countries*, Indo-British Symposium on Climate Change, 15-17 January 1992, New Delhi, pp. 202 -206.

Mitroff Ian I., R.O. Mason and V.B. Barabba (1983). *The 1980 census: Policymaking amid turbulence*, Heath and Co., Lexington.

Myrdal, G. (1944). *An American Dilemma. The negro problem and modern democracy*, New York.

Parikh, J. (1992). IPCC strategies unfair to the South, *Nature*, Vol. 360, 10 December 1992, pp. 507-508.

Pearce, D., WR Cline, AN Achanta, S Fankhauser, RK Pachauri, RSJ Tol, P Vellinga (1995). The Social Costs of Climate Change, in Bruce, J., Hoesung Lee and E. Haites (eds.) *Climate Change 1995: Economic and Social Dimensions of Climate Change; Contribution of Working Group III to the Second Assessment Report of the Intergovernmental Panel on Climate Change*, Cambridge University Press, 178-224.

Priem, H.N.A. (1995). The CO_2 ideologie, Wetenschap en Onderwijs, *NRC Handelsblad*, 6 Juli, pp.1-2.

Rijsberman, F and F.R. Swart (1990) (eds.). *Targets and Indicators of Climatic Change*, The Stockholm Environment Institute.

Rich, R. F. (1981). *The knowledge cycle*, SAGE Publications, Beverley Hills.

Rich, R.F. (1991). Knowledge creation, diffusion and utilization: perspectives of the founding editor of Knowledge in *Knowledge: Creation, diffusion and utilization* 12, pp. 319-337.

Segalstad, Tom V. The distribution of Co_2 between atmosphere, Hydrosphere and Lithosphere: Minimal influence from anthropogenic CO_2 on the Global Greenhouse Effect, in

John Emsley (ed.) *The Global Warming Debate: The Report of the European Science Forum*, Bournemouth Press Limited, Dorset. pp. 41-50.

Shiva, Vandana and J. Bandyopadhyay (1986). Environmental Conflicts and Public Interest Science (January 11, 1986) in *Economic and Political Weekly* Vol. XXI, No.2., p. 84-90.

Singer, S.Fred (1996): A preliminary critique of IPCC's Second Assessment of Climate Change, in John Emsley (ed.) *The Global Warming Debate: The Report of the European Science Forum*, Bournemouth Press Limited, Dorset. pp. 146-157.

Vall, Van de M (1988). De waarden-context van sociaal-beleidsonderzoek: een theoretisch model. In van de Vall, M. and Leeuw, F.L. (eds.) *Sociaal beleidsonderzoek*. VUGA, Den Haag.

Vellinga, P., J.H.G. Klabbers, R.J. Swart, A.P. van Ulden, M. Hisschemöller and M.M. Berk, (1995). Climate change, policy options and research implications, in S. Zwerver, R.S.A.R. van Rompaei, M.T.J. Kok en M.N. Berk, *Climate Change Research, Evaluation and Policy implications, proceedings of the international Climate Change Research Conference*, Maastricht, the Netherlands, 6-9 December 1994.

Verbruggen, H. (1995). *Global Sustainable Development: Efficiency and Distribution*, Annual Report, Institute for Environmental Studies.

Victor, D. (1995). Global warming: Avoid Illusory Goals, *International Herald Tribune*, 29 March 1995.

Victor, D. and J.E. Salt (1995). Keeping the Climate Treaty Relevant, commentary in *Nature*, Vol. 373, 26 January 1995.

Weiss, C.H. (1977). *Using Social Research in Public Policy Making*. Heath and Co. Lexington.

Winsemius P. (1986). *Gast in eigen huis*, Beschouwingen over milieumanagement. Alphen aan de Rijn.

Legal and other documents

Review of the Implementation of the Commitments and of other Provisions of the Convention, FCCC/CP/1998/11.

Kyoto Protocol to the United Nations Framework Convention on Climate Change, 37 I.L.M. 22; (however, the corrected text is available at the website of the climate secretariat, http://unfccc.de).

United Nations Framework Convention on Climate Change, (New York) 9 May 1992, in force 24 March 1994; 31 I.L.M. 1992, 822.

Rio Declaration and Agenda 21. Report on the UN Conference on Environment and Development, Rio de Janeiro, 3-14 June 1992, UN doc. A/CONF.151/26/Rev.1 (Vols.1-III).

Establishment of a single intergovernmental negotiating process under the auspices of the General Assembly, supported by UNEP and WMO, for the preparation by an Intergovernmental Negotiating Committee for a Framework Convention on Climate Change (INC/FCCC), UNGA res. 45/212 (1990).

Conference statements

1979 Declaration of the World Climate Conference, World Meteorological Organization, Geneva, 12-23 February.

1988 Conference Statement, Conference on the Changing Atmosphere: Implications for Global Security, June 27-30 1988.

1989 Meeting of Heads of State, Declaration of the Hague, 11 March 1989.

1989 Noordwijk Declaration on Climate Change, in Vellinga, P., Kendall, P. and Gupta J (ed.) Noordwijk Conference Report Volume I.

1989 Statement of the Heads of Government of the Ninth Conference of Heads of State or Government of Non-Aligned Countries, Beograd.

1990 ECE Bergen Ministerial Declaration on Sustainable Development in the ECE Region, Bergen.

1990 Ministerial Declaration of the Second World Climate Conference, Geneva.

1990 Scientific Declaration of the Second World Climate Conference, Geneva

Government/ EU documents

Council Resolution on the Greenhouse Effect and the Community, Council of the European Communities, June 1989.

European Commission (1996). Verslag van de Commissie: Tweede evaluatie van de nationale programma's in het kader van het bewakingsmechanisme voor de uitstoot van CO_2 en andere broeikasgassen in de gemeenschap.

Chapter 10

EPILOGUE: SCIENTIFIC ADVICE IN THE WORLD OF POWER POLITICS

S. Boehmer-Christiansen

Every need to which reality denies satisfaction compels to belief.'
J. W. von Goethe, 1809[i]

10.1 Introduction

Ministries are reporting that they are increasingly overwhelmed by the enormity of the implications of the Kyoto Protocol'
Reported from Bonn CoP meeting, 1998.

...the future of the democratic process is being shaped in risk-management decisions.
Slovic and Fischhoff, 1983.

The book so far has explained why a vague commitment to prevent climate change, made in 1992, led, in 1997, to the signing of the Kyoto Protocol (see Chapter 9). As amended there, Annex I countries - that is OECD (Organisation for Economic Cooperation and Development) member states and countries labelled as 'economies in transition', are to reduce their *net* emissions of six greenhouse gases (GHGs), converted to carbon dioxide equivalents, by *differentiated* amounts between 2008 and 2012. This scientifically impossible conversion to equivalents is estimated to cut carbon dioxide emissions by a global average of 5.2% of below 1990 levels by 2012. For the treaty to come into force, however, at least 55 governments must have ratified whose 1990 emissions amount to 55% of the world total. Ratification is therefore by no means assured.

A reduction of 5.2% would have little effect on atmospheric concentration of these gases, but is considered by believers in the carbon dioxide induced global warming threat to be a step in the right direction. Subsequent negotiations have remained largely indecisive and now concern the involvement of the developing countries which are expected to become the main emitters of greenhouse gases in coming decades.

Current negotiations are about how to implement the emission reductions accepted by developed countries, including through emission trading and joint implementation, and how to entice reluctant developing countries to join a proposed technology transfer regime that has been agreed in principle only. However, technology transfer is defined neither operationally nor theoretically, and is understood by business at least to mean aid. The United States are refusing to reopen their arrangements with Russia and seeks pure market solutions. Business and R&D interests are expecting opportunities to sell, if necessary with the aid of green subsidies, energy technologies and knowledge. Widely differing per capita emissions and abatement costs have introduced the concept of burden sharing at least inside the European Union; the main battle now being about who is to pay. No country wants to weaken its competitiveness in consequence of taking climate protection measure and the experts have long promised 'win-win' and 'no regret' strategies. The Parties to the FCCC (Framework Convention on Climate Change) met for the fourth time in Buenos Aires in 1998 when most important decisions were once again postponed (Rentz, 1999).

Three points must be stressed, or repeated, at the outset of this epilogue. Under the Convention many knowledge and information intensive tasks are already being undertaken, generally funded by the public purse. Many more will have to be attempted long before any impact on climate might be noticed. Administrations are already engaged in drawing up reduction and mitigation plans. They are to formulate, implement, publish and regularly update national and (where appropriate), regional programmes containing measures by 'addressing' anthropogenic emissions by sources and removals by sinks of all greenhouse gases. Measures to facilitate adequate adaptation to climate change will have to be agreed and governments have already been asked to promote and co-operate in the development, application and diffusion, including transfer, of technologies, practices and processes that control, reduce or prevent anthropogenic emissions.

Secondly, climate and climate change remain undefined in the legal texts and in the documents of the Intergovernmental Panel on Climate Change (IPCC). Is climate measured or calculated? The World Meteorological Organisation (WMO) and the United Nations Environment Programme (UNEP) in 1987, with the International Council of Scientific Unions (ICSU) in the background have been the main promoters of climate change research and policy developments. ICSU and WMO had very strong research interests in climate variability and climate change as scientific subjects, while the fledgling UNEP added a powerful policy dimension: planning for global sustainability. These bodies gave IPCC the task of advising mankind on the science and impacts of climate change, as well as on 'realistic' response strategies, a truly formidable task. Governments, and hence bureaucracies, have formed an essential component of the IPCC from its informal beginnings in 1987 shortly after the collapse of oil prices in 1986.

Thirdly the research community, today more accurately described as the research enterprise always seeking funding, is not a single political actor, but consist of networks of competing and co-operating organisations and types of knowledge that do not find collaboration easy. Most climate models discussed in this book (Chapter 3 and 7) are the creation of meteorologists, not biologists, geologists, space physicists or hydrologists who have become the main challengers of the current consensus. The battles between these branches of scientific research for status and funding under the global change or climate change labels is an important part of this story, though it is not an easy one to unravel (Agrawala, 1998). The research enterprise needs lobbies to defend itself and ensure its prosperity in what is often perceived to be an increasingly hostile world. Science is therefore under constant danger of being 'captured' by these vested interests, including environmentalists, and must itself attempt to influence its masters. Even natural scientists (supported by some globalising financial, commercial and environmental groups) have begun to talk about global environmental management.

For example, the Executive Director of the International Geosphere Biosphere Programme (IGBP) believes that Global Change science has the role of 'underpinning a sustainable global society' (IGBP, Newsletter 1996:1). Together with other global research programmes, which are largely managed from the political 'North', the IGPB and the WCRP aim to develop 'practical predictive capability' for the Earth System as expressed in mathematical formulae. Others question whether this is methodologically sound. (Wiin-Nielson, 1998). The place of such 'predictability' in real world policy-making requires examination. It is argued here that, in contrast to formalised decision-theory (see chapter 9) and apolitical integrated assessment models (see chapter 7), scientific advise is only one ingredient of policy once policy is viewed as the outcome of politics, not 'rational' expertise. Politics, as distinct from policy designed by experts, is based on knowledge that is selected and used not as the determinant of action but as weapon and tool to pursue pre-existing interests and values. Human groups use knowledge to support behaviour that is shaped by interest, prejudice, belief and culture, as well as knowledge about the physical world. Governments are expected to make many decisions on a large number of issues and do so rarely for only one reason, such as environmental protection. Academic policy models therefore tend to be flawed from the very start. Governments select policies, given a choice, according to their own priorities, priorities, which may change rapidly over space and time and are often deliberately not consistent. Many 'public' strategies and even laws remain unimplemented, though they may well serve important political, non-environmental purposes.

Given the enormous brief and implications of the FCCC, it is therefore necessary to ask whether non-scientific and non-environmental grounds for and against taking 'action' to reduce GHG-emissions, trade emissions to reduce costs, and provide aid to the weak, have played a role in policy-formation alongside scientific knowledge. I shall argue that these 'other grounds' have been more effective than scientific knowledge in constructing a problem and then persuading societies to fund available solutions. Environmental bureaucracies and experts, as well as organised commercial interests in the competitive world of fuel and energy technology supply, have a strong interest in the debate. Environmental bureaucracies are a relatively new species which has been multiplying fast at all level and is working hard to expand its competences and keep up the momentum gained at Rio in 1992. These more immediate interests have been supported by environmental non- or quasi-governmental organisations (NGOs) which they have carefully cultivated since Rio.

Robert Watson, who has replaced Bert Bolin as the chairman of the IPCC but now also a senior environment official at the World Bank and who started his professional life as atmospheric research scientist in the UK before moving to NASA - began his keynote address to the Uranium Institute, London 1998, with the words:

> *.... I would like to present the case of why climate change is a serious environmental issue, which presents the nuclear industry with a major opportunity to meet the growing energy demands of the world. There is absolutely no question we need to increase the amount of energy in the world. Today there are two billion people who do not have electricity...... I do believe climate change is an argument your industry can use, that we need to move away from the fossil era, to produce and use energy in very different ways from the ways we do it today* (Watson, 1998:19)

So the idea of helping developing countries (and the Bank) has also been added as more immediate benefit of 'addressing' climate change. 76% of the world's energy services are derived from fossil fuel and two billion people still have no access to commercial energy - hence gaining access to markets for energy technologies, while limiting the use of cheap fossil fuels, -

particularly of available and substitutable coal - is an opportunity, not only for gas and nuclear power. But it has also been transformed into a moral objective for seeking 'sustainability' and 'poverty alleviation'.

One claimant for this mission has already prepared itself. The World Bank and overseas aid administrations have responded positively to calls to invest in cost-effective global environmental benefits, meaning the 'transfer' of cleaner technologies and fuels to industrialising countries where energy demand is expected to increase sharply in the coming decades. A small range of projects is already being funded under the climate protection label by national or multi-lateral financial bodies, especially the Global Environment Facility (GEF), a small off-spring of the World Bank which also tried to market itself as the administrator of a carbon fund at Buenos Aires. (Boehmer-Christiansen, 1999). These projects and associated activities are heavily information and expertise intensive. Their very complexity makes intergovernmental administration not only slow and difficult, but also requires considerable pay-offs to environmental and financial consultants and researchers, the very people who have been the strongest advocates of climate protection policy. Both the World Bank and UN rely on experts to justify policy interventions.

What then remains of the function of scientific advice in the actual policy-making process? How did the experts - a scientist becomes an expert if selected by authority to give advice - shape policy in the real world of politics: a world of clashing interests, beliefs and values? This chapter is based on a political analysis of the actual role played by scientific advice in public policy (and hence law making) since the mid-1980s (Boehmer-Christiansen, 1993; 1994; Agrawala, 1998). It is argued that environmental and scientific institutions alone remained far too weak compared to economic and political bodies to shape (as distinct from justify) the observed policy developments. Science as research enterprise could not remain an objective judge because it depends on commerce and environmentalism for its own survival. Scientists as research managers are almost by definition opportunistic seekers of political attention, creators of public concerns and providers of 'policy-relevant' findings for their funders. Their autonomy and independence is always threatened and has sometimes to be forfeited, or is traded-off with impotence in the sense that supplied knowledge is not actually used (Mukerji, 1989).

In this chapter I question whether there is a global society that can be modelled as a single actor and ask how such a society really uses scientific advice as distinct from how scientists think it should. Given what we know about how existing societies and competing political regimes behave, what does 'underpinning' global policy for science mean in political as well as environmental terms? Who would gain or who would loose in this managed global society? One thing seems clear, even if a tonne of carbon emitted in the United States and a tonne of carbon emitted in China both have the same effect on the climate of Washington DC, the impacts are not likely to be same in these two places. Peking may even benefit from warming.[ii] Who would decide how human behaviour, including trade, is be to changed where governments use natural science predictions to develop and strengthen international environmental law and institutions?

People, who doubt that human induced global warming is a dangerous threat to all humanity and who distrust existing models and knowledge claims, tend to get a bad press in some countries, including the Netherlands and the UK.[iii] They might be accused of a limited sense of ethics. But is mankind politically and socially equipped to manage itself globally in response to an uncertain threat? The cure may be worse than the disease. An attempt to cure may also involve serious, if non-environmental, dangers, which need to be balanced against 'predicted' benefits. Adaptation may not be the best environmental response, though it may be the best political reply once the dangers of 'global management by uncertain science' are recognised. Raising such complex and controversial questions is attempted.

Section 10.2 discusses the issues that are raised when scientists claim that knowledge, and especially future knowledge, is needed by policy-makers. It does so with reference to the Framework Convention (FCCC) and from the perspective of government. Section 10.3 describes eleven different uses of science that have been observed in the policy making process, noting that only one of these is generally recognised in 'scientific' policy design. To support the argument that science is much more important to politics than policy, a brief review of the history of scientific advice on climate change follows in Section 10.4. In Section 10.5, the allies of the research enterprise, which helped to make the climate threat so politically potent, are described. Reflections on the roles of 'the environment' and scientific research in global politics, together with some suggestions for further thought and discussion, conclude the chapter (10.6).

10.2 The role of scientific advice and the climate treaty

10.2.1 Moving towards implementation?

When the Contracting Parties to the Framework Convention on Climate Change (FCCC) met in 1995 they were 'required to review the adequacy of the commitment of developed countries to bring their greenhouse gas emissions back to 1990 levels by the year 2000' (Estarda-Oyuela 1993). In Berlin they decided that adequacy had not been reached, but little else.

Negotiations would continue, and money would be spend on public planning and 'capacity building' in poor countries. In the Kyoto follow up to the 1997 Conference of the Parties (CoP) generous emission reduction targets were agreed among developed countries, with the buying 'hot air', that is emissions reduced by the economic collapse or restructuring of countries 'in transition', became a possibility. Implementing Kyoto to included developing countries, meant calls for governmental interventions in energy markets in the form of subsidies, concessional aid and policy pressure. Who will be in charge of, and who will benefit from, these interventions: the implementation of emission reductions, including emission trading in all its 'flexible' complexity, data collection and reporting, monitoring, the building of institutional capacities and the 'transferring' of technology to Group 77 countries.

In 1999, when once widely advocated carbon/energy taxes had been abandoned in the European Union on political grounds and rejected at the international level in favour of permits and trading, a number of countries has adopted tax measures and other voluntary national means to reduce the 'carbon intensity ' of their economies. This would make them less dependent on fossil fuels. Others have done so primarily to raise revenue. Since the late 1980s there have been many government interventions in fuel prices to 'internalise' alleged external costs of fossil fuel combustion with immediate benefits to their treasuries. On the other hand, the scientific debate about the hypothesis of significant man-induced climate change continues unabated though with little visibility in Europe (Michaels, 1992; Lindzen, 1992; Boettcher, 1996; Calder, 1998; 1999). [iv]

In mid-1994 one could only wonder what might be decided in the light of the inability of the Intergovernmental Panel on Climate Change (IPCC) to bring new decisive knowledge to bear on the subject of global warming in time for the 1995 Berlin Meeting. In 1996 the situation had changed but little as internal debates about the conclusiveness of evidence erupted inside the IPCC and important phrases were changed under pressure. The 1995 IPCC Executive Summary states that 'the balance of evidence suggests that there is a discernible human influence on global climate'[v] (note: not warming!), but also warns that

'....our ability to quantify the human influence...is currently limited because the expected signal is still emerging from the noise of natural variability, and because there are uncertainties in key factors.'

This is cautious language indeed; other key factors being extra-terrestrial impacts and the behaviour of water in the atmosphere, by far the most significant greenhouse gas, and emerging as a new candidate for warming in the late 1990s, and the influence of solar variability on the energy budget of the Earth. Regional predictions of climate change remained a major challenge for modellers.[vi]

Pleas from within scientific institutions for more research being essential, such as for a global observation system, continue. In mid-1996, the World Meteorological Organisation claimed that:

'The questions still remains as whether we can attribute this (warming) to accumulation of greenhouse gases. We can't be 100 per cent sure because we don't know how much climate variation is natural and how much is man-made.[vii]'

Relying on work done by Danish scientists after hours, Calder (1997) has collected not only much evidence for the argument that the 'increases in carbon dioxide in the air from year to year are a result, not a cause, of climate change' (Calder, 1999:16), but has described in detail the difficulties faced by scientists in Europe, challenging the IPCC consensus. He also makes a prediction, namely that:

'...carbon dioxide increments year-by-year should continue to obey temperature deviation, but the inexorable warming required by the enhanced greenhouse hypothesis should never occur.(ibid.)'

The precise mechanism of how changes in solar activity, including cosmic rays and the solar wind, influence climate remains unknown, but is now becoming a new research subject - not of meteorology, but of solar and space plasma physics (Taylor, 1997). Those who study the sun and its interactions with the earth, suggest that the current enhanced warming phase, if caused by an extra active sun, is likely to end early next century, around 2007 (Landscheidt, 1999). So the jury on the causation of the observed warming is still out. Should policy wait?

10.2.2 Early doubts: scientific uncertainty and interests

Senior research scientists, interviewed in confidence in 1994, doubted that it would be possible to 'know' for sure for another 20 years or so whether man was responsible for climate change. (If mankind is not the culprit, there is little point in 'mitigation' if this is affect future climate rather than justify aid and trade programmes, that is interests). Other scientists believed that predictions of climate change will never be precise and localised enough to allow rational policy responses at the level, where control policies can be implemented.[viii] In the opinion of some even more directly interested groups, for example the World Coal Council, the FCCC has already committed governments to too much. Greenpeace and the nuclear power industry, on the other hand, assert that far too little was required, both would prefer a 60% cut in global carbon dioxide emissions. All these groups try to justify their positions with reference to science, and if science does not deliver certainty, to the precautionary principle. Some governments now recognise, according to the coal industry, that to achieve mere global stabilisation of these emissions 'may require a 100% reduction in emissions in industrial countries'.[ix] What would be the societal consequences? At Buenos Aires OPEC countries asked in vain for compensation mechanisms to be set up for lost income to fossil fuel exporters; only those mechanisms that promised income to bureaucracies were accepted.

If the climate threat is real and man-made, one may now ask why the political responses of the perhaps best informed governments, the USA, Germany and UK, have been so tawdry with regard to actually doing anything beyond what their economies are delivering anyway and what their treasuries covet? If extra carbon dioxide emissions are absorbed by the biosphere (forests are said to be growing faster than ever in Europe), then the link between increased emissions and accumulation in the atmosphere is anything but linear and there may be no need to worry, nature will adjust.[x] Since uncertainty has been with us from the beginning of the debate, the political analyst needs to ask why so many political actors have accepted the subject as important enough for intergovernmental negotiations and committed themselves to stabilisation and joint implementation. The threat has clearly worked and provided many researchers, planners and corporations with tasks and incomes. However, most of them have asked government for subsidies.

10.2.3 Believing scientific advice on climate change

The dissemination of scientific warnings to the international public (as distinct from scholarly debates) dates back to the 1970s. The process has been studied, together with responses by other policy actors (Boehmer-Christiansen, 1993; 1994 a; b).[xi] Given a particular institutional role or mission, advocacy of a specific position on, or belief in, 'global warming' is not too difficult to predict. Table 10.1 reports empirical findings.

The World Coal Council, from observation, is attracted to views, supported by scientific evidence, that global greening is good for mankind because nature likes carbon dioxide, the nuclear industry and energy efficiency lobbies have every incentive to stress the damage that global warming might cause. So far, however, there can be no doubt that for immediate benefits, the main winners in the debates surrounding the Climate Change Convention have been the Global Change research enterprise, national and international environmental bureaucracies, the oil companies as the owners of natural gas, and the earth information industry (Boehmer-Christiansen, 1994 a, b).[xii] This invites questions about the role of science in public international policy, in particular about the apparently important role of scientific advice in specific cases, such as the stratospheric ozone depletion case (Maxwell and Weiner, 1993).[xiii] How, by whom and why was scientific advice funded and given, and by whom and for what purposes was it subsequently used? What was the nature of the advice in terms of commitment to change in policy?

Table 10.1 Likely institutional positions on global warming

Institution	**Belief in warming**
Research director of GC institute	strong (in public at least)
Director of Space Agency	strong (in public at least)
Chairman of Nuclear Company	strong
Director of Greenpeace/WWF	very strong
Chairman of Coal mining company	weak
Electricity consumers faced with green taxes	doubtful
UN/World Bank environmental administrator	very strong
Seller of gas turbines/energy efficiency	strong

Indeed, the number of vested interests and organised groups attracted to one side or the other of the climate change debate is now so large that the possibility of a neutral and objective assessment of scientific advice itself may be in doubt. Political and economic rationality are often at loggerheads. While large economic institutions promote 'globalisation', political ones tend to warn of its impacts on social cohesion. Underlying these broader questions is the need for deeper awareness of changing power relationships. How do the complexities of natural systems compare with those of human societies? How much of our imperfect knowledge of nature and even more so of societies does in fact influence the ways societies regulate themselves and interact with each other? Are 'predictions' about how societies ought to change trustworthy enough to be used as public policy instruments? Proposed global warming mitigation strategies, it should be noted, may not only be very costly in economic terms, but other threats may arise from their feared, political impacts. These impacts can also be described as costs.[xiv] Mitigation may be resisted because of socio-political problems, such as poverty, which would require resolution during or even before the implementation of mitigation strategies. Societies simply may not have the capacity - which is not quite the same as 'political will' - to change in the manner suggested by policy advocates and formulators. Judging the implementability of policy proposals, made by the academic community, luckily remains a function of political systems.

The processes by which governments have tried to cope with the issue of potential climate change, or rather global warming, need to be understood from observation in order to answer the question of what role scientific advice plays in the political process. Does expertise undermine democracy by promoting technocracy, a priesthood claiming to possess the facts or objective truths that cannot be challenged? Or does politics make such claims for science when certainty is desired? How easily is science corrupted when used for unintended purposes or in association with unacceptable means?

10.2.4 The need for transparency

Though the ultimate target of expert advice, policy-formulators and decision-makers are not easily discovered. They tend to hide in 'government' as expert advisors and members of 'policy-communities', which may include members from outside government. Policy-formulators in bureaucracies and final decision-takers usually also represent institutions with vested interests and values to protect, such as bureaucratic territories or 'turfs', ministerial pecking orders and roles, access to resources and the public, electability, legitimacy and, ultimately, national sovereignty. Important decision-takers are found not only in government, but also in industry and research, from where they try to influence or lobby 'government'. Do 'decision-makers' really need ever more knowledge, or do they fund research when they want to postpone decisions? Understanding policy-makers, that 'interactive' group of unelected bureaucrats and elected politicians, is important because only they possess the formal power to arbitrate in political battles and imprint their own perceptions and expectations on outcomes. They define the goals towards society is to strive and decide the rules or norms within which change is to take place. Only sometimes do they also possess the power to actually implement these norms. To do this, society must be persuaded to change.

Experts assist the policy-makers in making policy and they do so with persuasion. Politicians are often heard to have acted on the basis of technical advice. These technical experts, be they lawyers or scientists, are not formally accountable to any parliament or court, but they are clearly powerful. Who were the advisors on climate change policy? Who selected them and what institutional interests, if any, did they represent? The public should know whether and how the knowledge generated by researchers and academic policy analysts, for example by the designers of IMAGE and integrated assessment models (Chapter 7), is used by that ill-defined body of men and few women we call 'the decisions-makers'. Was knowledge used for the purposes for which it was funded? Were the experts the real decision-makers behind which politicians may hide?

These questions are empirical but raise the difficult problems: should expert advice be attributed to individuals and made public? When is the knowledge base, brought to 'the state' by the experts, considered 'sound' enough for policy purposes as distinct from merely being useful to politicians and commercial actors in power struggles or competition for market access? What are the roles of knowledge in the political process? Knowledge itself is selected early in problem identification that is at a time when undesirable knowledge may also be discarded or marginalised. Research objectives and priorities are therefore very important in shaping, or misshaping, the future. The political and commercial actors, the selection of knowledge is based on criteria, which promise the selector more rather than less negotiating strength and institutional power. Once knowledge is an instrument of negotiations behind closed doors (and in the end most effective negotiations tend to take place in secret), those without access to independent knowledge will not be able to participate. Formal transparency is of little help. Knowledge may include as well as exclude. Scientific knowledge tends to exclude those who do not know.

10.2.5 Nightmares of policy-makers

The complexities of man-nature relationships are a nightmare for policy-makers because nature, not being able to act for itself, unavoidably becomes a political football for all those human groups that are attracted to environmental problems. Many questions about climate change still worry them:

- What climatic changes can be expected in the area for which 'we' are responsible?
- Is global warming a myth created to benefit the research enterprise and several industries?
- What are the political implications if major emission reductions were implemented?
- Why do scientists continue to emphasise uncertainty?
- What is the link between low fossil fuel prices and the call for selective energy taxes?
- What will be the effect of emission trading on the economy?
- Where is the 'missing' carbon?

Almost by definition, policy proposals require government interventions in order to change 'business as usual'. Political actors will be drawn to proposals that are to their particular advantage. Governments therefore come under contradictory pressures and often, rather democratically, adopt contradictory measures. For example, to reduce CO_2, should governments expand the use of nuclear power or/and subsidise renewables? Both of these are higher cost options with very different market structures and degrees of political support. Should governments make laws to promote a 'dash' to natural gas at enormous costs to coal miners and the landscape, as happened in Britain? [xv] Or should they regulate only for energy conservation, or care more for longer-term technology change by subsidising R&D efforts, though this tends to mean higher levels of taxation or loss of public services somewhere else? In the end, these are societal and not governmental decisions for which 'science' has no answer.

Yet decision-makers would like to be 'told' what to do by experts and hence tend to shift responsibility and liability to 'science'. Answers to these policy questions must be decided, but they do not add to our store of facts and theories about how nature works or even how societies have changed nature. They are not attractive to the global research enterprise. Rather they raise the issue of whether and when it is wise for governments to attempt to change society, in order to prevent the consequences of mankind's interactions with their physical surroundings - such as burning fossil fuels, building ever faster transport and communication systems, and altering nature itself through genetic engineering.

While bureaucracies generally like to acquire new powers, they often lack the resources and social support for implementing new policies that may be devised under these powers. It was surely not by chancc that 'environmental demands' became so prominent during the 1980s at a time when ideological fashion, especially in Anglo-Saxon countries, was to 'shrink' government and hand many of its functions to 'the market' and 'civil society', that is to consumers and organised private interests. Demands for environmental planning and strategic thinking for 'sustainability' in the face of global change thus served as countervailing argument, at both the national and intergovernmental level, for bureaucracies threatened by redundancy and loss of competence. In political battles over fundamentals such as distribution of power, science tends to become a tool of many engaged in politics. The various uses of this tool are outlined below.

10.3 Eleven uses of science in politics

10.3.1 Concepts and definitions: what is politics?

To identify the uses of 'scientific consensus' (that is agreed, if still uncertain knowledge about nature and society) in the world of politics and policy, some definitions and concepts are needed at the outset. In the qualitative worlds of political analysis, bounded rationality and cultural diversity, these remain subjective and change with context and purpose.

10.3.2 Politics as purposeful activity involving the use of power by institutions

Politics has been defined as the *art of persuasion*. This implies that human beings find it difficult to agree on a lot of issues and channel their identities, disagreements, interests and concerns through institutions, that is organisations to represent them. Politics can also be viewed as a way by which societies individually and together manage disagreement and resolve conflicts with the help of institutions and rules. Parliaments and intergovernmental bodies may make rules, but other institutions (courts, inspectorates and the policy) are asked to implement and enforce. Note that intergovernmental bodies largely escape the oversight of parliaments and may therefore be attractive to politicians who want to 'escape' popular pressures. National institutions and legal regimes have developed over centuries, with the European nation state, legal traditions and democratic decision-making (in its various forms) probably as the crowning, if always vulnerable, achievement. Societies' search for methods of agreement, which minimise the destruction of institutions, serve the preservation of governability and could be identified with the process of civilisation.

Politics can also be conceived of as a *game of strategy*, largely played by institutions and some individuals with different roles and degrees of dependence on 'the state'. This may range from full independence to total dependency. The stakes are high even if individuals withdraw from politics or refuse to recognise that politics matters. Outside the family, politics is practised by institutions rather than individuals, which defines the most fundamental difference between economics and politics as theoretical approaches to the study of social behaviour. The assumption is made here that the search for resources, status and influence motivates all institutions, which participate in the political process. This applies as self-evident to institutions, which look after the interests of science, the research enterprise.

Outside the textbook or lecture room, therefore, institutional politics rather than expertise determine policy, change the law and decide what may be taught and done without attracting the attribute of criminality.

The best examples of rapid and thorough societal change are provided by dictatorships with the power to fully 'synchronise' the objectives of most institutions under their formal control. Their longer-term success is not, however, reassuring, but *the possibility of environmental threats being used to centralise government and weaken democratic institutions* must be recognised. Dictatorships are generally served by technocrats who make decisions in secret in the name of some alleged common interest, good or god; experts may equally hide behind masters and unstated assumptions.

In more democratic societies, the utility and application of knowledge is decided by *bargaining (also called negotiation)* in the world of practical politics, that is by a mix of commercial (corporations, banks), ideological (churches, political parties) and political institutions (governments, professional bodies) interacting in a shared search for resources. In this world of conflicting interests and values, of explicit and implicit alliances, bargaining and negotiation are the only ways of settling disagreements if violence and coercion are to be avoided. Measures of a society's success - power, access to resources (including knowledge), health and wealth - are distributed by the political process among its members. Complex ideologies are constructed, many with the help of science, to justify these distributions. Arising conflicts are managed through law and legalised coercion, and where this fails or does not exist, as in the international domain, by violence and threats of violence, including economic pressure. It would be foolish indeed to assume that politics ceases when an environmental problem is being identified or dealt with.

In this setting the *structures and rules, which structure decision-making processes* (often wrongly dismissed as 'bureaucracy'), become extremely important. They determine how and by whom policy is made, which experts are invited to advise and which are excluded. Policies are understood as measures and instruments associated with a programme or set of guidelines. They may include regulations or investments aimed at producing or resisting change. An environmental problem, once perceived and understood as such, is recognised during a political process in which science may serve many purposes and environmental policy is usually made by public institutions, which have been granted the right to do so and are therefore accepted as legitimate. Democratic decision-making, in its various forms, is not the only, but currently the most widely advocated system of providing legitimacy, though the desire of political actors to make policy more predictable and avoid too much democracy is also very strong.

10.3.3 The allocation of public resources and the research enterprise

The extraction and distribution of resources for public use is a major function of public policy-making. Both are highly political processes, which reflect the competing claims and priorities of many actors. The research enterprise, as the set of institutions, which looks after the interests of research, is a small but highly strategic actor in this game. Its power rests in its ability to make knowledge claims. Research bodies also pursue strategies motivated by existing commitments and competence, their own visions of the future and the search for researchable problems. The role of experts is seen to be quite limited compared to that of formulator of options for politicians. Other institutions and sometimes powerful individuals will select from these options, usually after they have undergone filtration and selection at a lower level, for example by the media and lower levels of administrations. In the end, however, politicians and their immediate advisors (most unelected and often unknown to the public) decide which type of research is funded on the basis of what criteria. To infiltrate the institutions of power must therefore be one aim of the research enterprise. The current interest in researching environmental problems surely relates to the declining status of 'pure' science and its declining utility to the 'defence' industry, where it may be said to have been too successful for its own good. What has emerged instead, is the claim that science is useful in defending mankind against future environmental threats.

Part of this 'infiltration' is the promise of research to provide solutions to existing, and even better, to future problems, that is to claim policy relevance. Individuals in government and industry will test this claim when taking advice from the research enterprise. They will have their own rationales for listening to scientists and for funding further research. Competition for resources within the research enterprise is generally fierce and is fought 'dirty' and behind closed doors, as well as in public. These battles provide the enterprise with the opportunity of bringing researchable problems, especially those that can be labelled 'environmental', to the attention of politics and the media. In this game of 'support me but not my opponent' threats and promises are freely used, but are effective only if they are believed. Credibility is essential, which means that even in the process of persuasion, rationality plays a role. An environmental problem, once perceived and understood as such, is the outcome of a process in which appeals to science may serve more than one purpose. All political actors strive to make credible knowledge claims based on some agreed authority.

10.3.4 The functions of science in politics

Science is used by most political actors that desire to be considered 'rational'. This attractiveness of science itself is a subject of much interest to philosophers and sociologists. In this chapter only personal observations, drawn from many years of research into environmental policy-making, are presented. I have identified eleven different uses:

1. Ultimate source of authority
2. Tool for persuasion and justification
3. Greenwash of unpopular policy
4. Scapegoat for policy failure
5. Mechanism for delaying or avoiding action
6. Maker of prophesy and hence tool to shape future
7. Clarifier of conflicting interests
8. Protector of sovereignty
9. Problem solver/ source of warning
10. Tool for monitoring and surveillance
11. Judge or arbiter in conflicts

(1) The political utility of a God, or The Law, as the ultimate *source of authority*, is obvious. Authorities other than science include taboos, experience, dogmas, ideologies, religion and the law, but these are not open to quantification and hence are often rejected as 'irrational'. Modern secular and pluralist societies, and therefore international bureaucracies and institutions cannot appeal to a single God or dogma, as there are so many competing ones. Weak institutions in particular are in great need of sources of authority. Only the natural sciences and international legal principles, which are often contradictory, make claims of global applicability. Why the natural sciences should have gained such authority and hence political utility cannot be explored here, but this surely relates to their claim to be the 'facts' or 'objective' knowledge. Given the complexity of the environmental sciences, this is of course very disputed. However, reference to 'science' as a source of authority provides legitimacy to positions on disputed policies. This faces scientists with a profound dilemma: to remain useful to politics without becoming too useful, they must also face the risk of loosing their integrity and hence weakening their authority in the future. For example, if the IPCC's scientific consensus - so widely disseminated as the voice of the scientific community -turns out to be wrong, the question is raised why many scientists, working at the frontiers of knowledge, seemed to agree in the mid-1980s that man-made warming was a serious threat requiring immediate action. While non-scientific sources of authority are usually more

powerful than science in politics, science has become the favourite source of authority for the environmental movement and at the intergovernmental level. The reasons cannot be explored here, but may reflect the difficulty of finding global sources of authority.

(2) In all political debates, scientific arguments, theories and evidence are used selectively with the purpose of changing an opponent's belief and hence his or her position in a debate. To work, both sides accept the authority of science, which then becomes a useful *tool for persuasion and justification*. Environmental debates and negotiations tend to be about the balance of scientific evidence, especially during their early stages, when the evidence for and against environmental damage, allegedly caused by pollution, is a major tool in intergovernmental negotiations. Negotiators tend to select the advice that best fits their existing interests and commitments, including consistency with non-environmental objectives. In fields where scientific disputes exist, partial evidence is useful politically, long before a problem is sufficiently understood for 'rational policy' as defined in earlier chapters. Rather, science is used to persuade the public, attract supporters and above all the media, and weaken opponents. Another illustration is the tendency to describe environmental standards as 'scientific' rather than as based on scientific evidence. Only the act of measurement is a truly scientific task, the setting of standards and thresholds involves judgements of costs and benefits broadly defined.

(3) *Greenwashing an unpopular policy* means giving an environmental justification for an unpopular policy that was adopted for non-environmental reasons, though it may have positive environmental side-effects. Behind most current environmental rhetoric, including many policies adopted by international and national bodies, one can discover 'non-environmental' aims, which, while usually quite legitimate, are 'dressed up' as environmental protection and hence as justified by 'science'. Non-environmental benefits may include the desire to create jobs and tasks for bureaucracies and experts, sell profitable new technologies or gain competitive advantages in new markets. *The 'greenwash' function* has provided the natural sciences with considerable influence and institutional power. Examples include the closure of coal mines, subsidising nuclear power and the 'dash' to gas in electricity generation. All of these have been explained to opponents as measures against global warming, though the same policies would have been pursued anyway. No-regret-policies against 'global warming' also tend to fit this category. The major concern of public authorities may well be to reduce dependence on imported fossil fuel rather than to reduce greenhouse gas emissions.

(4) Related to the roles of greenwash and providing authority is the ability of science to act as *scapegoat for policy failure*. Science has the advantage to be allowed to change its conclusions: to admit and correct its own errors. This virtue is useful not only to learning, but also to politics. Policy errors and failures can be explained with reference to new knowledge rather than to past mistakes or wrong judgements. Changing scientific advice, allegedly or real, allows policies and positions to be altered without loss of face or loss of credibility. Science may be used for covering up for poor judgements or errors. The argument is put as: 'since new knowledge has come to light', change is needed. For policy-makers this misuse of science is difficult to resist.

(5) When science remains uncertain, as is the case when evidence is abundant but contradictory, appeals to science become extremely useful in another sense again, *as mechanism for delaying or avoiding action, as antidote to precautionary action.* When faced with something difficult, government can always support more research. Inability to take action may be the outcome of a lack of institutional capacity, or can arise when the associated conflicts are too great. As both the factors are typical at the intergovernmental level, this level is the most appropriate one for research agendas to be presented. Scientific uncertainty, so plentiful in climate change, allows more research to be presented as a policy response. Uncertainty and ignorance are the incentives of research, and it will always be in the interests of scientific expertise, derived from research organisations to call for more research rather than to define implementable policy.[xvi] It is used for global warming within the EC and in the USA, i.e. by countries where rapid GHG reductions would cause serious political (and economic) difficulties. The counter argument also applies; experts may be used to urge action on uncertain scientific grounds (precaution), especially when non-environmental benefits accrue to the self-defined pollution victim. As economists and engineers have a greater interest in regulation than natural scientists, they tend not to be impressed by this use of science, and they prefer to opt for 'the precautionary principle'.

(6) The usefulness of science to politics because it helps in *shaping the future*, endows science with the function of *prophesy,* also called forecasting and scenario building. Prophecy has always been closely associated with politics: providing kings with reassurance and prophets, if wise enough, with considerable influence and a good life at court. A great deal of climate policy could be considered as shaping the future by planning and anticipation. But are the available climate and policy models predicting the future well enough to be used as policy instruments? Or are there rather heuristic devices to allow the human brain to think more clearly, collectively, and integratively? The latter use is far removed from that as 'authoritative' methodology for imposing solutions and institutions on the weak. This issue is of interest not only to sociologists, but also to policy-makers and scientists, because actions taken today clearly affect the options of future generations. Precautionary policies, in particular, provide those holding the reins of power with functions and resources to shape the future in the name of a common interest.

(7) Scientific debates also help policy-makers, if in a listening mode, with a means to which they can *clarify the interests of major participants* in a policy debate. The experts, speaking for conflicting interests, also provide information about likely political impacts of policy options. Without using science in this fashion, governments would be even more at the mercy of competing lobbies, be these industrial, research or environmental. WG-III of the IPCC was largely used for this purpose by governments.

(8) By combining several of the above uses, science may also become a *diplomatic tool for protecting sovereignty*. The use of scientific criteria as the bases for international agreements, apart from creating delays so that more research may be done, because that is all governments can agree on, tends to maximise sovereignty because scientific control instruments (bubbles, air quality standards) rarely affect technology and industrial policy. Directly, rather they add competence to national governments, which are then called upon to define more precise regulations.

(9) Scientists hope that they are most useful to politics when they provide *practical knowledge for problem understanding and solving*. Scientific knowledge can clearly help government to first warn that a problem does or may exist, and help with the design of solutions to those problems, which the political process (often as a result of pressure from scientific institutions) has identified as 'real'. Science acts as a provider of useful knowledge for designing targets, measures and instruments. Here at last is the 'rational' use of science. Expertise rather than bargaining or the irrational public is expected to set environmental standards. This is the use scientists themselves recognise most clearly. It means that their knowledge can be operationalised and assists directly in the definition of policy measures and of the instruments by which these can be implemented.[xvii] Here science becomes a part of the implementation process. However, scientific standards need to be translated into technological or behavioural change. Other forms of knowledge are therefore required as well. The inability of modellers to provide credible regional predictions for climate change, that is in particular earth bound knowledge, has reduced their usefulness.

(10) Related to above is the use of science for *monitoring environmental objectives as well as for policy compliance*. For environmental protection it is not the capacity of science to forecast that is really important, but whether the implementing institutions effectively solve environmental problems. Science here is useful in monitoring change and measuring actual performance. For effective implementation, policies must be much more than 'correct' in the sense of defining technical measures by which targets can be achieved, but must also be socially acceptable and politically feasible in the shorter run. Therefore, the knowledge base for predictable outcomes cannot only be the natural sciences. Global prescriptions of what ought to be done in terms of global emission reduction targets, for example, are useless to policy implementation if they cannot be translated into industrial sector targets and timetables, which existing institutions can understand, achieve and monitor. Emitters can only be effectively controlled at the national level, trees cannot be planted by UNEP, though UNEP may advise and the World Bank/GEF may fund afforestation. Because at the global level resources for such 'aid' remain extremely limited, the suspicion remains that monitoring data will serve modellers rather than policy implementation.

(11) Once implementation is seriously sought by the law-makers, science may be called upon again in its eleventh function, that of *judge or arbiter in disputes,* as an instrument for deciding whether a breach of regulations has occurred or whether a claim is true or not. In court science acts as one, but only one, source of evidence.

10.3.5 The gap between policy models and policy implementation

The aim of policy is often assumed to be effective implementation, but this is incorrect. Science clearly has many social and political uses in which knowledge as such, and implementation are of secondary importance, and policy is symbolic. Through these non-uses alone, scientific research may prosper while businessmen, politicians and administrators, responsible for outcomes of policies, have a far greater need for being right than scientists who have no responsibility for policy failure. Practical implementation, for which the natural sciences are not responsible, rarely concerns environmental scientists. It is therefore right that policy-makers should be cautious about believing scientists, and scientists should be cautious about making careless claims. As the utility of science to politics ultimately relies on the faith of non-scientists, that science will deliver what it has promised, scientists tend to reject this naive belief and are often prepared to admit that they deal with hypotheses, best guesses, uncertainties, informed opinion and negotiated conclusions. Yet the faith in science itself is the most formidable political tool scientific institutions possess. They know that politics is there as an ally only as long as they deliver practical knowledge. So scientists too have models of how policy ought to be made.

Three different policy models are very briefly used to summarise the tensions underlying the policy process compared with scientific policy options proposed by academe.

The simplest model, preferred by scientists, sees policy as the outcome of science (provided by research) and values (provided by policy-makers (note the avoidance of the term 'politicians')). The chairman of the IPCC has recently outlined such a view.

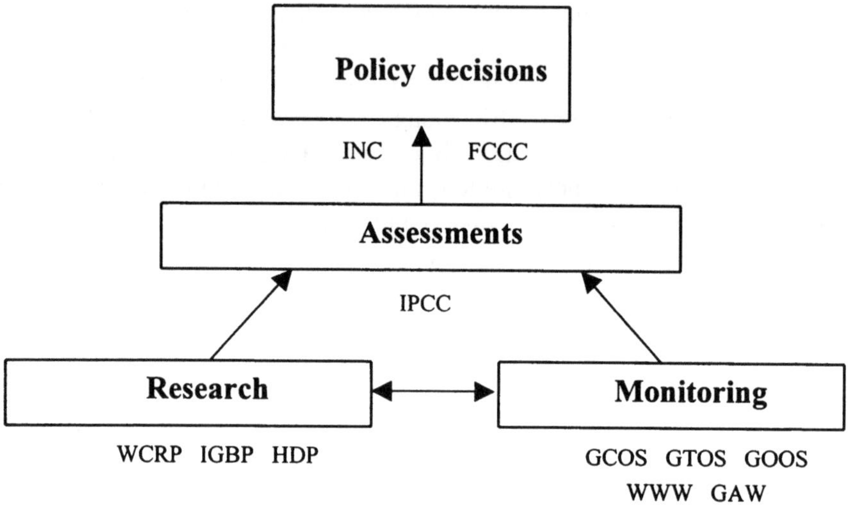

Figure 10.1 IPCC assessments provide the pathway for policy-makers to be made aware of the results of global change research programmes (such as the World Climate Research Prohramme, the International Geosphere-Biosphere Programme and the Human Dimensions of Global Environmental Change) and international monitoring stuides (such as the World Weather Watch, the Global Atmospheric Watch and the Global Climate Observing System, Global Terrestrial Observing System and Global Ocean Observing System).(Source: Bolin , 1994).

The figure suggests a policy process in which IPCC assessments provide the pathway for policy-makers to be made aware of the results of global change research programmes (such as the WCRP, the IGBP and the Human Dimensions of Global Environmental Change) and international monitoring studies (the World Weather Watch, the Global Atmospheric Watch and the proposed Global Climate Observing System, Global Terrestrial Observing System and Global Ocean Observing System). This 'science as policy foundation' model has the advantage of allowing scientists to present scientific knowledge not only as objective and neutral, but also as the very basis of policy in a world where knowledge can easily be divided into value and fact. The model excludes the institutional dimension entirely and with it power and power differentials. It is misleading because this conceptualisation explains neither process, behaviour nor national variations in policy response, and greatly exaggerates the role of knowledge, while those of interest, like the roles of dogma, culture and ignorance are ignored.

Policy implementation is not even considered. However, this science based approach, while unable to predict the barriers and issues, which invariably arise when human behaviour is to be changed and technology to be transformed, will lead to a clear problem definition problem. This serves the research enterprise.

A second type of policy model develops the Bolin approach to provide rational options for politicians to select from, but integrates a much wider range of knowledge. This approach is demonstrated elsewhere in this book (Chapter 7), especially by the IMAGE model, which is meant to aid formulation (Dowlatabadi and Morgan, 1993). This model significantly increases complexity and choice. However, it remains an input into policy-making and, as such, by-passes the political process and political competition. Reality remains simplified by assumed 'rationality'. Forecasts of measurable parameters are made on the basis of only those assumptions that can be measured. As such, the model is useful to those who share its assumptions or like what it predicts. The model is useful for justification, persuasion and negotiation, but can it define implementable policies? Integrated assessment (policy) models therefore face similar problems to those of GCMs. By combining uncertain predictions of climate change with socio-economic assumptions (upon which the former are based) they create a degree of 'cascading' uncertainty, which remains too large for the needs of implementable policy, though useful for many of the above mentioned purposes.[xviii] These policy design models are probably best viewed as academic experiments, which should become an issue of political concern only if taken too seriously, that is if they are used to impose global policies.[xix]

Another approach is to view policy empirically as the outcome of complex political interactions, decided by power differentials and strategies (Hart and Victor, 1993: 643-680). This is the approach favoured here and makes no claim to modelling the future, but is rather based on historical analysis. It postulates a desire for autonomy by the research sector, which must nevertheless participate in political battles to maintain status and gain funding. The interest of science is to influence research policy, which in turn is viewed as the product of political activity in which the environment becomes a justification for learning more, rather than regulating economic and social behaviour, which are issues outside the realm of natural science. This approach is not, however, particularly useful for policy-makers, maybe except for delivering warnings about the potential for science to be misused. It is proper, therefore, that this model does not dominate policy advice, but rather serves to point out the limitations of the others, thereby hopefully throwing 'the ball' of responsibility back into the court of governments and investors.

10.3.6 The ultimate irrelevance of the natural sciences?

Natural science is usually of little use in ensuring effective implementation, that is in managing 'society'. Other forms of expertise, especially technological knowledge and the arts of regulation and management, become more sought after. This makes implementation costly for politicians and taxpayers. Implementation may therefore never be sought in practice; more research is both cheaper and readily achievable; there is demand for 'knowledge creation'.

Giving new rights or duties to public institutions, such as designing climate policy, is sufficient for altering the distribution of bureaucratic power and resources. Calling for a reduction of GHG-emissions means demanding a change in industrial or commercial policy, which usually means intervention by the state in the workings of the economy by developing new framework laws, imposing emission rules or taxes, funding certain activities and withdrawing subsidies from others. In each case science stands to benefit long before implementation on the ground may be attempted.

While society has many mechanisms for making the unequal distribution of power acceptable to its members, resentment and political instability, even revolution, may threaten if too much change is demanded too quickly. Political parties do not like to face loss of power, economic elites even more so. Policy may therefore remain symbolic or an academic discipline in which its practitioners publish elegant solutions to problems they have defined for publication in learned journals. Political and not just economic costs are a part of the package of considerations that politicians and administrators must take into account and which are always likely to prevail over scientific advice, when it comes to practical implementation. Even then it is assumed that 'societies' have the capacity to effectively implement change, which may not be the case.

10.4 The origin of scientific advice on climate change and its linkage to energy policy

10.4.1 From weather modification to a New Ice Age and the limits of growth

As outlined in previous chapters, the IPCC has advised the world on climate change science, impacts and realistic response strategies since 1988. Its small secretariat is based in WMO headquarters in Geneva, but its activities are supported by the big research laboratories of the world in the

USA, Germany, UK, France, Japan, Netherlands and Canada. To understand why it 'emerged' in international relations in 1987 and rapidly achieved quite unexpected prominence, a short excursion into history and a look at the changing global policy context, especially the price of energy, are needed. The intellectual debate, underlying IPCC advice and concurrent political activities leading up to the ratification of the FCCC, can be traced back well over 100 years. Table 10.2 attempts a selective summary.

Table 10.2 Main events leading up to the FCCC

Year	Main event
1827	Fourier speculates about man-made climate change
1896	Arrhenius *calculates* warming between 4-6 degree C due to burning of fossil fuel
1938	Callender agrees with above but dismisses concern because of nuclear power
1957	Geophysical experiment with the atmosphere proclaimed by US scientists
1961	Keeling *measures* increase in carbon dioxide concentration in atmosphere
1971	Prior to Stockholm Conference worry about cooling due to aerosols (hence a New Ice Age scare), later used by IPCC to explain insufficient predicted warming
1970s	Limits to growth debate, fast breeder technology and the nuclear winter models influence perceptions and advance atmospheric and energy demand modelling
1979	UNEP declares that mankind is carrying out an uncontrolled experiment with the atmosphere and urges action
1981	Exaggerated predictions of energy demand growth (IIASA) fuel concern about warming hypothesis; high fossil fuel prices continue to encourage renewable and nuclear energy technologies, but WMO remains neutral on climate change theories
1985	AAAG declares warming to be a dangerous, if uncertain risk; calls for more research and a convention
1986	Oil price collapse begins and with it a return to cheap fossil energy; fast increase in computation power favours development of 'predictive', coupled climate models
1987	IPCC being formed from AAAG core and WMO scientists, backed by the International Council of Scientific Unions and UNEP
1990	WMO Second Climate Conference calls for more research and accepts the warming hypothesis as more likely and the World Bank 'green' prior to Rio

In 1827 Fourier raised the possibility of man-made climate change on the basis of physical calculations and in 1896 Arrhenius predicted an increase in air temperature between 4 and 6 degrees Centigrade from laboratory tests. In 1938 a British steam technologist called Callendar (who during the 1940s influenced Flohn in Germany) also calculated that the planet was getting warmer because of carbon dioxide emissions. He concluded that there was no need for concern, as another Ice Age would be prevented and nuclear power would reduce emissions anyway. Forecasting weather and climatology remained fairly lowly subjects for geographers, rather than 'real' science. The splitting of the nucleus and nuclear technology, rather than the complexity of nature, came to occupy the brightest minds. The climate change research did not begin in earnest until the late 1950s, with reference to *weather modification,* with the USA, Germany and Sweden deeply involved (Weart, 1992).

In 1957 American scientists pointed out that human beings were carrying out a large-scale geophysical experiment, which might yield far-reaching insights into the processes determining weather and climate. In 1961 Charles Keeling proved by measurement that the carbon dioxide concentration in the atmosphere was indeed increasing. Warnings in the media, made by the rising environmental movement, began shortly afterwards. The study of man's potential influence on climate advanced slowly during the 1960s, thanks to computers, but began to take-off properly when research institutions learnt to link their interests to fundable human concerns and political opportunities (Hart and Victor, 1993). In the 1970s the *limits to growth and nuclear winter* debates were strong stimuli, advancing computer-modelling capacities in the institutions, capable of participating in earth systems science. As early as 1965, the US President's Science Advisory Council argued vis-à-vis its funders that useful climate change predictions down to the regional level would be possible within two or three years, given sufficient computing power. Almost 30 years later, no such detail is possible and some doubt that it ever will (McCracken, 1992: 13).

10.4.2 Aggressive expansion of climate research

In 1970 the Massachusetts Institute of Technology, already at the forefront of research planning, convened a 'Study of Critical Environmental Problems', which concluded that global warming was a serious possibility and advocated the aggressive expansion of climate research, combined with population control and protection of the food system. Science was ready to solve these 'new' problems with the help of the pill and the Green Revolution. By 1972, the Stockholm Conference recommended more climate change research at the suggestion of the *First International Conference on Environmental Futures,* at which an UN institute for planetary survival had been proposed. In a keynote paper on 'Climatic modification by air pollution', the American climatologist discussed the role of aerosols and the ozone layer. Atmospheric dust was suspected to be the cause of global cooling, though in discussion he admitted to a 'sneaking' suspicion that the loud demands for more monitoring were 'mostly for the care and feeding of big computers', rather than the welfare of man (Bryson, 1972: 165-7). Others argued that the development of numerical general circulation models, which successfully simulate the present climate and the behaviour of the atmosphere in long time runs, was the first step in attempts to predict what happens to the atmosphere as a result of man's activities, then considered to be the ultimate goal of the Global Atmospheric Research Programme (GARP), the project managed by Bert Bolin as a young man and considered to be the predecessor to the IGBP which was planned in the early 1980s and is now being implemented.

The Executive of the WMO began to include climate change into its research portfolio and by the end of the 1970s UNEP (a product of Stockholm and a rather weak intergovernmental body in search of a function), made climate both a research and development and hence aid issue. Important new 'allies' to the climate change cause had been attracted and UNEP's director soon mentioned energy and the need for better climate forecasting in 1974, though still without any reference to global warming. When addressing the First World Climate Conference in 1979, however, he referred to climate change as the process of carrying out an uncontrolled experiment in the earth's atmosphere and called for preventative action. The worlds of politics and energy remained unimpressed, as energy prices soared thanks to OPEC policies of limiting oil supplies. A rosy future for nuclear power and other new, low carbon fuels was still widely envisaged. Energy demand forecasters became interested in the subject of climate change and research groups began to collaborate on energy forecasting, linked to climate change, caused by increased emissions from the combustion of fossil fuel, a 'given' assumption because of the expected industrialisation and population growth in many countries outside the still existing 'West'.

The designers *of fast breeder reactors* in particular were particularly attracted to the subject, as only a vast nuclear programme would be able to cope both with the enormous predicted growth of energy demand and global warming. In Germany this led to a productive collaboration between Herman Flohn, a famous German climatologist, and Wolf Hafele, the German father of the breeder reactor, at IIASA in the late 1970s (Cavender and Jager, 1993). Estimated temperature rises then put forward for a doubling of carbon dioxide concentration were slightly lower than those put forward in the 19th century (Kellogg and Schware, 1981), but differed very little from those proposed a decade later by the IPCC.[xx]

10.4.3 The Advisory Group on Greenhouse Gases: 'independent science' warns

As mentioned earlier, this small group of environmental scientists and research managers, working on energy and climate, met again in 1985 at a conference, revealingly called 'the second joint UNEP/ICSU/WMO international assessment of the role of carbon dioxide and other greenhouse gases in climate variations and associated impacts'. They set up the Advisory Group on Greenhouse Gases (AGGG), which successfully launched global warming into world politics, one year before the oil prices collapsed, surely a major reason for the wide appeal of its message to certain energy supply interests.

The 1985 Conference had been organised by the Swedish International Meteorological Institute, the home of the IPCC chairman, and the Stockholm Environment Institute. Participants concluded that 'it is now believed that in the first half of the next century a rise of global mean temperature could occur, which is greater than any in man's history'. Furthermore, they recommended that science based emission or concentration targets should be worked out to limit the rate of change in global mean temperature to a maximum of 0.1 degree Celsius, (WMO, 1986). The scientific papers read at Villach, commissioned and peer-reviewed by its organisers, were published jointly by WMO/ICSU and UNEP (Bolin *et al.*, 1988). The conference was attended mainly by non-governmental researchers. There was no need for consensus generating procedures as only those in agreement with the aims of the group had been invited. Another 10-20 years of observation would be needed before the detection of global warming was likely to occur and ecologist William Clark from Harvard stated that uncertainties, from emission rates through environmental consequences to socio-economic impacts, dominated the greenhouse gas question (WMO, 1986: 24). The Conference felt that refining estimates was 'a matter of urgency 'and recommended a list of actions, which remained vague with respect to policy. The IGBP and the WCRP were recommended to governments. Decision-making rules under specific kinds of risks, the determination of damage costs from greenhouse warming, as well as the behaviour of policy-makers were to be researched. An action plan for global environmental management was clearly in the making.

10.4.4 A call for a global convention and policy advocacy turn against fossil fuels

Villach also approved the broader, political brief of the UNEP, which urged delegates to support the setting up of a non-governmental International Greenhouse Gas Co-ordinating Committee to:

- promote and co-ordinate research, monitoring and assessment; promote the exchange of information related to climate warming;
- prepare and disseminate educational material;
- approve the possible advantages of an intergovernmental agreement on global convention.

The change from climate variation to warming had been made on the basis of a specially requested modelling exercise, linking energy demand predictions with temperate rises. AGGG members subsequently organised the 1988 Toronto NGO Conference, which called for a 20% reduction of CO_2 emissions, which caused unease among governments, industry and other scientists. The AGGG also organised the Second World Climate Conference in 1990 (which failed to agree on CO_2 reduction targets and opted for stabilisation) and wrote up the results (Jager and Ferguson, 1991). It prepared the Meeting of Legal and Policy Experts, held in February 1989 in Ottawa, which recommended an 'umbrella' consortium to protect the atmosphere, which was to be implemented through subsequent protocols and proposed a World Atmosphere Trust Fund, as well as a Convention that should be served by a panel of independent experts. There was therefore considerable bitterness when the AGGG was replaced by the IPCC under pressure from the US.[xxi]

By 1985, therefore, a network of science leaders (the managers of the research enterprise) had formed, which included people deeply involved in energy and policy research and determined to initiate a dialogue with 'policy-makers', a selected few of whom they invited to a 1986 meeting in Italy. The AGGG had succeeded in taking the policy debate into the world of politics, but its institutional base proved too weak to keep the issue out of the hands of big, accountable institutions: governments and the WMO, which had the ability to fund the global change experiment. The IPCC is therefore best understood as a response to the AGGG and its policy advocacy. Informal links between the IPCC and the IGBP are strong (IGBP, 1994). There are programmes to strengthen the scientific capacity of developing countries and they constitute a common interest with the IPCC and the WMO. Space-science and -technology are deeply involved in this global research effort.[xxii]

The research enterprise spoke with more than one voice. For example, two climatologists, who attended the 1985 Villach Conference and contributed to the IPCC , and who therefore appeared to have added their voices to calls for immediate action, a little later argued, vis-à-vis the science policy-makers, that the range of scientific uncertainties was still so large that neither 'do nothing' nor 'prevent emissions' could be excluded from consideration (Warrick and Jones, 1988: 48-62). It was imperative, they felt, that full support was extended to a research effort that would narrow the range of scientific uncertainty concerning the greenhouse effect. While the IPCC had 'calculated with confidence' that an immediate reduction of 60% of emissions would have to take place if the stabilisation of current concentrations was to be the goal. The goal itself was not advocated and the difficulties of linking calculated temperature rises to real climate were pointed out.

The US Committee on Global Change and the US National Committee for the IGBP have stated that the IGBP was developed as a step in the evolving process of defining the scientific needs for understanding changes in the global environment, which, it is asserted, are of great concern to the public. Here it is argued that this concern was largely created by the research enterprise to market its agenda. Human beings, or human welfare, hardly entered this agenda, except as justification for its funding. One chapter out of nine of its recent report deals with humanity (US National Research Council, 1990). The small section of the Global Change research agenda, devoted to the Human Dimension, deals mainly with the physical impacts of landuse changes. The research agenda of the European Union also reflects this priority, responding probably to the wishes of its own space technology and information lobby.[xxiii] The importance of 'Global Change' (as label earth systems science and remote sensing) is great indeed. It would not gain from an advisory body that hurried policy-formation along, by recommending action, rather than decades of further research. To achieve such an outcome, science would need allies that would 'balance' the environmental lobby'. For itself, the strategy was creative ambiguity.

10.4.5 From non-governmental to intergovernmental science: ambiguity prevails

The IPCC is largely a representative of governmental scientific institutions, especially meteorological offices and atmospheric physics. It has repeatedly been called upon by the UN General Assembly for advice (Churchill and Freestone, 1992), but its role in the FCCC remains weak, as the very size of the Panel today makes its unmanageable for governments.

The Panel is nevertheless able to use the good offices and machinery of the WMO and the UNEP, and increasingly the WHO, to reach the heart of government with the assistance of environmental bureaucracies. By absorption and rejection of the AGGG network, IPCC lead authors were assembled in the late 1980s by science leaders, meeting at the WMO, and set to work to write carefully planned chapters, summarising available knowledge and pointing out uncertainties, as well as research needs.

The Dutch and American environmental bureaucracies collaborated in creating a strong global warming threat (in 1990), by prescribed emission scenarios for IPCC WG-I.[xxiv] These assumed a rapid doubling of CO_2 and rapid global economic growth. IPCC scientific research needs, on the other hand, became closely integrated in UK environmental policy formulation in order to advance British science policy. The threat of global warming was disseminated by the IPCC in a highly condensed form through its 'policy-makers' summaries. While much has been made of the IPCC scientific consensus, especially in Europe, its essence is agreement on uncertainty. An impression of certainty and authority was, however, carefully crafted by the ambiguous use of language (Boehmer-Christiansen, in Boetcher and Emsley, 1996). Instead of more neutral terms, such as numerical experimentation and scenario building, the terms 'model' and 'prediction' were used.

IPCC lead authors on the 1990 scientific assessment claimed that they could calculate with confidence average global temperature increases from models, which a few pages later were admitted to be 'comparatively crude' and from which two important greenhouse gases, water vapour and ozone, as well as aerosols and processes involving the biosphere, had been omitted. The combined 'will result' with 'a likely increase' in the same sentence is surely a grammatical device, which gives the impression of both certainty and uncertainty (Houghton *et al.*, 1990). The predicted rate of temperature increase was likely to come true only if assumed emission scenarios, doubling the concentration of carbon dioxide by prescribed dates, would take place and provided that there would be no surprises in the understanding of the carbon cycle.[xxv] The scientific essence of the 1995 IPCC science report to policy-makers has been cited above. Another report is being prepared.

While policy-makers might well have decided to adopt mitigation- and adaptation-strategies immediately, because of undoubted increases in GHG-concentrations, the IPCC has never advised this, though many commentators have done so. Rather, its policy relevant formulations have allowed various interpretations, except for the recommendations that additional research was needed. Here the number of recommendations actually increased between 1990 and 1992. Environmentalists and the promoters of advanced energy technologies could therefore select from IPCC statements those which allowed them to assert IPCC-support for their particular positions. [xxvi] When addressing the mentioned World Energy Council, John Houghton, then still chairman of IPCC WG-1, discarded the idea of certainty altogether and stated that 'IPCC publications explain the degree of scientific uncertainty, regarding future climate change...Research therefore needs to be undertaken urgently in order to reduce scientific uncertainty' (WEC, 1993: 47). The chairman of the IPCC supported him by arguing that 'as long as we do not know the natural carbon cycle adequately, the prediction of atmospheric concentrations due to future emissions remains uncertain' (WEC, 1993: 43). While uncertainty is worrying scientists, Greenpeace continues to campaign with reference to the view 'that climate is in danger of catastrophic destabilisation' (Greenpeace, undated), a view which the current British government appears to share.

10.5 The research enterprise attracts powerful allies

10.5.1 The United Nations seek an environmental role

As outlined, global warming as a threat to mankind made its way to the top of the international political agenda during the second half of the 1980s, a period when the bi-polar political system began to disintegrate and the UN lost some of its status and began to seek new 'tasks'. The global warming threat was disseminated most effectively by northern environmental lobbies and the UN bureaucracy (Boehmer-Christiansen, 1994 b), which suggests that both found the issue attractive politically. It is argued that the success of the threat in activating intergovernmental politics was due not so much to the arrival of new knowledge, but to the dissemination of powerful images - the global hothouse and rising oceans - by lobbies not only from within environmental lobbies (and some researchers), but also from policy-making sections of the UN.

The bare details of these developments are not described here except for noting that a new role for the UN, as an intergovernmental body, is now more urgently needed than ever before. One hope is that by linking 'aid' to 'environment', mutually beneficial links between rich and poor areas and groups can be forged. To justify these links 'scientifically', credible global environmental threats seem to have become a necessary condition.

10.5.2 Energy lobbies seek opportunities

Environmentalists were supported not only by the 'technocratic' UN bureaucracy, but most importantly, from the political perspective, also by advocates of policy measures, which promised to increase the price of fossil fuels, especially coal and oil. Figure 10.2 illustrates the persistent belief in, or hope for, higher oil prices, while reality has turned out very different. By the late 1990s, crude oil prices were at their lowest levels since the 1960s, though very little of this was passed on to consumers. World coal prices are also very low today and related to oil price fluctuations. Indeed, by the end of the millennium, the world was awash with cheap fossil energy, not an incentive to energy efficiency and renewables.

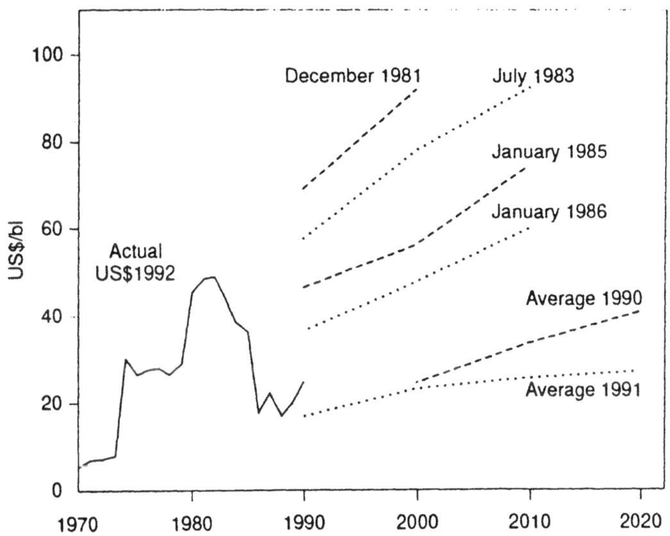

Figure 10.2 Oil price expectations and real oil prices 1970-1991
(Source: Stevens, 1996)

Falling fossil fuel prices at a time of global energy glut and stagnating demand meant that by 1987, at the latest, it became obvious that a period of high energy prices had ended; indeed that was abnormal rather than normal. This created many real or potential losers eager to attract subsidies or capture regulatory debates by becoming 'environmentally aware' about energy. The warming threat also gained credibility because of a series of long, hot summers in the Northern Hemisphere. The Chernobyl nuclear accident, however, weakened the alliance for rapid abatement action as the hope of some that nuclear power would be part of this solution evaporated, at least for the time being.

Hopes for higher fossil fuel prices were continuously nourished during the late 1980s and early 1990s by 'scientific' proposals for greenhouse gas emission reduction measures, that were marketed as technical solutions to global warming. For example, both IPCC leaders cited in 10.4.5 immediately above, spoke out in favour of energy efficiency and higher fossil fuel prices. Bolin in particular has pronounced on energy policy matters since the 1970s, and Houghton remains a strong advocate for 'pricing' cars off British roads. The links between the AGGG and nuclear interests have been mentioned. Science became the language of these energy policy related disputes and negotiations. The research enterprise prospered, serving all combatants and continuously 'disseminating' not its knowledge, but its ignorance, the uncertainties. Each party would soon use IPCC advice for justifying its position. This encouraged political activity and ensured that global warming remains 'an issue', but also discouraged effective policy-making and even more so, implementation.

Advocates of policies that promised to increase the prices of fossil fuels or to reduce their markets, therefore became the most effective and enduring political agents, supporting the serious development of climate policies by governments. Very powerful energy interests (nuclear, renewables, energy efficiency, natural gas) entered debates and as yet largely inconclusive negotiations, at many national levels. In the new energy markets of industrialising countries, like China, Indonesia and Brazil, rapid growth in energy demand was expected and is indeed taking place. Future profits from energy investments are most likely here, and these countries therefore need to be opened up for international capital, for example by World Bank conditionalities. All these developments give a commercial and financial impetus to 'joint implementation' policies, because these policies also imply 'inward' investments and sharing of profits. Global and national energy interests therefore possessed powerful motives to become politically active, either in support of intervention or, if based on coal or oil, in opposition to intervention.

10.5.3 Threatened national bureaucracies also seek sustainability

Governments, or rather administrations, also seek knowledge they can use legitimately and, hopefully, in the end effectively. Research groups in government want to keep their jobs and, like everybody else, prefer to expand their brief, influence and resource. Research inside government is an important and highly competitive activity, even if research is increasingly done by outside bodies, whose services are bought under contract. Such research inside and outside government is managed by individuals, who may themselves seek power and influence, as members of policy-makers for science (Weale, 1992: 195). When they do discover common interests with government, this elite acts for 'the scientific community' inside government, which mainly means attracting research funding by whatever arguments government considers legitimate or 'politically correct' at the time. In the 1990s this meant that science needed to define the problems it wishes to solve, and environmental problems were popular and would serve the interests of an environmental civil service and their expert advisors, which had developed since the late 1960s. A closer alliance between research science and bureaucracy, also sometimes referred to as 'regulatory science', developed and stood to benefit from a serious approach to climate change.

This raises two issues for environmental administrators: that of the quality of knowledge produced under conditions of contractual obligations, and that of its ownership. It is generally true that the producers of knowledge lose all rights over the fate of 'their' products, once these have been published, but a great deal of research done for government is never published and even IPCC publications have undergone very thorough governmental 'review'. The political implications are serious, when research output remains the property of its funders, be these government departments or industries. Most global environmental research is funded by governments, but does this mean that governments own the results? In this dawning age of information, such questions will surely come to haunt us. Once knowledge is in the open, as governments know only too well, other actors on the political stage may use it as they see fit. This may also cause much concern to scientists, who often feel misunderstood and misused, but is here supported as the only way (apart from education) we can protect society against the misuse of knowledge by policy-makers. Knowledge that is not tested in the world of open politics is always, by its very nature, in danger of leading to 'command and control' decision-making that is likely to face social acceptability and hence implementation problems.

For example, official British responses to climate change included vigorous participation in global environmental diplomacy after 1987, a flowering of bureaucratic interest and commitment. The use of global change rhetoric even spread to politicians, who used 'green' arguments to justify unpopular official policies (Boehmer-Christiansen, 1995). A sudden and significant increase in funding for British global environmental research after 1988 has been a major outcome and is defined as a symbolic policy response. An explanation for this success postulates that:

- environmental policy-formulators in the Department for the Environment used research for policy design purposes. This increased their budgets and influence inside government, but did not have much impact on 'real' policy, where 'the market' was still considered the best institution to determine environmental outcomes;

- energy policy-makers used science as justification for changes adopted or desired for political and economic reasons, such as the rapid switching from coal to natural gas and the imposition of taxes on energy, both highly unpopular policies, which attracted an official 'greenwash';
- environmental diplomats used science both as legitimisation and policy instrument in Britain's struggle against encroachment by the European Community legislation into the energy-environment policy space, which the UK wanted to keep separate as domestic matters.

A convergence of political, bureaucratic and commercial interests in climate change also initiated a new phase in the science policies of most OECD countries, with the natural science- and space-technology as the main beneficiaries. This, in my view, has been the major policy response to the climate issue so far. This has been an intellectual debate and planning exercise, rather than a policy requiring practical action. After 1988, for example, much of British research - environmental, technological, even social and space-related - began to define itself in relation to the IPCC and global environmental change. The leaders of science had succeeded in capturing the political imagination and interest of the Conservative Party, as well as environmental groups, which were increasingly led by natural scientists. Together, these groups persuaded politicians to direct public funding towards climate change, with the IPCC as the flagship of British environmental diplomacy. British scientists thus enjoyed advantages, which were shared with their Dutch colleagues, but not with those in Germany or France. Science policy allowed Britain, as well as the Netherlands, to continue to play an active role in global change research, in spite of severe cutbacks in research elsewhere. Many other countries, without such expertise, could not participate in the climate negotiations as 'equals'. Science therefore has another function not even mentioned above: it excludes many 'stakeholders'.

10.6 Conclusions: the environment in global politics[xxvii]

The question of why modern societies create rather than use so much new knowledge about the physical world is perhaps the most difficult question raised by 'global change' science. Is understanding of environmental change largely *technology-driven*? Without growth in computing power and earth observation, science-based scenario building and often dubious claims to 'rigorous' quantification and predictability, would not be possible. By going beyond explanation and analysis of what is to be the design of options, how society ought to be in the future, producers of knowledge enter the political arena very directly indeed.

The question of why global warming was picked up by world politics as a threat in 1986/1987 can now be tentatively answered. It was brought about by an alliance of the natural science research enterprise with assorted allies in bureaucracies and energy industries. All of these discovered that aspects of environmentalist arguments, served their particular institutional interest. A global research agenda was ready when fossil fuel prices collapsed. This created an energy glut, which included the growing availability in regional markets of natural gas. The energy winners of the 1970s had become the losers of the 1980s and demanded government intervention in energy markets and energy related R&D. Global warming provided an attractive greenwash for a most powerful, if tacit, alliance in which environmentalism acted as a catalyst. The prophecy of cooling would not have created such an alliance.

In the ensuing political battles, science and its uncertainties became tools of persuasion, prophecy, justification and potential conflict resolution, rather than providers of practical knowledge. The Framework Convention now codifies a global research agenda, which may prove useful in drawing attention to the plight of many poor countries. It is far less clear now where the interests of Europe lie. By transferring technology, via the FCCC, to developing countries, competitors will be strengthened. Political rather than environmental problems, may be creating the grave situations of 'degradation' the environmental prophets of doom have predicted. But for the opposition of US/UK policy advisors, fully aware of the uncertain scientific base of climate change, an international climate regime reflecting the demands of European environmentalists might have been 'sold' to the world, which would have committed all countries to energy policy changes, which might have created disaster for the poorest groups everywhere, as well as for countries most dependent on cheap energy. If the prophets of doom turn out to be 'right' and the causality of change is due to human activity, there will have to be very painful adjustments by future generations.

Natural science research, information intensive policies and natural gas have emerged as major 'winners' of climate change responses so far. To achieve this, science (as an institution, not a body of knowledge) had to market itself with reference to a threat, which provided it with policy relevance in the live-now, pay-later consumer society of today. It did not, however, act alone, but 'in partnership' with institutions that also stood to benefit from the threat of global warming. However, political scientists should not assume the existence of environmental threats on the basis of scientific claims. They must ask how and by whom such threats were constructed, who placed them on political agendas and who acted in response. Threats as well as policies come and go, only some lead to change 'on the ground'. With sociologists of science they investigate the social construction of environmental problems and ask not only how the arts of modelling and academic decision-making theory influence policy, but also how scientific knowledge is used by political and commercial interests.

The argument that 'scientific consensus' is about truth is dangerous, because truth, from the political perspective, varies not only with the availability of knowledge, but also over time and between places and institutions. Empirical study of how scientific institutions and governments deal with the climate threat, demonstrates that 'science' as knowledge plays a much more ambivalent and complex role than tends to be assumed by scientists themselves. While it may be a chief ingredient of policy justification, other forces also shape policy and are likely to determine how much of the 'plan' or 'strategy' is actually implemented in practice. Chief among these forces are vested interests, visions of the future and existing patterns of power, law, wealth and commitments. These interests, visions and identities initiate political power games with research agendas and the promise of new knowledge. New threats and promises are constructed from limited knowledge about our environment.

Most environmental regulations change power relationships. Usually they do so in favour of the State, as the representative of the public interest and of its professional advisors; the 'experts' as the providers and users of the types of knowledge. Making plans and strategies in response to global change implies controlling 'the market' as well as intervening in 'business as usual' in general, all in the name of environmental protection. This is not to say that environmental protection is not needed, only that on the ground, protection efforts have political implications as well as motivations. The existing patterns, or contexts, will select and filter the new knowledge and theories to minimise the threat of disturbance and to ensure consistency with existing patterns. The global warming hypothesis (rather than 'neutral' climate change) was therefore quickly accepted in the late 1980s, by those political actors that stood to benefit directly or indirectly from the desire for more knowledge, from regulations that promised higher fossil fuel prices, or from subsidies and public investment in alternative energy technologies. The link between energy and climate interests is therefore strong, not only at the modelling level, but also in the world of commerce, regulation and resource competition.

As the overall costs of mitigation became apparent to governments, however the idea that the threat of global warming might be exaggerated, in order to attract public 'intervention' also occurred to governments. Given precedents like Star Wars and fusion research, this is not surprising. As an argument against 'precautionary action', this suspicion may well benefit the poor and weak of today. Much wealth would have to be withdrawn from current investment and consumption via taxes and higher prices, had the world responded as Greenpeace and some scientists recommended. Only world politics made sure that all knowledge claims were fully tested, but this too is no guarantee for 'truth'.

For action to take place, the worlds of knowledge must become tied to the worlds of interest, indeed the latter may select and construct knowledge to suit itself. Persuasion, not rationality, is the primary political process in areas where knowledge is insufficient and yet claimed to be the major instrument for policy design. Uncertainty becomes a part of two distinct realms of politics: science policy as well as environmental policy formation. Both, in turn, have major impacts on the real policy target, in our case a complex mix of interventions in energy markets, through which an emission reduction policy, allegedly designed to protect the climate, is to be implemented. Climate policy will fail if it does not serve concrete interests and pursue several goals; it will also fail, in a different way, if it solves a problem that does not exist or cannot be solved by the measures taken. But would this matter? Such a rather negative conclusion invites a number questions, which may serve both as revision and stimuli for further thoughts and discussion about the linkages between politics and science.

10.7 Questions for further thought and discussion

1. Why has scientific research been so successful globally? Can societies cope with the growing flood of knowledge that is being produced?
2. How important is expert advice in policy-making in your country or institution? What role does it play compared to what it could play? Were it to play a greater role, would policies become more or less acceptable to a majority of people?
3. Why did climate change become such a hot issue during the mid-1980s and why has it remained so until today? You might consider the outcome if oil prices had indeed continued to rise.
4. Have you observed any differences in the use of scientific advice between public and private institutions? Are there implications for environmental policy?
5. Under what condition may it be true that a better understanding of an environmental problem leads not to a remedial or precautionary policy, but to it being ignored?

6. Before solutions can be attempted, information that there might be a problem, needs to be translated into practical knowledge. Why are scientists rather bad at providing this?

References

Agrawala, S.(1998) 'Context and Early Origins of the Intergovernmental Panel on Climate Change' *Climatic Change:* 39:605-620.

Boehmer-Christiansen, S.A. (1999) 'Climate Change and the World Bank: Opportunity for Global Governance? *Energy and Environment:* 10:1:27-50.

Boehmer-Christiansen, S.A. (1995) 'Britain's role in the politics of global warming', *Environmental Politics* 4, 1.

Boehmer-Christiansen, S.A. (1994a) 'Global climate protection policy: the limits of scientific advice - Part I', *Global Environmental Change* 4, 2: 140-159.

Boehmer-Christiansen, S.A. (1994b) 'Global climate protection policy: the limits of scientific advice - Part II', *Global Environmental Change* 4, 3: 185-200.

Boehmer-Christiansen, S.A. (1993) 'Science policy, the IPCC and the Climate Convention: the codification of a global research agenda', *Energy and Environment* 4, 4: 362-406.

Boehmer-Christiansen, S.A. (1990) 'Energy policy and public opinion: manipulation of environmental threats by vested interests in the UK and West Germany', *Energy Policy* 18, 9.

Boetcher, F. and Emsley, J. eds. (1996) Global Warming: A Report from the European Science Foundation, ESF, London.

Bryson, R. A. (1972) 'Climatic modification by air pollution', in N Polunin, *The Environmental Future*, London: Macmillan, 133-167.

Bolin, B. (1994) 'Science and policy-making', *AMBIO* 23, 1: 25-29.

Bolin, B. *et al.* (eds) (1988) *The Greenhouse Effect: Climatic Change and Ecosystems*, SCOPE 29, Chichester: Wiley & Sons.

Calder, N., *The Manic Sun*, Pilkington, 1997.

Calder, N. The Carbon Dioxide Thermometer and the Cause of Global Warming, Energy & Environment, 10,1:1-18.

Cavender, J. and Jager, J. (1993) *International Environmental Affairs* 5,1:3-18

Churchill, R. and Freestone D. (1992) *International Law and Global Climate Change*, London: Graham & Trotman/Nijhoff.

Dowlatabadi, H. and Morgan G. M. *Science* 259,26 March 1993:1813-4.

Estarda-Oyuela, R. 1993, UN/UNEP/WMO *Climate Change Bulletin,* 1.1 3rd quarter 1993:1.

Hart and Victor (1993) 'Scientific elites and the making of US policy for climate change research 1957-1974', *Social Studies of Science* 23, 4: 643-680.

UK, Her Majesty's Treasury (1992). *The economics of man-made climate change,* London:HMSO.

Houghton, J. T., Jenkins, G. J., and Ephraums, J. J. eds. (1990) *Climate Change: The IPCC Scientific Assessment*; Cambridge/New York: Cambridge University Press, published for the IPCC. Also 1992 Update by same publisher and main author.

ICSU/IGBP (1992) *Reducing Uncertainties*, Stockholm: Royal Swedish Academy of Sciences.

IGBP (1994) *Global Change Newsletter* 17, March and 25, March 1996.

Jäger, J. and Ferguson, H.L. eds (1991) Climate Change: Science, Impacts, Policy: Proceedings of the Second World Climate Conference, London: Cambridge University Press for WMO.

Kellogg, W. W. and Schware, R. (1981) Climate Change and Society: Consequences of Increasing Carbon Dioxide, New York: Westview.

Lindzen, R. (1992) 'Global warming: The Origin and Nature of the Alleged Scientific Consensus', *Special Issue of Energy and Environment* of Proceedings of OPEC Seminar, Vienna April 1992, Brentwood: Multi-Science.

Maxwell, J.H and Weiner S.L. (1993) 'Green Consciousness of Dollar Diplomacy? The British Response to the Threat of Ozone Depletion, *International Environmental Affairs*, 5,1 Winter :19:41.

McCracken, M. (1992) in S. Veggeberg 'Global Warming Researchers say they need breathing room', *The Scientist* 6, 2 (New York)

Michaels, Patrick J. (1992) *Sound and Fury: The Science and Politics of Global Warming,* Washington: Cato Institute.

Mukerji, C. (1989) *A Fragile Power: Scientists and the State*, Oxford: Princeton.

Rentz, H. Outcomes of the Fourth Conference of the parties to the Convention on Climate Change, Buenos Aires 1998, *Energy & Environment* :10:2 :157-190.

Slovic and Fischhoff (1983) 'How safe is safe enough?' in Walker C A, Gould L C and E J Woodhouse, *Too hot to handle*, Yale University Press, New York.

Tayler, R. J, *The Sun as a Star*, Cambridge University Press, 1997.
US National Research Council (1990) *Research Strategies for the US Global Change Research Program*, National Academy Press: Washington.

Warrick, R.A. and Jones, P. D. (1988) 'The greenhouse effect: impacts and policies', *Forum for Applied Research and Public Policy* 3, 3: 48-62.

Watson, R .T. Climate Change: the Challenge for Energy Supply' *Energy & Environment*, 10: 1:19-26.

Weart, S. (1992) 'From the nuclear frying pan into the global fire', *The Bulletin of Atomic Scientists* 48, 5: 18-2

Wiin-Nielson (1999), 'Limited Predictability and the Greenhouse Effect -A Scientific Review', *Energy &Environment: Special Issue -The Misinterpreted Greenhouse Effect* 9, 6 : 633-646.

World Energy Council (1993) *Journal*, July: 26.

World Meteorological Organisation (WMO) (1986) 'Report of the International Conference on the assessment of carbon dioxide and other greenhouse gases in climate variations and associated impacts, Villach 9-15 October 1985, WMO no. 661, Geneva: WMO.

World Meteorological Organisation (WMO) (1992) *The World Climate Programme 1992 - 2001*, WMO publication No. 762 : Geneva.

Wynne, B. (1992) 'Global Environmental Change: Human and Policy Dimensions, *Global Environmental Change*, Vol. 2, No. 2. pp. 11-127.

Endnotes

i Cited by R. Shroeder, 'Cyberculture, cyborg post-modernism and the sociology of virtual reality, *Futures*, 1994, 26, 5: 519

[ii] The idea that global warming is good for the planet, a Russian belief since the 1970s, led to their exclusion from IPCC WG-1 and the marginalisation of paleoclimatological studies inside the IPCC, though work is continuing under the IGBP. Russian academics have since devoted much effort to challenging IPCC predictions on empirical grounds.

[iii] Publication of criticisms of IPCC views is difficult, because so much official research funding is currently based on the assumption of anthropogenic warming. For reports of the scientific challenge to the IPCC based consensus, see, for example Calder cited here and the Special Issue of *Energy & Environment (*9, 6), which published all the papers given at a conference held in Bonn 1997 and organised by the European Academy for Environmental Affairs, Academiua Scientiarum et Artium Europaea, and The Science and Environmental Policy Project, and edited by Professor Helmut Metzner, of the University of Tübingen.

[iv] The writings of opponents of the IPCC scientific assessment were studied by the author. It was concluded that this opposition is less about science, than the use to which it is being put, e.g. with respect to energy policy, and including the careless incorporation of emission scenarios based on assumptions that were not widely discussed or even revealed in 1990.

[v] IPCC Climate Change 1995, Report of Working Group I, p.11

[vi].Senior UK MET Office administrators have been reported to me as being opposed to the development of regional models, as these could destroy the picture of a uniform threat and hence the incentive for 'global' action

[vii] Peter Scholfield, WMO's head of Climate Monitoring, quoted in Daily Mail, 2.2.1996, p.28; the ambivalent most recent IPCC statement has been noted above.

[viii] The distinctions between policy formulation, where expertise is very important, and policy-making, where politicians matter most, and policy-implementation, when administrators, industries and the public matter, are fundamental to this discussion. The danger is that administrators and politicians will hand decision-making powers to experts, who may have their own agendas.

[ix] ECOcoal, Newsletter of the World Coal Institute, April 1994. To find the flaws in the arguments of one group, look at those of its rival or commercial opponent.

[x] James Lovelock , famous for the Gaia hypothesis, is reported to have held a private meeting in Oxford for invited natural scientists and mathematicians to discuss this possibility.

[xi] The project Y320 25 3030 was funded by the British ESRC under its Global Environmental Change Initiative, as part of its contribution to Global Change research.

[xii] Gas is rapidly replacing coal in several industrial countries with weak nuclear sectors (USA, UK). It does so because it is cheap. This strategy allows CO_2-emission stabilisation by 2000, but not beyond. Climate change related natural science research has benefited greatly in countries with strong IPCC commitments (USA, UK, Germany, Netherlands, France, Australia, Canada, Japan, and India).

[xiii] These authors show that far from providing a blueprint for future environmental institutions, developed on the basis of scientific consensus and leadership (the conventional view), the ozone treaty was achieved by industrial competition in the environmental arena with science, serving not the environment, but ICI and Du Pont, the joint winners with smaller producers as losers. This is a typical tale of how international environmental agreements work in practice.

[xiv] As far as I am aware, all policy models ignore power relationships, because these are extraordinarily subtle, variable and impossible to measure quantitatively - hence the tendency to replace them with economic indicators or ignore them as 'values', as in the policy model put forward by Bolin.

[xv] Natural gas generates less carbon dioxide than coal or oil per unit electricity, because of its chemistry and because it can be burned more efficiently by combined cycle technology.

[xvi] The counter argument to the one put here, that uncertainty is emphasised by scientific advisors (or sought by policy-makers) in order to delay or avoid policy in order to promote research, is put by W. Rudig, 'Sources of Technological Controversy', in A. Barker and B. Peters (eds.), *The Politics of Expert Advice*, Edinburgh Uni Press, 1993. He calls especially for cutting edge science to be included in policy advice to reduce over-confidence in policy-makers and thereby generate a precautionary response. This does not follow, however. While this argument generally appeals to environmentalists, it tends to ignore the immediate political threats, arising from actions that are based on poorly understood problems and are advocated by ideologues or enthusiasts, or parties standing to gain from precaution, e.g. nuclear power in the global warming debate. In both cases, factors other than scientific ones are likely to determine policy.

[xvii] Global averages are useful for global prescription, but cannot be implemented by a single actor. This is why regional predictions of climate change are needed by policy-makers, yet science may well remain unable to deliver them, irrespective of the question of whether policy prescriptions derived from mathematical models can be trusted as 'true'.

[xviii] The most advanced assessment models of climate change policy appear to require almost divine abilities from their makers . They claim to be able not only to judge benefits, as well as costs, for the whole of mankind in future (there is no consultation, but numbers are taken from existing 'authoritative' sources, such as the World Bank), but also whether societies will be able to adjust.

[xix] In each case I know of, the results of such exercises tend to confirm the view that experts tend to conclude what their funders or governing bodies want to hear. Nordhaus concluded that early CO_2 abatement was not cost-effective, at a time when the Bush/Reagan administrations were totally opposed to carbon taxes and even scientists like Schlesinger calculated that there was still time; the Hope's PAGE model advises the EU to adopt stringent measures, and surprise, surprise, ICAM-0 'finds' that 'subjective perceptions of different actors are more important in determining policy objectives than scientific uncertainty'. Are such models a substitute for empirical research, or a toy?

[xx] This book is based on Aspen Institute workshops held in the USA and West Germany. Major participants were the US National Corporation for Atmospheric Physics (NCAR), based in Boulder, and the Austrian International Institute for Applied Systems Analysis (IIASA), where the climate threat was explored by two Germans, H. Flohn (climatologist, WMO) and Wolf Hafele (inventor of the breeder reactor) during the 1970s.

[xxi] This had the support of the US Department of Energy, a major sponsor of carbon dioxide research. The US State Department wanted scientific assessment to stay in governmental hands, not in those 'of free-wheeling academics'. In contrast to the UK, American negotiators did not want to let their own Environmental Protection Agency get deeply involved.

[xxii] A 1992 German IGBP newsletter (*Global Change Prisma*) lists 56 satellite and 4 space shuttle launches between 1992 and 1998 for remote sensing of the earth. The countries involved are USA (NASA and NOAA), Japan, Russia, Germany, India, UK and France: all major global change research nations.

[xxiii] Opposition by social scientists working for the EC Commission was overruled. OECD is now considering Global Change research mega-science. Space observation is its most costly sector; this information must find markets as well as justification. Science funding decisions, supporting these efforts, are not made by parliaments.

[xxiv] Such scenarios have been found unsuitable for policy-making, though useful for the testing of climate models.

[xxv] These dates were prescribed by the USA and the Netherlands. This methodology has since been abandoned.

[xxvi] A former director of UK Friends of the Earth, for example, argued in 1993 that a 60% reduction in CO_2 needed to stabilise the climate at an acceptable temperature, which was the consensus view of the IPCC. The UK Treasury, basing itself on the same 'authoritative IPCC assessment by several hundred scientists', on the other hand, emphasised 'huge uncertainties' and accepted private advice that another decade of research would be needed before policy could be made in response to man-made changes, provided that there should not be excessive costs for the world economy (Her Majesty's Treasury 1992).

[xxvii] Readers, who may be offended by the argument of this chapter, may note that the book by Roger Tayler, describing recent developments in our understanding of the sun and its relationship with the earth, is dedicated to Dr. Peter James Christiansen. Peter died in 1992, long before he could complete his work on the magnetosphere. He was my husband and encouraged me to question the IPCC consensus. I dedicate this chapter to him.

Index

MIX
Papier aus verantwortungsvollen Quellen
Paper from responsible sources
FSC® C105338

If you have any concerns about our products,
you can contact us on
ProductSafety@springernature.com

In case Publisher is established outside the EU,
the EU authorized representative is:
Springer Nature Customer Service Center GmbH
Europaplatz 3, 69115 Heidelberg, Germany

Printed by Libri Plureos GmbH
in Hamburg, Germany